Analog and Digital Communication Systems

Analog and Digital Communication Systems

SECOND EDITION

Martin S. Roden

DEPARTMENT OF ELECTRICAL AND COMPUTER ENGINEERING
CALIFORNIA STATE UNIVERSITY, LOS ANGELES

PRENTICE-HALL., Englewood Cliffs, N.J. 07632

059856669

Library of Congress Cataloging in Publication Data

RODEN, MARTIN S.

Analog and digital communication systems.

Bibliography: p. 439
Includes index.
1. Telecommunication.   2. Digital communications.   I. Title.
TK5105.R64 1985   621.38′0413   84-18114
ISBN 0-13-032822-7

Editorial/production supervision and
   interior design: Tom Aloisi
Cover design: 20/20 Services, Inc.
Manufacturing buyer: Tony Caruso

Printed in the United States of America

10 9 8 7 6 5 4 3 2 1

ISBN: 0-13-032822-7   01

Prentice-Hall International, Inc., *London*
Prentice-Hall of Australia Pty. Limited, *Sydney*
Editora Prentice-Hall do Brasil, Ltda., *Rio de Janeiro*
Prentice-Hall Canada Inc., *Toronto*
Prentice-Hall Hispanoamericana, S.A., *Mexico*
Prentice-Hall of India Private Limited, *New Delhi*
Prentice-Hall of Japan, Inc., *Tokyo*
Prentice-Hall of Southeast Asia Pte. Ltd., *Singapore*
Whitehall Books Limited, *Wellington, New Zealand*

Contents

Preface

Analog and Digital Communication Systems, Second Edition, represents a greatly modified and enhanced version of the earlier edition. Particular attention has been given to digital communication systems. The book is intended as an introductory text for the study of analog and/or digital communication systems, with or without noise. Although all necessary background material has been included, a prerequisite course in linear systems analysis is helpful.

The text stresses a mathematical systems approach to all phases of the subject matter. The mathematics used throughout is as elementary as possible but is carefully chosen so as not to contradict any more sophisticated approach which may eventually be required. An attempt is made to apply intuitive tehniques prior to grinding through the mathematics. The style is informal, and the text has been thoroughly tested in the classroom with excellent success.

Chapter 1 lays the mathematical groundwork of signal analysis, including a study of discrete and continuous signals. Chapter 2 applies this analysis to linear systems with emphasis upon analog and digital filters. Chapter 3 introduces probability and random analysis and applies these to the study of narrowband noise. If it is desired to omit noise in a first approach to the subject, this chapter may be skipped. The factor of noise has been introduced at the end of each of the remaining chapters. The student may therefore eliminate noise analysis without adversely affecting continuity. Chapter 4 is a thorough treatment of AM, including applications to broadcast radio, television, and AM stereo. Chapter 5 parallels Chapter 4, but for angle instead of amplitude modulation. A section on broadcast FM and FM stereo is included. Chapter 6 explores analog forms of pulse modulation and introduces the concepts of time division multiplexing and cross talk. Chapter 7 treats digital communication with emphasis upon PCM and delta modulation. An extensive discussion of quantization noise, both for uniform and for non-uniform quantizing, is included. The final chapter investigates ASK, FSK, and PSK. The matched filter is introduced, and the error performance of detectors is discussed. Appendices cover a list of references, a summary of symbols used in the text, a brief table of Fourier Transforms, a table of error functions, and answers to selected problems.

Problems are presented at the end of each chapter. Numerous solved illustrative examples are given within each chapter.

It gives me a great deal of pleasure to acknowledge the assistance and support of the following people, without whom this text would not have been possible:

- To the many classes of students who were responsive during lectures and helped indicate the clearest approach to each topic.
- To my colleagues at Bell Telephone Labs, Hughes Aircraft Ground Systems, and California State University, Los Angeles.
- To all who adopted the earlier text and submitted evaluations and comments.
- To Mr. William Gehr of the Trident Shop, without whom this text would have been delayed several years.
- To Mr. Dennis J. E. Ross for guidance and for assistance with the diagrams.
- To Professor A. Papoulis, who played a key role during the formative years of my education.

Martin S. Roden

Introduction for the Student

Communications is perhaps the oldest applied area within electrical engineering. As is the case in many technical disciplines, the field of communications is experiencing a revolution several times each decade. Some important milestones of the past include

- *The television revolution*: After only four decades, television has become a way of life. Wrist watch television is now available, as is wall size television.
- *The space revolution*: This has been the catalyst for many innovations in long distance communication. Satellite communication is now permitting universal access.
- *The digital revolution*: The tremendous emphasis upon digital electronics and processing has resulted in a rapid change in direction from analog to digital communication.
- *The microprocessor revolution*: This is changing the shape of everything related to computing and control, including many phases of communication. Home computers with modems will soon be in almost every household. This will permit direct data communications by the general public.
- *The consumer revolution*: Consumers first discovered calculators and digital watches. They then expanded into sophisticated TV recording systems, CB radio, and video games. When digital watches were not enough, numerous other functions were added to the wrist device, such as calculators, TV, and pulse and temperature indicators.

There is every reason to expect this revolution rate to continue or, indeed, accelerate. Two-way interactive cable and videotext will greatly reduce our need for hard copies of literature and for travelling. People are already juggling bank accounts, ordering groceries, sending electronic mail, and holding business-related conferences without leaving their homes.

The problems of urban centers are being attacked through communications. Indeed, adequate communication can make cities unnecessary. The potential impact of these developments is mind boggling, and the possibilities involving communications are most exciting.

All of the above should excite you to consider the field of communications as a career. But even if you decide not to approach the field any more deeply than in this text, you will be a far more aware person having experienced even the small amount contained herein. If nothing more, the questions arising in modern communications as it relates to everything from the space program to home entertainment will take on a new meaning for you. The devices around you will no longer by mysterious. You will learn what magic force lights the "stereo" indicator on your receiver, the basics of how a video game works, and the mechanism by which your voice can travel to any part of the globe in a moment's time.

Enjoy the book. Enjoy the subject. And please, after studying this text, communicate any comments that you may have (positive, negative, or neutral) to me at California State University, Los Angeles, California 90032. Thank you.

Martin S. Roden

Analog and Digital Communication Systems

Chapter 1
Signal Analysis

By this stage of your education, you have probably heard of Fourier Series expansions. You have probably also learned that the earlier part of any course of study tends to be the least interesting. Certain groundwork must be laid prior to getting into the interesting applications.

We could start by presenting the rules for finding Fourier Series expansions and from there go directly into the business of signal analysis. We choose instead to take a more general approach. Since the principles introduced in this approach are not crucial to the understanding of *elementary* communication systems, their presentation at this time must be somehow justified.

The purpose of this section is to put the study of signals into proper perspective in the much broader area of applied mathematics. Signal analysis, and indeed most of communication theory, is a mathematical science. Probability theory and transform analysis techniques (a term which will have to be defined later) form the backbone of all communication theory. Both of these disciplines fall clearly within the realm of mathematics.

It is quite possible to study communication systems without relating the results to more general mathematical concepts, but this seems to be quite narrow-minded and tends to downgrade engineers (a phenomenon to which we need not contribute). More important than that, the narrow-minded approach reduces a person's ability to extend or modify existing results in order to apply them to new problems.

We shall therefore begin by examining some of the more common properties of orthogonal vector spaces. We do this with the intention of generalizing the results to apply to orthogonal function spaces. The generation of the Fourier Series representation of a function (our immediate goal) will prove to be a trivial application of the principles of orthogonal function spaces.

As an added bonus, the techniques and principles we develop will prove crucial to later studies in more advanced communication theory. For example, orthogonality is basic to detection problems (e.g., radar).

If words such as "space" and "orthogonal" make you uncomfortable, skip to Section 1.3 and start with the definition of a Fourier Series. This will cost nothing as far as basic communication theory is concerned.

1.1   ORTHOGONAL VECTOR SPACES

We shall use the word *orthogonality* (Webster defines this as relating to right angles) of vectors as it applies to the dot (or inner) product operation. That is, two vectors are said to be orthogonal to each other if their dot product is zero. This is equivalent to saying that one vector has a zero component in the direction of the other; they have nothing in common. A set of vectors is called an orthogonal set of vectors if each member of the set is orthogonal to every other member of the set. An often used set of orthogonal vectors in three-dimensional Euclidean space is that set composed of the three rectangular unit vectors,

$$\bar{a}_x, \ \bar{a}_y, \ \bar{a}_z$$

The bar above a quantity is the notation which we have adopted for a vector. The three unit vectors shown above form an orthogonal set since

$$\bar{a}_x \cdot \bar{a}_y = \bar{a}_x \cdot \bar{a}_z = \bar{a}_y \cdot \bar{a}_z = 0 \qquad (1.1)$$

Recall that in three-dimensional space, the dot product of two vectors is given by the product of their lengths multiplied by the cosine of the angle between them.

We know that any three-dimensional vector can be written as a sum of its three rectangular components. This is equivalent to saying that it can be written as a "linear combination" of the three orthogonal unit vectors presented above. If $\bar{A}$ is a general three-dimensional vector, then

$$\bar{A} = c_1 \bar{a}_x + c_2 \bar{a}_y + c_3 \bar{a}_z \qquad (1.2)$$

where c_1, c_2, and c_3 are scalar constants. Suppose we take the dot product of each side of Eq. (1.2) with $\bar{a}_x$. Since each side of this equation is certainly a vector, we are justified in doing this.

$$\bar{A} \cdot \bar{a}_x = c_1 \bar{a}_x \cdot \bar{a}_x + c_2 \bar{a}_y \cdot \bar{a}_x + c_3 \bar{a}_z \cdot \bar{a}_x \qquad (1.3)$$

By virtue of the orthogonality of the unit vectors and the fact that their lengths are unity, we have

$$\bar{a}_x \cdot \bar{a}_x = 1 \qquad \bar{a}_y \cdot \bar{a}_x = 0 \qquad \bar{a}_z \cdot \bar{a}_x = 0 \qquad (1.4)$$

and the right side of Eq. (1.3) reduces to just c_1. Similarly, we could have taken the dot product of both sides of Eq. (1.2) with $\bar{a}_y$ or $\bar{a}_z$ in order to find c_2 or c_3. The results would be

$$c_1 = \bar{A} \cdot \bar{a}_x \qquad c_2 = \bar{A} \cdot \bar{a}_y \qquad c_3 = \bar{A} \cdot \bar{a}_z \qquad (1.5)$$

Thus, each weighting constant, c_n, corresponds to the component of the vector, $\bar{A}$, in the direction of the associated unit vector. Note how simple it was to find the values of the c_n's. This would have been far more complicated if the component unit vectors had not been orthogonal to each other.

In the case of a three-dimensional vector, it is obvious that if we break it into its components, we need three component vectors in most cases. Suppose that we were not endowed with such intuitive powers, and wished to express a general three-dimensional vector, $\bar{A}$, as a linear combination of x and y unit vectors only.

$$\bar{A} \approx c_1 \bar{a}_x + c_2 \bar{a}_y \qquad (1.6)$$

Equation (1.6) represents an approximation, since one cannot be sure that the vector $\bar{A}$ can be expressed exactly as the sum shown in Eq. (1.6). Indeed, the reader with intuition knows that, in general, this cannot be made into an exact equality. For example, suppose that $\bar{A}$ had a non-zero component in the z-direction. Equation (1.6) could then never be an exact equality.

In many cases of interest, we either cannot use as many approximating vectors as we know are required, or we are not even sure of how many are necessary. A significant question would therefore be, "How does one choose c_1 and c_2 in Eq. (1.6) in order to make this approximation as *good* as is possible?" (whatever "good" means). The standard way of defining goodness of an approximation is by defining an *error* term. The next step is to try to minimize this error term.

We shall define the error as the difference between the actual $\bar{A}$ and the approximation to $\bar{A}$. This difference will be a vector, and since defining error as a vector is not too satisfying (i.e., What does "minimum" mean for a vector?), we shall use the magnitude, or length of this vector as the definition of error. Since the magnitude of a vector is a term which is virtually impossible to work with mathematically, we shall modify this one more time by working with the square of this magnitude. Therefore,

$$e = \text{Error} = |\bar{A} - c_1 \bar{a}_x - c_2 \bar{a}_y|^2 \qquad (1.7)$$

If we minimize the square of the magnitude, we have simultaneously minimized the magnitude. This is true since the magnitude is always positive, and the square increases as the magnitude itself increases (i.e., a monotonic function). Differentiating the error, e, with respect to the variables which we wish to find, c_1 and c_2, and then setting these derivatives equal to zero in order to minimize the quantity, we have[1]

$$e = [\bar{A} - c_1 \bar{a}_x - c_2 \bar{a}_y] \cdot [\bar{A} - c_1 \bar{a}_x - c_2 \bar{a}_y]$$

$$\frac{\partial e}{\partial c_1} = -2[\bar{A} - c_1 \bar{a}_x - c_2 \bar{a}_y] \cdot \bar{a}_x = 0 \qquad (1.8)$$

---

[1]Since the error, as defined, is quadratic, we are assured of a nice minimum. One can easily be convinced that the zero-derivative point corresponds to a minimum and not a maximum of the error.

and

$$\frac{\partial e}{\partial c_2} = -2[\bar{A} - c_1\bar{a}_x - c_2\bar{a}_y]\cdot\bar{a}_y = 0 \qquad (1.9)$$

from which we get, after expanding and using the orthogonality property,

$$c_1 = \bar{A}\cdot\bar{a}_x \qquad c_2 = \bar{A}\cdot\bar{a}_y \qquad (1.10)$$

Equation (1.10) is a very significant result. It says that, even though the orthogonal set used to approximate a general vector does not contain a sufficient number of elements to describe the vector exactly, the weighting coefficients are chosen as if the set were sufficient. In vector terms, even though the unit vector in the z-direction was missing, c_1 was still chosen as the component of $\bar{A}$ in the x-direction, and c_2 as the component of $\bar{A}$ in the y-direction. This indicates a complete independence between the various pairs of components. The z-direction component has nothing to do with that in the x- or y-direction. This could have been expected due to the orthogonality (nothing in common) property of the original set of vectors. It probably appears obvious in this case. However, it will be generalized and used in the following section where it will certainly no longer be obvious.

1.2   ORTHOGONAL FUNCTION SPACES

In communication systems, we deal primarily with time functions. In order to apply the above results to functions, the concept of vectors will now be generalized with an appropriate modification of the dot product operation. We will consider a set of real time functions, $g_n(t)$, for n between 1 and N. That is, we have N functions of time. This set will be defined to be an orthogonal set of functions if each member is orthogonal to every other member of the set. Two time functions are defined to be orthogonal over the interval between t_1 and t_2 if the integral of the product of one with the complex conjugate of the other is zero. This integral therefore corresponds to the inner, or dot, product operation. That is, $f(t)$ is orthogonal to $h(t)$ if

$$\int_{t_1}^{t_2} f(t)h^*(t)\,dt = 0 \qquad (1.11)$$

Therefore, the set of time functions, $g_n(t)$, forms an orthogonal set over the interval between t_1 and t_2 if

$$\int_{t_1}^{t_2} g_j(t)g_k^*(t)\,dt = 0 \qquad (\text{for all } j \neq k) \qquad (1.12)$$

The restriction $j \neq k$ assures that we integrate the product of two different members of the set.

Suppose that we now wish to *approximate* any time function, $s(t)$, by a linear combination of the $g_n(t)$ in the interval between t_1 and t_2, where the $g_n(t)$ are orthogonal over this interval. Notice that we use the word "approximate" since there is no assurance that an exact equality can be obtained no matter what values are chosen for the weighting coefficients. The number of g_n's necessary to attain an exact equality is not at all obvious. Compare this with the three-dimensional vector case where we knew that three unit vectors were sufficient to express any general three-dimensional vector.

The approximation of $s(t)$ is then of the form

$$s(t) \approx c_1 g_1(t) + c_2 g_2(t) + \cdots + c_N g_N(t) \tag{1.13}$$

or in more compact form,

$$s(t) \approx \sum_{n=1}^{N} c_n g_n(t) \tag{1.14}$$

Recalling our previous result [Eq. (1.10)], we conjecture that the c_n's are chosen as if Eq. (1.13) were indeed an exact equality. If it were an equality, we would find the c_n's as we did in Eq. (1.3). That is, take the inner product of both sides of the equation with one member of the orthogonal set. In this case, multiply both sides of Eq. (1.14) by one of the $g_n(t)$, say $g_{17}(t)$ [assuming, of course, the $N \geq 17$ in Eq. (1.13)] and integrate both sides between t_1 and t_2. By virtue of the orthogonality of the $g_n(t)$, each term on the right side would integrate to zero except for the 17th term. That is,

$$\int_{t_1}^{t_2} s(t) g_{17}^*(t)\, dt = c_1 \int_{t_1}^{t_2} g_1(t) g_{17}^*(t)\, dt + \cdots$$

$$+ c_{17} \int_{t_1}^{t_2} |g_{17}^2(t)|\, dt + \cdots + c_N \int_{t_1}^{t_2} g_N(t) g_{17}^*(t)\, dt$$

Identifying the terms which are zero yields

$$\int_{t_1}^{t_2} s(t) g_{17}^*(t)\, dt = c_{17} \int_{t_1}^{t_2} |g_{17}^2(t)|\, dt \tag{1.15}$$

Solving for c_{17}, we get

$$c_{17} = \frac{\displaystyle\int_{t_1}^{t_2} s(t) g_{17}^*(t)\, dt}{\displaystyle\int_{t_1}^{t_2} |g_{17}^2(t)|\, dt}$$

Being extremely clever, we can generalize the above to find any c_n.

$$c_n = \frac{\displaystyle\int_{t_1}^{t_2} s(t) g_n^*(t)\, dt}{\displaystyle\int_{t_1}^{t_2} |g_n^2(t)|\, dt}, \quad 1 \leq n \leq N \tag{1.16}$$

Note that the numerator of Eq. (1.16) is essentially the component of $s(t)$ in the $g_n(t)$ "direction" (one must be pretty open-minded to believe this at this point). The denominator of Eq. (1.16) normalizes the result since the $g_n(t)$, unlike the unit vectors, do not necessarily have a "length" of unity.

1.3   FOURIER SERIES

We have shown that, over a given interval, a function can be represented by a linear combination of members of an orthogonal set of functions. There are many possible orthogonal sets of functions, just as there are many possible orthogonal sets of three-dimensional vectors (e.g., consider any rotation of the three rectangular unit vectors). One such possible set of functions is the set of harmonically related sines and cosines. That is, the functions $\sin 2\pi f_0 t$, $\sin 4\pi f_0 t$, $\sin 6\pi f_0 t, \ldots$, $\cos 2\pi f_0 t$, $\cos 4\pi f_0 t$, $\cos 6\pi f_0 t, \ldots$, for any f_0, form an orthogonal set. These functions are orthogonal over the interval between any starting point, t_0, and $t_0 + 1/f_0$. That is,

$$\int_{t_0}^{t_0+1/f_0} g_n(t)g_m(t)\, dt = 0 \qquad \text{for all } n \neq m \tag{1.17}$$

where $g_n(t)$ is any member of the set of functions, and $g_m(t)$ is any other member. This can be verified by a simple integration using the "cosine of sum" and "cosine of difference" trigonometric identities.

Illustrative Example 1.1

Show that the set made up of the functions $\cos 2\pi n f_0 t$ and $\sin 2\pi n f_0 t$ is an orthogonal set over the interval $t_0 < t \leq t_0 + 1/f_0$ for any choice of t_0.

Solution

We must show that

$$\int_{t_0}^{t_0+1/f_0} g_n(t)g_m(t)\, dt = 0$$

for any two distinct members of the set. There are three cases which must be considered:

(a) Both $g_n(t)$ and $g_m(t)$ are sine waves.
(b) Both $g_n(t)$ and $g_m(t)$ are cosine waves.
(c) One of the two is a sine wave and the other, a cosine wave.

Considering each of these cases, we have
Case (a)

$$\int_{t_0}^{t_0+1/f_0} \sin 2\pi n f_0 t \, \sin 2\pi m f_0 t \, dt = \tfrac{1}{2}\int_{t_0}^{t_0+1/f_0} \cos(n-m)2\pi f_0 t \, dt$$

$$-\tfrac{1}{2}\int_{t_0}^{t_0+1/f_0} \cos(n+m)2\pi f_0 t \, dt \tag{1.18}$$

If $n \neq m$, both $n - m$ and $n + m$ are non-zero integers. We note that, for the function $\cos 2\pi k f_0 t$ the interval, $t_0 < t \leq t_0 + 1/f_0$ represents exactly k periods. The integral of a cosine function over any whole number of periods is zero, so we have completed this case. Note that it is important that $n \neq m$ since, if $n = m$, we have

$$\frac{1}{2} \int_{t_0}^{t_0 + 1/f_0} \cos(n - m) 2\pi f_0 t \, dt = \frac{1}{2} \int_{t_0}^{t_0 + 1/f_0} 1 \, dt = \frac{1}{2 f_0} \neq 0$$

Case (b)

$$\int_{t_0}^{t_0 + 1/f_0} \cos 2\pi n f_0 t \cos 2\pi m f_0 t \, dt = \frac{1}{2} \int_{t_0}^{t_0 + 1/f_0} \cos(n - m) 2\pi f_0 t \, dt$$
$$+ \frac{1}{2} \int_{t_0}^{t_0 + 1/f_0} \cos(n + m) 2\pi f_0 t \, dt \qquad (1.19)$$

This is equal to zero by the same reasoning applied to Case (a).

Case (c)

$$\int_{t_0}^{t_0 + 1/f_0} \sin 2\pi n f_0 t \cos 2\pi m f_0 t \, dt = \frac{1}{2} \int_{t_0}^{t_0 + 1/f_0} \sin(n + m) 2\pi f_0 t \, dt$$
$$+ \frac{1}{2} \int_{t_0}^{t_0 + 1/f_0} \sin(n - m) 2\pi f_0 t \, dt \qquad (1.20)$$

To verify this case, we note that

$$\int_{t_0}^{t_0 + 1/f_0} \sin k 2\pi f_0 t \, dt = 0$$

for all integer values of k. This is true since the integral of a sine function over any whole number of periods is zero. (There is no difference between a sine function and a cosine function other than a shift.) Each term present in Case (c) is therefore equal to zero.

The given set is therefore an orthogonal set of time functions over the interval $t_0 < t \leq t_0 + 1/f_0$.

We have talked about *approximating* a time function with a linear combination of members of an orthogonal set of time functions. We now present a definition. An orthogonal set of time functions is said to be a *complete* set if the approximation of Eq. (1.14) can be made into an equality (with the word "equality" being interpreted in some special sense) by properly choosing the c_n weighting factors, and for $s(t)$ being any member of a certain class of functions. The three rectangular unit vectors form a complete orthogonal set in three-dimensional space, while the vectors $\bar{a}_x$ and $\bar{a}_y$ by themselves form an orthogonal set which is not complete. We state without proof that the set of harmonic time functions,

$$\cos 2\pi n f_0 t, \ \sin 2\pi n f_0 t$$

where n can take on any integer value between zero and infinity, is an orthogonal complete set in the space of time functions defined in the interval between t_0 and $t_0 + 1/f_0$. Therefore any time function[2] can be expressed, in the interval between t_0 and $t_0 + 1/f_0$, by a linear combination of sines and cosines. In this case, the word equality is interpreted not as a pointwise equality, but in the sense that the "distance" between $s(t)$ and the series representation, given by

$$\int_0^{1/f_0} | s(t) - \sum_{n=1}^{N} c_n g_n(t) |^2 \, dt$$

approaches zero as more and more terms are included in the sum. This is what we will mean when we talk of equality of two time functions. This type of equality will be sufficient for all of our applications.

For convenience, we define

$$T \overset{\Delta}{=} \frac{1}{f_0}$$

Therefore, any time function, $s(t)$, can be written as

$$s(t) = a_0 \cos(0) + \sum_{n=1}^{\infty} [a_n \cos 2\pi n f_0 t + b_n \sin 2\pi n f_0 t] \qquad (1.21)$$

for

$$t_0 < t \le t_0 + T$$

An expansion of this type is known as a *Fourier Series*. We note that the first term in Eq. (1.21) is simply "a_0" since $\cos(0) = 1$. The proper choice of the constants, a_n and b_n is indicated by Eq. (1.16).

$$a_0 = \frac{\int_{t_0}^{t_0+T} s(t) \, dt}{\int_{t_0}^{t_0+T} 1^2 \, dt} = \frac{1}{T} \int_{t_0}^{t_0+T} s(t) \, dt \qquad (1.22)$$

and for $n \ne 0$,

$$a_n = \frac{\int_{t_0}^{t_0+T} s(t) \cos 2\pi n f_0 t \, dt}{\int_{t_0}^{t_0+T} \cos^2 2\pi n f_0 t \, dt} = \frac{2}{T} \int_{t_0}^{t_0+T} s(t) \cos 2\pi n f_0 t \, dt \qquad (1.23)$$

[2]In the case of Fourier Series, the class of time functions is restricted to be that class which has a finite number of discontinuities and a finite number of maxima and minima in any one period. Also the integral of the magnitude of the function over one period must exist (i.e., not be infinite).

$$b_n = \frac{\displaystyle\int_{t_0}^{t_0+T} s(t)\sin 2\pi nf_0 t\, dt}{\displaystyle\int_{t_0}^{t_0+T} \sin^2 2\pi nf_0 t\, dt} = \frac{2}{T}\int_{t_0}^{t_0+T} s(t)\sin 2\pi nf_0 t\, dt \qquad (1.24)$$

Note that a_0 is the average of the time function, $s(t)$. It is reasonable to expect this term to appear by itself in Eq. (1.21) since the average value of the sines or cosines is zero. In any equality such as Eq. (1.21), the time average of the left side must equal the time average of the right side.

A more compact form of the Fourier Series described above is obtained if one considers the orthogonal, complete set of complex harmonic exponentials, that is, the set made up of the time functions

$$\exp(j2\pi nf_0 t)$$

where n is any integer, positive or negative, is orthogonal over a period of $1/f_0$ sec.

The student should recall that the complex exponential can be viewed as (actually, it is *defined* as) a vector of length "1" and angle, "$n2\pi f_0 t$" in the complex two-dimensional plane. Thus,

$$\exp(j2\pi nf_0 t) \overset{\Delta}{=} \cos 2\pi nf_0 t + j \sin 2\pi nf_0 t$$

As before, the series expansion will apply in the time interval between t_0 and $t_0 + 1/f_0$. Therefore, any time function, $s(t)$, can be expressed as a linear combination of these exponentials in the interval between t_0 and $t_0 + T(T = 1/f_0$ as before).

$$s(t) = \sum_{n=-\infty}^{\infty} c_n e^{j2\pi nf_0 t} \qquad (1.25)$$

The c_n are given by

$$c_n = \frac{\displaystyle\int_{t_0}^{t_0+T} s(t)e^{-j2\pi nf_0 t}\, dt}{\displaystyle\int_{t_0}^{t_0+T} |e^{j2\pi nf_0 t}|^2\, dt} = \frac{1}{T}\int_{t_0}^{t_0+T} s(t)e^{-j2\pi nf_0 t}\, dt \qquad (1.26)$$

This is easily verified by multiplying both sides of Eq. (1.25) by $e^{-j2\pi nf_0 t}$ and integrating both sides. The negative sign in the exponent results from taking the complex conjugate as in Eq. (1.16).

The basic results are summed up in Eqs. (1.21) and (1.25). Any time function can be expressed by a weighted sum of sines and cosines or a weighted sum of complex exponentials in an interval. The rules for finding the weighting factors are given in Eqs. (1.22-1.24) and (1.26).

Equations (1.21) through (1.26) could have formed the starting point of our discussion of communication theory. For some students, this might still be the best place to begin.

Examination of the right side of Eq. (1.21) discloses that it is a periodic function outside of the interval $t_0 < t \leq t_0 + T$. In fact, its period is T. Therefore, if $s(t)$ happened to be periodic with period T, even though the equality in Eq. (1.21) was written to apply only within the interval $t_0 < t \leq t_0 + T$, it does indeed apply for all time! (Please give this statement some thought!)

In other words, if $s(t)$ is periodic, and we write a Fourier Series for $s(t)$ that applies over one complete period of $s(t)$, the series is equivalent to $s(t)$ for all time.

Illustrative Example 1.2

Evaluate the trigonometric Fourier Series expansion of $s(t)$ as shown in Fig. 1.1. This series should apply in the interval $-\pi/2 < t \leq \pi/2$.

Solution

Using the trigonometric Fourier Series form, $T = \pi$, and $f_0 = 1/T = 1/\pi$. The series is therefore of the form

$$s(t) = a_0 + \sum_{n=1}^{\infty} [a_n \cos 2nt + b_n \sin 2nt] \tag{1.27}$$

where

$$a_0 = \frac{1}{T} \int_{-T/2}^{T/2} s(t)\, dt = \frac{1}{\pi} \int_{-\pi/2}^{\pi/2} \cos t\, dt = \frac{1}{\pi} \sin t \Big|_{-\pi/2}^{\pi/2} \tag{1.28}$$

$$a_0 = 2/\pi$$

and

$$a_n = \frac{2}{T} \int_{-T/2}^{T/2} s(t)\cos 2nt\, dt = \frac{2}{\pi} \int_{-\pi/2}^{\pi/2} \cos t \cos 2nt\, dt$$

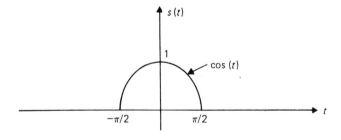

FIGURE 1.1 $s(t)$ for Illustrative Example 1.2.

$$a_n = \frac{1}{\pi} \int_{-\pi/2}^{\pi/2} [\cos(2n-1)t + \cos(2n+1)t]\,dt$$

$$= \frac{1}{\pi} \frac{\sin(2n-1)t}{(2n-1)} \Big|_{-\pi/2}^{\pi/2} + \frac{1}{\pi} \frac{\sin(2n+1)t}{(2n+1)} \Big|_{-\pi/2}^{\pi/2} \tag{1.29}$$

We note that

$$\sin\left\{ (2n-1) \left[\frac{\pi}{2} \right] \right\} = (-1)^{n+1}$$

$$\sin\left\{ (2n+1) \left[\frac{\pi}{2} \right] \right\} = (-1)^{n}$$

Therefore,

$$a_n = \frac{2}{\pi} \left[\frac{(-1)^{n+1}}{2n-1} + \frac{(-1)^{n}}{2n+1} \right] \tag{1.30}$$

Proceeding now to the evaluation of b_n, we find

$$b_n = \frac{2}{T} \int_{-T/2}^{T/2} s(t)\sin 2nt\,dt \tag{1.31}$$

Since $s(t)$ is an even function of time (i.e., $s(t) = s(-t)$), $s(t) \sin 2nt$ is an odd function (i.e., $s(t) \sin 2\pi nt = -s(t) \sin 2\pi n(-t)$), and the integral from $-T/2$ to $+T/2$ will be zero.

$$b_n = 0$$

Finally,

$$s(t) = \frac{2}{\pi} + \sum_{n=1}^{\infty} \left\{ \frac{2}{\pi} \left[\frac{(-1)^{n+1}}{2n-1} + \frac{(-1)^{n}}{2n+1} \right] \cos 2nt \right\} \tag{1.32}$$

Writing out the first few terms of this series, we have

$$s(t) = \frac{2}{\pi} \left[1 + \frac{2}{3} \cos 2t - \frac{2}{15} \cos 4t + \frac{2}{35} \cos 6t - \cdots \right] \tag{1.33}$$

We note that the Fourier Series in Eq. (1.33) is also the expansion of the periodic function, $s_p(t)$ shown in Fig. 1.2. That is, $s_p(t)$ is a periodic function which is equal to $s(t)$ over one period.

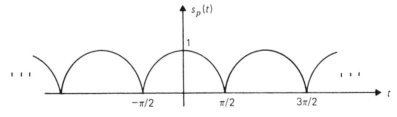

FIGURE 1.2 $s_p(t)$ representing Eq. (1.33).

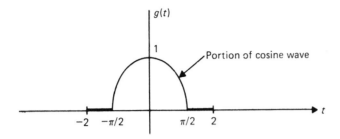

FIGURE 1.3 $g(t)$ similar to $s(t)$ of
Illustrative Example 1.2.

Suppose we now calculated the Fourier Series of $g(t)$ shown in Fig. 1.3, to apply in the interval $-2 < t \leq +2$.

The result will clearly be different from that of Example 1.2. One is readily convinced of this difference once it is noted that the frequencies of the various sines and cosines will be different from those of Example 1.2. However, for t between $-\pi/2$ and $+\pi/2$, both series represent the same function. Both series do not, however, represent $s_p(t)$ of Fig. 1.2. The periodic function corresponding to $g(t)$, denoted as $g_p(t)$, is sketched in Fig. 1.4.

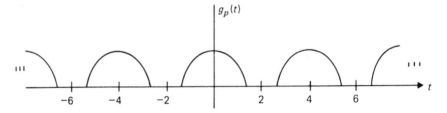

FIGURE 1.4   Periodic repetition of $g(t)$ of Fig. 1.3.

The series expansion of a function in a finite interval is therefore *not* unique. This should not upset anybody. In fact, there are situations in which one actually takes advantage of this fact in order to choose the type series which simplifies the results. (Solution of partial differential equations by separation of variables is one example.)

Illustrative Example 1.3

It is desired to approximate the time function,

$$s(t) = |\cos t|$$

by a constant. This constant is to be chosen so as to minimize the error. The error is defined as the average of the square difference between $s(t)$ and the approximating constant. Find the constant.

Solution

$$s(t) = |\cos t|$$

Approximation to $s(t) \overset{\Delta}{=} C$

Square error, $e^2(t) = [\,|\cos t| - C\,]^2$

Find the average square error by integrating,

$$\{e^2(t)\}_{\text{avg}} = \lim_{T \to \infty} \frac{1}{T} \int_{-T/2}^{T/2} [\,|\cos t| - C\,]^2 \, dt \qquad (1.34)$$

We could evaluate this integral and differentiate with respect to C in order to minimize the quantity, or we can be more devious and borrow a result from orthogonal vector spaces. Recall that, "Even though the orthogonal set used to approximate a general vector does not contain a sufficient number of elements to describe the vector exactly, the weighting coefficients are chosen as if the set were sufficient!"

The above problem is simply just such a case where we must approximate $s(t)$ by the first term in its Fourier Series expansion. The best value to choose for the constant, C, is the a_0, or constant term in the Fourier Series expansion. This expansion was found in Illustrative Example 1.2, where the value of a_0 was $2/\pi$. The function, $s(t)$, and its approximation are shown in Fig. 1.5.

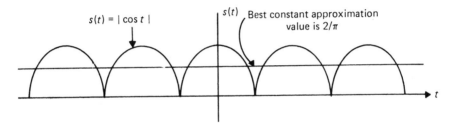

FIGURE 1.5   Result of Illustrative Example 1.3.

In summary, if we wish to approximate $|\cos t|$ by a constant so as to minimize the mean square error, the best value of the constant to choose is $2/\pi$.

1.4   COMPLEX FOURIER SPECTRUM (LINE SPECTRUM)

To each value of n in the complex Fourier Series representation of a time function, we have assigned a complex weighting factor, c_n. We can plot these c_n as a function of "n." Note that this really requires two graphs since the c_n are, in general, complex numbers. That is, we can have one plot of the magnitude of c_n and one plot of the phase. Alternatively, we could graph real and imaginary parts. We further note that this graph would be discrete. That is, it only has non-zero value for discrete values of the abscissa (e.g., $c_{1/2}$ has no meaning).

A more meaningful quantity to plot as the abscissa would be n times f_0, a quantity corresponding to the frequency of the complex exponential for which c_n is a weighting coefficient. This plot of c_n vs. $n f_0$ is called the "Complex Fourier Spectrum."

___

Illustrative Example 1.4

Find the Complex Fourier Spectrum of a full wave rectified (absolute value of) cosine wave,

$$s(t) = |\cos t|$$

as shown in Fig. 1.6.

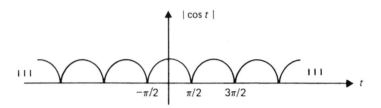

FIGURE 1.6 $s(t)$ for Illustrative Example 1.4.

Solution

In order to find the Fourier Spectrum, we must first find the exponential (complex) Fourier Series expansion of this waveform.

As in Illustrative Example 1.2, $f_0 = 1/\pi$. Instead of evaluating the c_n from

$$c_n = \frac{1}{T} \int_{-T/2}^{T/2} s(t) e^{-j2nt} \, dt$$

we can use the results of Example 1.2 to find the Complex Fourier Series directly. Recall that the trigonometric Fourier Series expansion was found to be

$$s(t) = \frac{2}{\pi} + \sum_{n=1}^{\infty} a_n \cos 2nt$$

where

$$a_n = \frac{2}{\pi} \left[\frac{(-1)^{n+1}}{2n-1} + \frac{(-1)^n}{2n+1} \right]$$

We note that Euler's identity for the cosine,

$$\cos 2nt = \frac{e^{j2nt} + e^{-j2nt}}{2} \tag{1.35}$$

can be used to go from the trigonometric to the exponential form of the expansion,

$$s(t) = \frac{2}{\pi} + \sum_{n=1}^{\infty} \frac{a_n}{2} e^{j2nt} + \sum_{n=1}^{\infty} \frac{a_n}{2} e^{-j2nt}$$

$$= \frac{2}{\pi} + \sum_{n=1}^{\infty} \frac{a_n}{2} e^{j2nt} + \sum_{n=-\infty}^{-1} \frac{a_{-n}}{2} e^{+j2nt} \qquad (1.36)$$

where, in the second line, we have simply made a change of variables. Comparing this to Eq. (1.25), we see that

$$c_0 = \frac{2}{\pi}$$

$$c_n = \frac{a_n}{2} \qquad (n > 0)$$

$$c_n = \frac{a_{-n}}{2} \qquad (n < 0)$$

Writing several terms, we have

$$c_1 = c_{-1} = \tfrac{1}{2}a_1 = \frac{2}{3\pi}$$

$$c_2 = c_{-2} = \tfrac{1}{2}a_2 = -\frac{2}{15\pi}$$

The series is of the form

$$s(t) = \frac{2}{\pi} + \frac{2}{3\pi} e^{j2t} + \frac{2}{3\pi} e^{-j2t} - \frac{2}{15\pi} e^{j4t} - \frac{2}{15\pi} e^{-j4t}, \dots \qquad (1.37)$$

The Fourier Spectrum is simply a sketch of c_n vs. nf_0, as shown in Fig. 1.7 for this example.

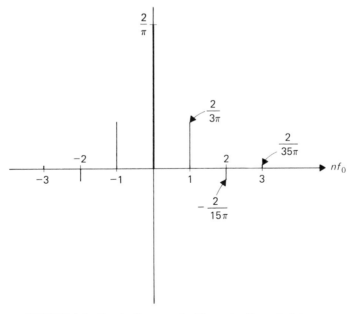

FIGURE 1.7   Fourier Spectrum for Illustrative Example 1.4.

Note that only one plot is necessary since, in this particular example, the c_n were all *real* numbers.

1.5   THE FOURIER TRANSFORM

The vast majority of interesting signals extend for all time and are nonperiodic. One would certainly not go through any great effort to transmit a periodic wave since all of the information is contained in one period. Instead, one could either transmit the signal over one period only, or transmit the values of the Fourier Series coefficients in the form of a list. The question therefore arises as to whether or not we can write a "Fourier-type" series for a nonperiodic signal.

A non-periodic signal can be viewed as a limiting case of a periodic signal, where the period of the signal approaches infinity. If the period, T, approaches infinity, the Fourier Series summation representation of $s(t)$ becomes an integral. In this manner, we could develop the Fourier Integral theory.

In order to avoid the limiting processes required to go from Fourier Series to Fourier Integral, this text will take an axiomatic approach. That is, we will *define* the Fourier Transform, and then show that it is extremely useful. There need be no loss in motivation by approaching the transform in this "pull out of a hat" manner, since its extreme versatility will become rapidly obvious.

You will recall that a common everyday *function* is actually a set of rules which substitutes one number for another number. That is, $s(t)$ is a set of rules which assigns a number, $s(t)$, to any number, t, in the domain. In a similar manner, a *transform* is a set of rules which substitutes one function for another function. We *define* one particular transform as follows:

$$S(f) \stackrel{\Delta}{=} \int_{-\infty}^{\infty} s(t)e^{-j2\pi ft} \, dt \qquad (1.38)$$

Since "t" is a dummy variable in integration, the result of the integral evaluation (after the limits are plugged in) is not a function of t, but only a function of the frequency, f. We have therefore given a rule that assigns to every function of t (with some broad restrictions required to make the integral of Eq. (1.38) converge) a function of f.

The extremely significant Fourier Transform Theorem (which we shall not prove since it involves limit arguments, as may have been expected . . . conservation of difficulty) states that, given the Fourier Transform of a time function, the original time function can always be uniquely recovered. The transform is *unique*! Either $s(t)$ or its transform, $S(f)$, uniquely characterizes a function. This is very crucial!!! Were it not true, the transform would be useless.

An example of a useless transform (the Roden Transform) follows:

To every function, s(t), assign the function

$$R(f) = f^2 + 1.3$$

This transform defines a function of "f" for every function of "t". The reason it has not become famous is that, among other factors, it is not unique. Given that the Roden Transform of a time function is $f^2 + 1.3$, you have not got a prayer of finding the $s(t)$ which led to that transform.

Actually, the Fourier Transform Theorem goes one step further than stating uniqueness. It gives the rule for recovering $s(t)$ from its Fourier Transform. The rule exhibits itself as an integral, and is almost of the same form as the original transform rule. That is, given $S(f)$, one can recover $s(t)$ by evaluating the following integral,

$$s(t) = \int_{-\infty}^{\infty} S(f)e^{j2\pi ft}\, dt \tag{1.39}$$

Equation (1.39) is sometimes referred to as "the inverse transform of $S(f)$." It follows that this is also unique.

We will usually use the same letter for the time function and its corresponding transform, the capital version being used for the transform. That is, if we have a time function called $g(t)$, we shall call its transform, $G(f)$. In cases where this is not possible, we find it necessary to adopt some alternative notational forms to associate a time function with its transform. The script capital "$\mathscr{F}$" and "$\mathscr{F}^{-1}$" are often used to denote taking the transform, or the inverse transform respectively. Thus, if $S(f)$ is the transform of $s(t)$, we can write

$$\mathscr{F}[s(t)] = S(f)$$

$$\mathscr{F}^{-1}[S(f)] = s(t)$$

A double ended arrow is also often used to relate a time function to its transform, sometimes known as a transform pair. Thus we would write

$$s(t) \leftrightarrow S(f)$$

or

$$S(f) \leftrightarrow s(t)$$

There are infinitely many unique transforms.[3] Why then has the Fourier Transform achieved such widespread fame and use? Certainly it must possess properties which make it far more useful than any other transform.

Indeed, we shall presently discover that the Fourier Transform is useful in a way which is analogous to the common logarithm. (Remember them from high school?) You may recall that, in order to multiply two numbers together, one often found the logarithm of each of the numbers, added the logs, and then found the number corresponding to the resulting logarithm. One goes through all of this trouble in order to avoid multiplication (a frightening prospect to high school students).

---

[3] As two examples, consider either time scaling or multiplication by a constant. That is, define $S_1(f) = s(2f)$ or $S_2(f) = 2s(f)$. For example, if $s(t) = \sin t$, $S_1(f) = \sin 2f$ and $S_2(f) = 2 \sin f$. The extension to an infinity of possible transform rules should be obvious.

$$a \quad \times \quad b \quad = \quad c$$
$$\downarrow \quad \downarrow \quad \downarrow \quad\quad \uparrow$$
$$\log(a) + \log(b) = \log(c)$$

There is an operation which must often be performed between two time functions which we call *convolution*. It is enough to scare even those few who are not frightened by multiplication. It will now be shown that, if the Fourier Transform of each of the two time functions is first found, an operation can be performed upon the transforms which corresponds to convolution of the time functions, but is considerably simpler. That operation, which corresponds to convolution of the two time functions, is multiplication of the two transforms. (Multiplication is no longer difficult once one graduates from high school.) Thus, we will multiply the two transforms together, and then find the time function corresponding to the resulting transform. It is now time for us to determine what this horrible operation of convolution involves.

1.6   CONVOLUTION

The *convolution* of two time functions, $r(t)$ and $s(t)$, is defined by the following integral operation,

$$r(t)*s(t) \overset{\Delta}{=} \int_{-\infty}^{\infty} r(\tau)s(t-\tau)\, dt = \int_{-\infty}^{\infty} s(\tau)r(t-\tau)\, dt \qquad (1.40)$$

The asterisk notation is conventional and is read, "$r(t)$ convolved with $s(t)$." The second integral in Eq. (1.40) results from a change of variables and states that convolution is commutative. That is, $r(t)*s(t) = s(t)*r(t)$.

The reason that one would ever want to perform such an operation will be deferred to Chapter 2. Suffice it to say that this operation is basic to almost any linear system we will study.

Note that the convolution of two functions of "t" is, itself, a function of "t" since "τ" is a dummy variable which is integrated out. The integral of Eq. (1.40) is, in general, very difficult to evaluate in closed form. An apparently simple example will back up this statement.

Illustrative Example 1.5

Evaluate the convolution of $r(t)$ with $s(t)$, where $r(t)$ and $s(t)$ are the square pulses shown in Fig. 1.8.

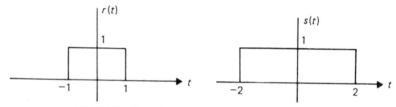

FIGURE 1.8 $r(t)$ and $s(t)$ for Illustrative Example 1.5.

Solution

We note that the functions can be written in the form

$$r(t) = U(t+1) - U(t-1)$$
$$s(t) = U(t+2) - U(t-2)$$

where $U(t)$ is the unit step function defined by

$$U(t) = \begin{cases} 1 & t > 0 \\ 0 & t < 0 \end{cases}$$

The convolution is defined by

$$r(t)*s(t) \stackrel{\Delta}{=} \int_{-\infty}^{\infty} r(\tau)\, s(t-\tau)\, d\tau$$

We see that

$$r(\tau) = U(\tau+1) - U(\tau-1) \qquad (1.41)$$

and

$$s(t-\tau) = U(t-\tau+2) - U(t-\tau-2)$$
$$r(\tau)s(t-\tau) = U(\tau+1)U(t-\tau+2) - U(\tau+1)U(t-\tau-2)$$
$$-U(\tau-1)U(t-\tau+2) + U(\tau-1)U(t-\tau-2) \qquad (1.42)$$

Therefore, breaking the integral into parts,

$$r(t)*s(t) = \int_{-\infty}^{\infty} U(\tau+1)U(t-\tau+2)\, d\tau$$
$$- \int_{-\infty}^{\infty} U(\tau+1)U(t-\tau-2)\, d\tau$$
$$- \int_{-\infty}^{\infty} U(\tau-1)U(t-\tau+2)\, d\tau$$
$$+ \int_{-\infty}^{\infty} U(\tau-1)U(t-\tau-2)\, d\tau \qquad (1.43)$$

We now note that $U(\tau+1)$ is equal to zero for $\tau < -1$, and $U(\tau-1)$ is zero for $\tau < 1$. Taking this into account, the limits of integration in Eq. (1.43) can be reduced to yield

$$r(t)*s(t) = \int_{-1}^{\infty} U(t-\tau+2)\, d\tau - \int_{-1}^{\infty} U(t-\tau-2)\, d\tau$$
$$- \int_{1}^{\infty} U(t-\tau+2)\, d\tau + \int_{1}^{\infty} U(t-\tau-2)\, d\tau \qquad (1.44)$$

To derive this, we have replaced one of the step functions by its value, unity, in the range in which this applies. We now try to evaluate each integral separately. Note that

$$U(t-\tau+2) = 0 \qquad (\tau > t+2)$$

and

$$u(t-\tau+2)=0 \qquad (\tau > t+2)$$

$$U(t-\tau-2)=0 \qquad (\tau > t-2)$$

Using these facts, we have

$$\int_{-1}^{\infty} U(t-\tau+2)\,d\tau = \int_{-1}^{t+2} d\tau = t+3 \tag{1.45}$$

provided that $t+2 > -1$, or equivalently, $t > -3$. Otherwise, the integral evaluates to zero. Likewise, if $t-2 > -1$, that is, $t > 1$,

$$\int_{-1}^{\infty} U(t-\tau-2)\,d\tau = \int_{-1}^{t-2} d\tau = t-1 \tag{1.46}$$

If $t+2 > 1$, that is, $t > -1$,

$$\int_{1}^{\infty} U(t-\tau+2)\,d\tau = \int_{1}^{t+2} d\tau = t+1 \tag{1.47}$$

If $t-2 > 1$, that is, $t > 3$,

$$\int_{1}^{\infty} U(t-\tau-2)\,d\tau = \int_{1}^{t-2} d\tau = t-3 \tag{1.48}$$

Using these four results, we find that Eq. (1.44) becomes

$$r(t)*s(t) = (t+3)U(t+3) - (t-1)U(t-1) - (t+1)U(t+1)$$
$$+ (t-3)U(t-3) \tag{1.49}$$

These four terms, together with their sum, are sketched in Fig. 1.9. From this modest example,

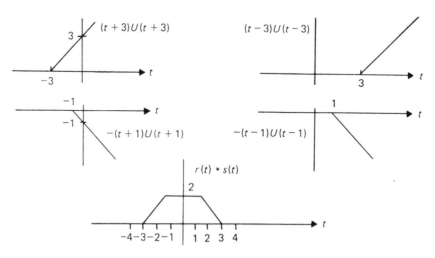

FIGURE 1.9   Result of Illustrative Example 1.5.

we can see that, if either $r(t)$ or $s(t)$ contains step functions, the evaluation of the convolution becomes quite involved.

Graphical Convolution

The claim is made that, for simple (what we mean by simple should be clear at the end of this section) $r(t)$ and $s(t)$, the result of the convolution can be obtained *almost* by inspection. Even in cases where $r(t)$ and $s(t)$ are quite complex, or even not precisely known, certain observations can be made about their convolution without actually performing the detailed calculations. In many communications applications, these general observations will be sufficient, and the exact convolution will not be required.

This inspection procedure is known as *graphical convolution*. We shall arrive at the technique by examining the definition of convolution.

$$r(t)*s(t) \stackrel{\Delta}{=} \int_{-\infty}^{\infty} r(\tau)s(t - \tau)\,d\tau$$

One of the original functions is $r(\tau)$, where the independent variable is now called "τ."

The mirror image of $s(\tau)$ is represented by $s(-\tau)$, that is, $s(\tau)$ flipped around the vertical axis.

The definition now tells us that for a given "t," we form $s(t - \tau)$, which represents the function $s(-\tau)$ shifted to the right by "t." We then take the product,

$$r(\tau)s(t - \tau)$$

and integrate this product (i.e., find the area under it) in order to find the value of the convolution for that particle value of t.

This procedure is illustrated in Fig. 1.10 for the two functions of Fig. 1.8, Illustrative Example 1.5. The ideal way to demonstrate graphical convolution is with an animated motion picture which shows the two functions, one of which is moving across the τ-axis. The motion picture would show the two functions overlapping by varying amounts as t changes.

Unfortunately, the constraints of book publishing do not allow us to present a motion picture herein. Instead, we illustrate a stroboscopic view of such a motion picture, that is, we show a number of frames.

Figure 1.10 shows twelve separate frames of the motion picture. For this particular example, it is not obvious that $s(t)$ was flipped to form the mirror image, since the original $s(t)$ was an even function of t.

It is necessary to interpolate between each pair of values of t shown in the figure. Note that it is the area under the product which represents the value of the convolution. Figure 1.11 plots this area as a series of points, with the straight line interpolation between these pairs of points. It should not be surprising to note that the result is piecewise linear. That is, the interpolation results in straight lines. This is true since the convolution becomes an integration of a constant, which results in a ramp function. With practice, only a few points are needed to plot the resulting convolution. It is comforting to compare Fig. 1.11 with Fig. 1.9 and to note that we obtained the same answer in both cases.

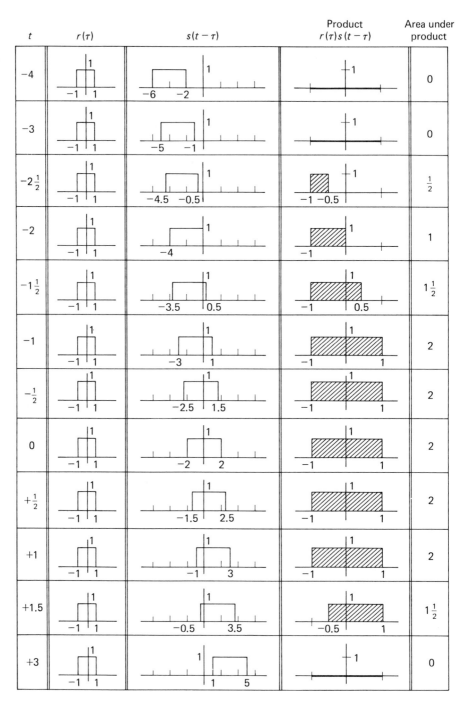

FIGURE 1.10 Illustration of graphical convolution of Illustrative Example 1.5.

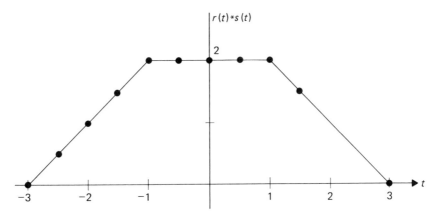

FIGURE 1.11 Convolution result from Fig. 1.10.

Illustrative Example 1.6

We wish to find $r(t)*r(t)$, where the only information we have about $r(t)$ is that it is zero for $|t| > \alpha$. That is, $r(t)$ is limited to the range between $t = -\alpha$ and $t = +\alpha$. A typical, but not exclusive, $r(t)$ is sketched in Fig. 1.12.

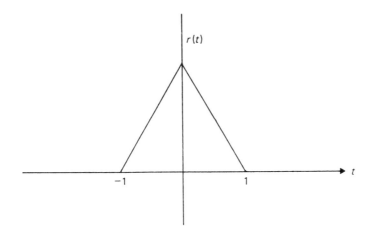

FIGURE 1.12 Typical $r(t)$ for Illustrative Example 1.6.

Solution

Not knowing $r(t)$ exactly, we certainly cannot find $r(t)*r(t)$ exactly. To see how much information we can obtain concerning this convolution, we attempt graphical convolution. Figure 1.13 is a sketch of $r(\tau)$ and $r(t - \tau)$.

FIGURE 1.13 $r(\tau)$ and $r(t - \tau)$ for representative $r(t)$ of Fig. 1.12.

We note that, as "t" increases from zero, the two functions illustrated have less and less of the τ-axis in common. When "t" reaches $+2\alpha$, the two functions separate. That is, at $t = 2\alpha$, we have

$$r(\tau)r(2\alpha - \tau) = 0$$

as sketched in Fig. 1.14.

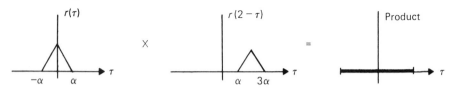

FIGURE 1.14 Convolution product of Illustrative Example 1.6 goes to zero at $t = 2\alpha$.

The two functions continue to have nothing in common for all $t > 2\alpha$. Likewise, for negative "t" one can see that the product of the two functions is zero as long as $t < -2\alpha$. That is, $r(-\tau)$ can move to the left by a distance of 2α units before it separates from $r(\tau)$.

Therefore, although we don't know $r(t) * r(t)$ exactly, we do know what range it occupies along the t-axis. We find that it is zero for $|t| > 2\alpha$. A possible form of this is sketched in Fig. 1.15. We again emphasize that the sketch does not intend to indicate the shape of this function.

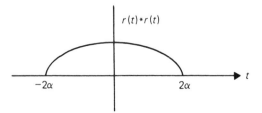

FIGURE 1.15 Typical result of the convolution in Illustrative Example 1.6.

It only shows what portion of the t-axis it occupies. The results of this particular example will be used later in the study of modulation systems.

Illustrative Example 1.7

Use graphical techniques to convolve the two functions shown in Fig. 1.16.

Solution

We again show a stroboscopic picture of eleven frames of the motion picture of convolution. This sampled view is shown in Fig. 1.17.

FIGURE 1.16 Two functions for convolution in Illustrative Example 1.7.

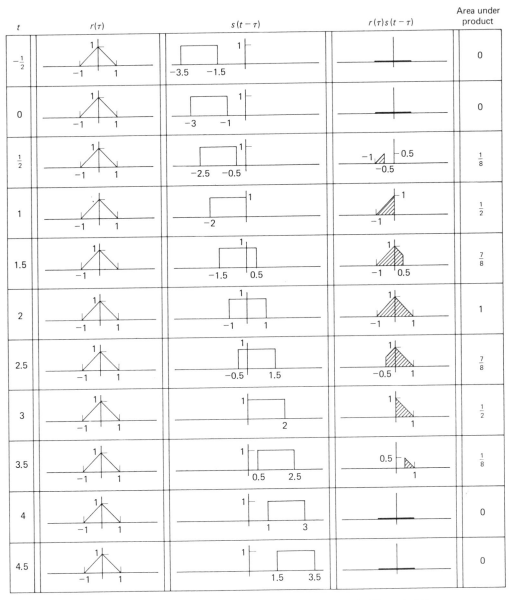

t	$r(\tau)$	$s(t-\tau)$	$r(\tau)s(t-\tau)$	Area under product
$-\frac{1}{2}$				0
0				0
$\frac{1}{2}$				$\frac{1}{8}$
1				$\frac{1}{2}$
1.5				$\frac{7}{8}$
2				1
2.5				$\frac{7}{8}$
3				$\frac{1}{2}$
3.5				$\frac{1}{8}$
4				0
4.5				0

FIGURE 1.17 Graphical convolution for Illustrative Example 1.8.

The samples of the convolution, together with the interpolation between these samples, are shown in Fig. 1.18. The result is parabolic, since the function being integrated can be

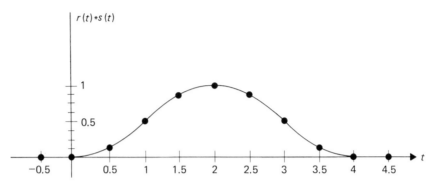

FIGURE 1.18   Result of convolution for Illustrative Example 1.7.

thought of as a ramp function. Again, with practice, this result could be sketched with only a few key points. For example, observing the shaded region of Fig. 1.17, we can see that the result increases slowly at first and then accelerates until the right edge of the square reaches the origin. The rate of increase then starts to get smaller again, as each incremental move to the right adds a shorter and shorter strip to the product.

Now that we have a feel for the operation known as convolution, let us return to our study of the Fourier Transform.

As promised, the *convolution theorem* states that the Fourier Transform of a time function which is the convolution of two time functions is equal to the product of the two corresponding Fourier Transforms. That is, if

$$r(t) \leftrightarrow R(f)$$

and

$$s(t) \leftrightarrow S(f)$$

then

$$r(t)*s(t) \leftrightarrow R(f)S(f) \qquad (1.50)$$

The proof of this is rather direct. Simply evaluate the Fourier Transform of $r(t)*s(t)$.

$$\mathscr{F}\left[r(t)*s(t)\right] \overset{\Delta}{=} \int_{-\infty}^{\infty} e^{-j2\pi ft} \left[\int_{-\infty}^{\infty} r(\tau)s(t-\tau)\, d\tau\right] dt$$

$$= \int_{-\infty}^{\infty} r(\tau) \left[\int_{-\infty}^{\infty} e^{-j2\pi ft}s(t-\tau)\, dt\right] d\tau \qquad (1.51)$$

If we now make a change of variables in the inner integral and let $t - \tau = k$, we have

$$\mathscr{F}[r(t)*s(t)] = \int_{-\infty}^{\infty} r(\tau) \left[\int_{-\infty}^{\infty} e^{-j2\pi ft} s(k) e^{-j2\pi fk} dk \right] d\tau$$

$$= \int_{-\infty}^{\infty} r(\tau) e^{-j2\pi f\tau} \left[\int_{-\infty}^{\infty} s(k) e^{-j2\pi fk} dk \right] d\tau \quad (1.52)$$

The integral in the square brackets is simply $S(f)$. Since $S(f)$ is not a function of "τ," it can be pulled out of the outer integral. This yields

$$\mathscr{F}[r(t)*s(t)] = S(f) \int_{-\infty}^{\infty} r(\tau) e^{-j2\pi f\tau} d\tau$$

$$= S(f)R(f) \quad \text{[q.e.d.]} \quad (1.53)$$

Convolution is an operation performed between two functions. They need not be functions of the independent variable "t." We could just as easily have convolved two Fourier Transforms together to get a third function of "f."

$$H(f) = R(f)*S(f) = \int_{-\infty}^{\infty} R(k)S(f-k) \, dk \quad (1.54)$$

Since the integral defining the Fourier Transform and that yielding the inverse transform are quite similar, one might guess that convolution of two transforms corresponds to multiplication of the two corresponding time functions. Indeed, one can prove, exactly in an analogous way to the above proof, that

$$R(f)*S(f) \leftrightarrow r(t)s(t) \quad (1.55)$$

In order to prove this, simply calculate the inverse Fourier Transform of $R(f)*S(f)$.

Equation (1.50) is sometimes called the "time convolution theorem" and Eq. (1.55), the "frequency convolution theorem."

Illustrative Example 1.8

Use the convolution theorem in order to evaluate the following integral,

$$\int_{-\infty}^{\infty} \frac{\sin 3\tau}{\tau} \frac{\sin (t - \tau)}{t - \tau} \, d\tau$$

Solution

We recognize that the above integral represents the convolution of the following two time functions,

$$\frac{\sin 3t}{t} * \frac{\sin t}{t}$$

The transform of the integral is therefore the product of the transforms of the above two time functions. These two transforms may be found in Appendix II. The appropriate transforms and their product are sketched in Fig. 1.19.

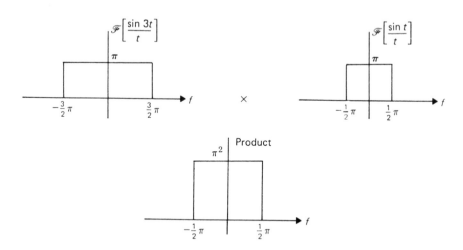

FIGURE 1.19   Transforms and product for Illustrative Example 1.8.

The time function corresponding to the convolution is simply the inverse transform of this product. This is easily seen to be

$$\frac{\pi \sin t}{t}$$

Note that when $(\sin t)/t$ is convolved with $(\sin 3t)/t$, the only change that takes place is the addition of a scale factor, π. In fact, if $(\sin t)/t$ had been convolved with $(\sin 3t)/\pi t$, it would not have changed at all. This surprising result is no accident. We will see, in the next chapter, that there are entire classes of functions which remain unchanged after convolution with $(\sin t)/\pi t$. If this were not true, many of the most basic communication systems could never function.

Parseval's Theorem

There is little similarity between a function and its Fourier Transform. That is, the waveshapes of the two functions are, in general, completely different. Certain relationships do exist, however, between the "energy" of a time function and the "energy" of its transform. Here, we use *energy* to denote the integral of the square of the function. We will explore this concept in detail in Sect. 2.9. Thus, if we knew the transform of a time function, and only wished to know the energy of the time function, such relationships would eliminate the need to evaluate the inverse transform.

Parseval's Theorem is such a relationship. It is easily derived from the frequency convolution theorem. Starting with that theorem, we have

$$r(t)s(t) \leftrightarrow R(f)*S(f)$$

$$\mathscr{F}[r(t)s(t)] = \int_{-\infty}^{\infty} r(t)s(t)e^{-j2\pi ft}\, dt$$

$$= \int_{-\infty}^{\infty} R(k)S(f-k)\, dk \tag{1.56}$$

Since the above equality holds for all values of f, we can take the special case of $f = 0$. For this value of f, Eq. (1.56) becomes

$$\int_{-\infty}^{\infty} r(t)s(t)\, dt = \int_{-\infty}^{\infty} R(k)S(-k)\, dk \tag{1.57}$$

Equation (1.57) is one form of Parseval's formula. It can be made to relate to energy by further taking the special case of

$$s(t) = r^*(t)$$

where the asterisk (*) represents complex conjugate.

We can show that, if $s(t) = r^*(t)$, then

$$S(f) = R^*(-f)$$

We illustrate this last relationship by finding the transform of $r(t)$, evaluating it at "$-f$," and taking the conjugate,

$$R(f) \overset{\Delta}{=} \int_{-\infty}^{\infty} r(t)e^{-j2\pi ft}\, dt$$

$$R(-f) = \int_{-\infty}^{\infty} r(t)e^{j2\pi ft}\, dt$$

$$R^*(-f) = \int_{-\infty}^{\infty} r^*(t)e^{-j2\pi ft}\, dt \overset{\Delta}{=} \mathscr{F}[r^*(t)] \tag{1.58}$$

Plugging these relationships into Parseval's formula (Eq. 1.57), we get

$$\int_{-\infty}^{\infty} |r^2(t)|\, dt = \int_{-\infty}^{\infty} |R^2(f)|\, df \tag{1.59}$$

We have used the fact that the product of a function with its complex conjugate is equal to the magnitude of the function, squared.

Equation (1.59) shows that the energy under the time function is equal to the energy under its Fourier Transform. The integral of the magnitude squared is often called the energy of the signal since this would be the amount of energy, in watt-seconds, dissipated in a 1-ohm resistor if the signal represented the voltage across the resistor (or the current through it).

1.7   PROPERTIES OF THE FOURIER TRANSFORM

We shall now illustrate some of the more important properties of the Fourier Transform. One can certainly go through technical life without making use of any of these properties, but to do so would involve considerable repetition and extra work. The properties allow us to prove something once and then to use the result for a variety of applications. They also allow us to predict the behavior of various systems.

Real/Imaginary—Even/Odd

The following table summarizes properties of the Fourier Transform based upon observations made upon the time function.

Property	Time Function	Fourier Transform
A	Real	Real part even; imaginary part odd
B	Real and even	Real and even
C	Real and odd	Imaginary and odd
D	Imaginary	Real part odd; imaginary part even
E	Imaginary and even	Imaginary and even
F	Imaginary and odd	Real and odd

Proof

The Fourier Transform defining integral can be expanded using Euler's Identity as follows:

$$S(f) = \int_{-\infty}^{\infty} s(t) e^{-j2\pi ft} \, dt$$

$$= \int_{-\infty}^{\infty} s(t) \cos(2\pi ft) \, dt - j \int_{-\infty}^{\infty} s(t) \sin(2\pi ft) \, dt$$

$$= I_1 - jI_2$$

I_1 is an even function of f, since when f is replaced with $-f$, the function does not change. Similarly, I_2 is an odd function of f.

If $s(t)$ is first assumed to be real, I_1 becomes the real part of the transform, and $-I_2$ is the imaginary part. Thus, property A is proven.

If in addition to being real, $s(t)$ is even, then $I_2 = 0$. This is true since the integrand in I_2 would be odd (the product of even and odd) and would integrate to zero. Thus, property B is proven.

If $s(t)$ is now real and odd, the same argument applies, but $I_1 = 0$. This proves property C.

Now we let $s(t)$ be imaginary. I_1 now becomes the imaginary part of the transform, and I_2 is the real part. From this simple observation, properties D, E, and F are easily verified.

Illustrative Example 1.9

Evaluate the Fourier Transform of $s(t)$, where

$$s(t) = \begin{cases} A & \text{for } |t| < \alpha \\ 0 & \text{otherwise} \end{cases}$$

as shown in Fig. 1.20.

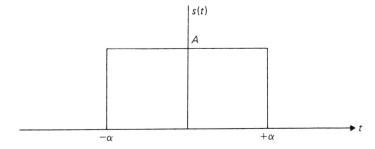

FIGURE 1.20 $s(t)$ for Illustrative Example 1.9.

Solution

From the definition of the Fourier Transform,

$$S(f) \overset{\Delta}{=} \int_{-\infty}^{\infty} s(t)e^{-j2\pi dt}\, dt$$

$$= \int_{-\alpha}^{\alpha} A e^{-j2\pi ft}\, dt$$

$$= \frac{-e^{-j2\pi ft}}{j2\pi f}\Bigg|_{-\alpha}^{\alpha}$$

$$= A\,\frac{e^{j2\pi f\alpha} - e^{-j2\pi f\alpha}}{j2\pi f}$$

$$= A\,\frac{\sin 2\pi f\alpha}{\pi f} \tag{1.60}$$

This transform is sketched in Fig. 1.21.

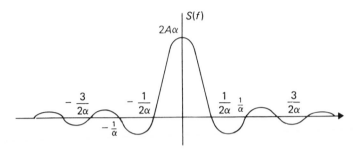

FIGURE 1.21 Transform of $s(t)$ of Illustrative Example 1.9.

Functions of the type illustrated in Fig. 1.21 are common in communication studies. To avoid having to constantly rewrite this type of function, the *sinc* function is defined as

$$\text{sinc}\,(x) \overset{\Delta}{=} \frac{\sin x}{x}$$

In terms of this, Eq. (1.60) becomes

$$S(f) = 2A\alpha\,\frac{\sin 2\pi f\alpha}{2\pi f\alpha} = 2A\alpha\,\text{sinc}(2\pi f\alpha)$$

The given time function in Illustrative Example 1.9 was real and even. Thus, it is not surprising that (see Property B) the resulting transform was real and even.

Time Shift

The Fourier Transform of a shifted time function is equal to the transform of the original time function multiplied by a complex exponential.

$$e^{-j2\pi f t_0}S(f) \leftrightarrow s(t - t_0) \tag{1.61}$$

Proof

The proof follows directly from evaluation of the transform of $s(t - t_0)$.

$$\mathscr{F}\,[s(t - t_0)] \overset{\Delta}{=} \int_{-\infty}^{\infty} s(t - t_0)e^{-j2\pi f t}\,dt \tag{1.62}$$

Changing variables, we let $t - t_0 = \tau$. For finite t_0, the transform becomes

$$\mathscr{F}\,[s(t - t_0)] = \int_{-\infty}^{\infty} s(\tau)e^{-j2\pi f(\tau + t_0)}\,d\tau$$

$$= e^{-j2\pi f t_0} \int_{-\infty}^{\infty} s(\tau)e^{-j2\pi f \tau} \, d\tau$$

$$= e^{-j2\pi f t_0} S(f) \tag{1.63}$$

Illustrative Example 1.10

Find the Fourier Transform of $s(t)$,

$$s(t) = \begin{cases} 1 & \text{for } 0 < t \le 2 \\ 0 & \text{otherwise} \end{cases}$$

as sketched in Fig. 1.22.

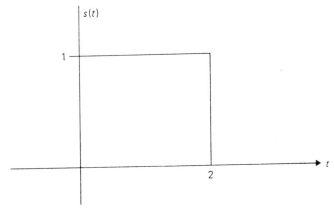

FIGURE 1.22 $s(t)$ for Illustrative Example 1.10.

Solution

From the definition of the Fourier Transform,

$$S(f) \overset{\Delta}{=} \int_{-\infty}^{\infty} s(t)e^{-j2\pi f t} \, dt$$

$$= \int_{0}^{2} e^{-j2\pi f t} \, dt$$

$$= \frac{-e^{-j2\pi f t}}{j2\pi f} \Big|_{0}^{2} = \frac{1}{j2\pi f}[1 - e^{-j4\pi f}]$$

$$= \frac{e^{-j2\pi f}}{j2\pi f} [e^{j2\pi f} - e^{-j2\pi f}]$$

$$S(f) = e^{-j2\pi f} \left[\frac{\sin 2\pi f}{\pi f} \right]$$

We first note that this $S(f)$ is complex as expected, since $s(t)$ is neither even nor odd.

The result of Illustrative Example 1.10 could have been derived in one step using the answer from Illustrative Example 1.9 and the time shift property. The $s(t)$ of Example 1.10 is the same as that of Example 1.9 except for a time shift of 1 sec.

Frequency Shift

The time function corresponding to a shifted Fourier Transform is equal to the time function corresponding to the unshifted transform multiplied by a complex exponential.

$$S(f - f_0) \leftrightarrow e^{j2\pi f_0 s}(t) \tag{1.64}$$

Proof

The proof follows directly from evaluation of the inverse transform of $S(f - f_0)$.

$$\mathscr{F}^{-1}[S(f - f_0)] = \int_{-\infty}^{\infty} S(f - f_0)e^{j2\pi ft} \, df \tag{1.65}$$

Changing variables, we let

$$f - f_0 = k$$

$$\mathscr{F}^{-1}[S(f - f_0)] = \int_{-\infty}^{\infty} S(k)e^{j2\pi t(k+f_0)} \, dk$$

$$= e^{j2\pi f_0 t} \int_{-\infty}^{\infty} S(k)e^{j2\pi kt} \, dk$$

$$= e^{j2\pi f_0 t} s(t) \tag{1.66}$$

Illustrative Example 1.11

Find the Fourier Transform of $s(t)$, where

$$s(t) = \begin{cases} e^{j2\pi t} & |t| < 1 \\ 0 & \text{otherwise} \end{cases}$$

Solution

This $s(t)$ is the same as that of Illustrative Example 1.9, except for a multiplying factor of $e^{j2\pi t}$. The frequency shift theorem indicates that if a function of time is multiplied by $e^{j2\pi t}$, its transform is shifted by 1 unit of frequency. We therefore take the transform found in Illustrative Example 1.9 and substitute $(f-1)$ for (f).

$$S(f) = \frac{\sin 2\pi(f-1)}{\pi(f-1)}$$

This transform is sketched in Fig. 1.23.

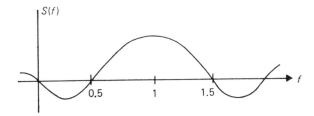

FIGURE 1.23 $S(t)$ of Illustrative Example 1.11.

Note that in this case the time function is neither even nor odd. However, the Fourier Transform turned out to be a real function. Such a situation can arise only because the time function is not real.

Modulation Theorem

The modulation theorem is very closely related to the frequency shift theorem. We treat it as a separate entity, since it forms the basis of the entire study of amplitude modulation.

We start by assuming that an $s(t)$ is given, with its associated Fourier Transform. $s(t)$ is then multiplied by a cosine waveform to yield

$$s(t)\cos 2\pi f_0 t$$

where the frequency of the cosine is f_0. The Fourier Transform of this waveform is given by

$$\mathscr{F}[s(t)\cos 2\pi f_0 t] = \tfrac{1}{2}S(f-f_0) + \tfrac{1}{2}S(f+f_0) \qquad (1.67)$$

Thus, the result of multiplying a time function by a pure sinusoid is to shift the original transform both up and down by the frequency of the sinusoid, and to cut the amplitude in half.

Proof

The proof of the modulation theorem follows directly from the frequency shift theorem. We split $\cos 2\pi f_0 t$ into its two complex exponential components (Euler's identity) to yield

$$s(t)\cos 2\pi f_0 t = \tfrac{1}{2}s(t)e^{j2\pi f_0 t} + \tfrac{1}{2}s(t)e^{-j2\pi f_0 t}$$

We can transform each of these terms individually, and Eq. (1.67) results from the frequency shift theorem.

Scaling in Time and Frequency

We are rapidly approaching the point of diminishing returns in presenting properties of the Fourier Transform. At some point, it is worth dealing with the individual properties as they arise. We shall terminate this exploration with a companion set of two properties referred to as time and frequency scaling. The usefulness of these properties arises when you take a time function or a Fourier Transform and either stretch or compress it along the horizontal axis. Thus, for example, if you already know the Fourier Transform of a pulse with width two units (as was shown in Fig. 1.20), you need not do any further calculations to find the Fourier Transform of a pulse of any other width. We will thus be able to handle both shifts and scaling, which should considerably decrease the amount of calculations necessary in later work.

Time Scaling

Suppose we already know that the Fourier Transform of $s(t)$ is $S(f)$. We wish to find the Fourier Transform of $s(at)$, where a is the real scaling factor. Thus, for example, if $a = 2$, we are compressing the function by a factor of 2 along the t-axis, and if $a = \frac{1}{2}$, we are expanding it by a factor of 2. The result can be derived directly from the Fourier Transform integral definition as follows:

$$\mathscr{F}[s(at)] = \int_{-\infty}^{\infty} s(at)e^{-j2\pi ft}\, dt$$

Now letting $at = \tau$, we have

$$\mathscr{F}[s(at)] = \int_{-\infty}^{\infty} s(\tau)e^{-j2\pi f\tau/a}\, \frac{d\tau}{a} = \frac{1}{a}\, S\left(\frac{f}{a}\right) \tag{1.68}$$

This result represents a complementary operation on the frequency axis and an amplitude scaling. Thus, for example, if $a = 2$, the time axis is being compressed. In finding the transform, we would expand the frequency axis by a factor of 2, and scale the transform by dividing by 2.

By the result of Illustrative Example 1.9, the transform of the pulse of width 2 was $(\sin 2\pi f)/\pi f$ as repeated in Fig. 1.24. Now the transform of the unit width pulse will be $\sin(\pi f)/(\pi f)$, as sketched in Fig. 1.25.

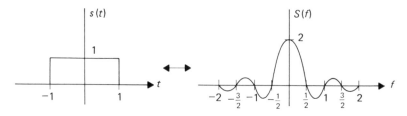

FIGURE 1.24   Fourier Transform of pulse of width 2.

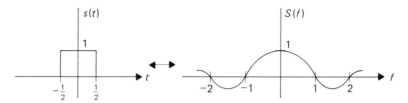

FIGURE 1.25   Fourier Transform of pulse of width 1.

Frequency Scaling

If it is already known that the Fourier Transform of $s(t)$ is $S(f)$, then the time signal which has $S(af)$ as its transform, where a is a real scaling factor, is given by

$$\mathcal{F}^{-1}[S(af)] = \frac{1}{a} s\left(\frac{t}{a}\right) \tag{1.69}$$

This is proven from the inverse Fourier Transform integral.

$$\int_{-\infty}^{\infty} S(af)e^{j2\pi ft}\, df = \int_{-\infty}^{\infty} S(\Omega)e^{j2\pi\Omega t/a}\, \frac{d\Omega}{a}$$

$$= \frac{1}{a} s\left(\frac{t}{a}\right)$$

In the second equation, a change of variables has been made with

$$af = \Omega$$

Think about how you might have proven this from the time scale property given in Eq. (1.68). (Hint: What if $a' = 1/a$ in Eq. (1.68)?).

Linearity

The final property that we wish to explore is undoubtedly the most important. We have saved it for last since it is probably the most obvious, and we can use a breathing spell before getting to the next section.

The Fourier Transform of a linear combination of time functions is a linear combination of the corresponding Fourier Transforms. That is,

$$as_1(t) + bs_2(t) \rightarrow aS_1(f) + bS_2(f) \tag{1.70}$$

where a and b are any constants.

Proof

The proof follows directly from the definition of the Fourier Transform, and from the fact that integration is a linear operation.

$$\int_{-\infty}^{\infty} [as_1(t) + bs_2(t)]e^{-j2\pi ft}\, dt = a \int_{-\infty}^{\infty} s_1(t)e^{-j2\pi ft}\, dt$$

$$+ b \int_{-\infty}^{\infty} s_2(t)e^{-j2\pi ft}\, dt$$

$$= aS_1(f) + bS_2(f)$$

Illustrative Example 1.12

Find the Fourier Transform of $s(t)$, where

$$s(t) = \begin{cases} 1 & -1 < t \le 0 \\ 2 & 0 < t \le 1 \\ 1 & 1 < t \le 2 \\ 0 & \text{otherwise} \end{cases}$$

as sketched in Fig. 1.26.

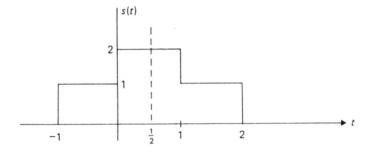

FIGURE 1.26 $s(t)$ for Illustrative Example 1.12.

Solution

Using the linearity property, we observe that $s(t)$ given above is simply the sum of the time function of Illustrative Example 1.9 with that of Illustrative Example 1.10. Therefore, the transform is given by the sum of the two transforms.

$$S(f) = \frac{\sin 2\pi f}{\pi f}[1 + e^{-j2\pi f}] \tag{1.71}$$

This transform is complete in the above form. However, we shall perform a slight manipulation to illustrate a point.

$$S(f) = \frac{\sin 2\pi f}{\pi f} e^{-j\pi f}[e^{j\pi f} + e^{-j\pi f}]$$

$$= 2\frac{\sin 2\pi f \cos(\pi f)}{\pi f}[e^{-j\pi f}] \tag{1.72}$$

We now recognize that $S(f)$ is in the form of a real, even function of "f" multiplied by $e^{-j\pi f}$.

This indicates that the corresponding time function must be the result of shifting a real, even time function by $t_0 = \frac{1}{2}$. This is indeed the case as can be easily observed from Fig. 1.26.

1.8   SINGULARITY FUNCTIONS

Suppose that we wished to evaluate the transform of a function, $g(t)$, which was equal to 1 for $|t| < a$, that is, a pulse of width $2a$. We would proceed exactly as we did in Illustrative Example 1.9, where "a" was given as "1." Alternatively, the time scaling property could be used. Either way, we would find that

$$G(f) = \frac{\sin a 2\pi f}{\pi f} \tag{1.73}$$

As a check, for $a = 1$, this reduces to the $S(f)$ found in Example 1.9. The functions $g(t)$ and $G(f)$ are sketched in Fig. 1.27.

FIGURE 1.27   A pulse and its transform.

Suppose that we now wanted to find the Fourier Transform of a constant, say $s(t) = 1$ for all t. We could consider this as the limit of the above $g(t)$ as "a" approaches infinity. We attempt this roundabout approach since the straightforward technique fails in this case. That is, plugging $s(t) = 1$ into the defining integral for the Fourier Transform yields

$$S(f) = \int_{-\infty}^{\infty} e^{-j2\pi ft}\, dt \tag{1.74}$$

This integral does not converge (i.e., it oscillates). If the limit argument is used, we find that unfortunately, the limit of the transform in Eq. (1.73) as "a" approaches infinity does not exist. We can make the general observations that, as "a" increases, the height of the main peak of $G(f)$ increases while its width decreases. Thus, it can only be timidly suggested that, in the limit, the height goes to infinity and the width to zero. This sounds like a pretty ridiculous function, and, indeed, it is not a function at all, since it is not defined at $f = 0$. If we insist upon saying anything about the Fourier Transform of a constant, we must now restructure our way of thinking.

This restructuring is begun by defining a new function which is not really a function at all. This new "function" shall be given the name, *impulse*. We will get around the fact that it is not really a function by also *defining* its behavior in every common situation. The symbol for this function is the Greek letter *delta* (δ), and the function is denoted as $\delta(t)$.

The usual, though non-rigorous, definition of the impulse function is formed by making three simple observations, two of which have already been hinted at. That is,

$$\delta(t) = 0 \qquad (t \neq 0)$$
$$\delta(t) = \infty \qquad (t = 0) \qquad\qquad (1.75\,a)$$

The third property is that the total area under the impulse function is equal to unity.

$$\int_{-\infty}^{\infty} \delta(t)\, dt = 1 \qquad\qquad (1.75\,b)$$

Since all of the area of $\delta(t)$ is concentrated at one point, the limits on the above integral can be moved in toward $t = 0$ without changing the value of the definite integral. Thus,

$$\int_{a}^{b} \delta(t)\, dt = 1$$

as long as $a < 0$ and $b > 0$.

One can also observe that the integral of $\delta(t)$ is $U(t)$, the unit step function. That is,

$$\int_{-\infty}^{t} \delta(\tau)\, d\tau = \begin{cases} 1 & t > 0 \\ 0 & t < 0 \end{cases} \triangleq U(t)$$

We called the definition comprised of the above three observations "non-rigorous." Some elementary study of singularity functions shows that these properties do not uniquely define a delta function. That is, there are other functions in addition to the impulse which satisfy the above three conditions. However, these three conditions can be used to indicate (not prove) a fourth. This fourth property serves as a unique definition of the delta function, and will represent the only property of the impulse which we ever make use of. We shall integrate the product of an arbitrary time function with $\delta(t)$.

$$\int_{-\infty}^{\infty} f(t)\delta(t)\, dt = \int_{-\infty}^{\infty} f(0)\delta(t)\, dt \qquad\qquad (1.76)$$

In Eq. (1.76) we have claimed that we could replace $f(t)$ by a constant function equal to $f(0)$ without changing the value of the integral. This requires some justification.

Suppose that we have two functions, $g_1(t)$ and $g_2(t)$, and we consider the product of each of these with a third function, $h(t)$. As long as $g_1(t) = g_2(t)$ at all values of "t" for which $h(t)$ is non-zero, then $g_1(t)h(t) = g_2(t)h(t)$. At those values of "t" for which $h(t) = 0$, the values of $g_1(t)$ and $g_2(t)$ have no effect on the product. One possible example is sketched in Fig. 1.28. For $g_1(t)$, $g_2(t)$ and $h(t)$ as shown in Fig. 1.28,

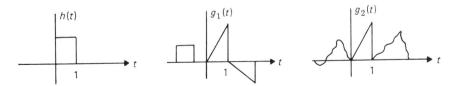

FIGURE 1.28   Example of $h(t)$, $g_1(t)$, and $g_2(t)$.

$$g_1(t)h(t) = g_2(t)h(t)$$

Returning to Eq. (1.76), we note that $\delta(t)$ is zero for all $t \neq 0$. Therefore, the product of $\delta(t)$ with any time function only depends upon the value of the time function at $t = 0$. Figure 1.29 illustrates several possible functions which will have the same product with $\delta(t)$ as does $s(t)$.

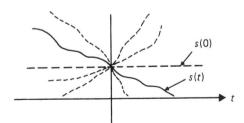

FIGURE 1.29   Several possible functions all having the same product with $\delta(t)$.

Out of the infinity of possibilities, the constant time function is a wise choice, since we can now factor it out of the integral to get

$$\int_{-\infty}^{\infty} s(t)\delta(t)\,dt = s(0)\int_{-\infty}^{\infty} \delta(t)\,dt = s(0) \qquad (1.77)$$

This is a very significant result, and is sometimes appropriately given the name of "the sifting or sampling property of the delta (impulse) function," since the impulse sifts out the value of $s(t)$ at the time, $t = 0$. Note that a great deal of information has been lost since the result depends only upon the value of $s(t)$ at one point.

In a similar manner, a change of variables yields a shifted impulse with the analogous sifting property.

$$\int_{-\infty}^{\infty} s(t)\delta(t - t_0)\, dt = \int_{-\infty}^{\infty} s(k + t_0)\delta(k)\, dk$$

$$= s(k + t_0)|_{k=0} = s(t_0) \qquad (1.78)$$

Figure 1.30 shows a sketch of $\delta(t)$ and $\delta(t - t_0)$. The upward pointing arrow is a generally accepted technique to indicate an actual value of infinity. The number next to the arrow indicates the total area under the impulse, sometimes known as its "strength." In sketching the impulse, the height of the arrow is often made equal to the strength of the impulse.

FIGURE 1.30   Pictorial representation of delta function.

Equations (1.77) and (1.78) are the only things that one must know about the impulse function. Indeed, either of these can be treated as the definition of the impulse. Note that $s(t)$ can be any time function for which $s(t_0)$ can be defined.

Illustrative Example 1.13

Evaluate the following integrals:

(a) $\displaystyle\int_{-\infty}^{\infty} \delta(t)[t^2 + 1]\, dt$

(b) $\displaystyle\int_{-1}^{2} \delta(t - 1)[t^2 + 1]\, dt$

(c) $\displaystyle\int_{3}^{5} \delta(t - 1)[t^3 + 4t + 2]\, dt$

(d) $\displaystyle\int_{-\infty}^{\infty} \delta(1 - t)[t^4 + 2]\, dt$

Solution

(a) Straightforward application of the sifting property yields

$$\int_{-\infty}^{\infty} \delta(t)[t^2 + 1]\, dt = 0^2 + 1 = 1 \qquad (1.79)$$

(b) Since the impulse falls within the range of integration,

$$\int_{-1}^{2} \delta(t-1)[t^2+1]\, dt = 1^2 + 1 = 2 \tag{1.80}$$

(c) The impulse occurs at $t = 1$, which is outside the range of integration. Therefore,

$$\int_{3}^{5} \delta(t-1)[t^3+4t+2]\, dt = 0 \tag{1.81}$$

(d) $\delta(1-t)$ falls at $t = 1$ since this is the value of "t" which makes the argument equal to zero. Therefore,[4]

$$\delta(1-t) = \delta(t-1)$$

and

$$\int_{-\infty}^{\infty} \delta(1-t)[t^4+2]\, dt = 1^4 + 2 = 3 \tag{1.82}$$

From Eq. (1.77) it is a simple matter to find the Fourier Transform of the impulse function.

$$\delta(t) \leftrightarrow \int_{-\infty}^{\infty} \delta(t)e^{-j2\pi ft}\, dt = e^0 = 1 \tag{1.83}$$

This is indeed a very nice Fourier Transform for a function to have. One can guess that it may prove significant since it is the unity, or identity, multiplicative element. That is, anything multiplied by 1 is left unchanged.

One manifestation of this occurs when one convolves an arbitrary time function with $\delta(t)$.

$$\delta(t) * s(t) \overset{\Delta}{=} \int_{-\infty}^{\infty} \delta(\tau)s(t-\tau)\, d\tau = s(t-0) = s(t) \tag{1.84}$$

That is, any function convolved with an impulse remains unchanged. The proof of this is equally trivial in the frequency domain. Recall that convolution in the time domain corresponds to multiplication in the frequency domain. Since the transform of $\delta(t)$ is unity, the product of this transform with the transform of $s(t)$ is simply the transform of $s(t)$.

---

[4]Many arguments can be advanced for regarding the impulse as an even function. For one thing, it can be considered as the derivative of an odd function. For another, we will see that its transform is real and even.

$$s(t)*\delta(t) \leftrightarrow S(f).\mathscr{F}[\delta(t)] = S(f) \leftrightarrow s(t) \qquad (1.85)$$

Thus, convolution with $\delta(t)$ does not change the function.

If we convolve $f(t)$ with the shifted impulse, $\delta(t - t_0)$, we find

$$\delta(t - t_0)*s(t) = \int_{-\infty}^{\infty} \delta(\tau - t_0)s(t - \tau)\,d\tau = s(t - t_0) \qquad (1.86)$$

The same derivation in the frequency domain involves the time shift theorem. The transform of the shifted impulse is $e^{-j2\pi f t_0}$ by virtue of the time shift theorem.

$$\delta(t - t_0) \leftrightarrow e^{-j2\pi f t_0}$$

Therefore, by the time convolution theorem, the transform of the convolution of $s(t)$ with $\delta(t - t_0)$ is the product of the two transforms.

$$\delta(t - t_0)*s(t) \leftrightarrow e^{-j2\pi f t_0} S(f) \qquad (1.87)$$

$e^{-j2\pi f t_0} S(f)$ is recognized as the transform of $s(t - t_0)$ by virtue of the time shift theorem.

In summation, convolution of $s(t)$ with an impulse function does not change the functional form of $s(t)$. It may only cause a time shift in $s(t)$ if the impulse does not occur at $t = 0$.

It is almost time to apply all of the theory we have been laboriously developing to a practical problem. Before doing that, we need simply evaluate several transforms which involve impulse functions.

Let us now return to the evaluation of the transform of a constant, $s(t) = A$. If we applied the definition and tried to evaluate the integral, we would find that it does not converge as already stated,

$$A \leftrightarrow \int_{-\infty}^{\infty} Ae^{-j2\pi f t}\,dt \qquad (1.88)$$

For $f \neq 0$, the above integral is bounded by $A/\pi f$ but does not converge. A function need not approach infinity if it does not converge, but may oscillate instead. This case is analogous to finding the limit of $\cos t$ as "t" approaches infinity. For $f = 0$, the above integral clearly blows up. If we are to find any representation of the transform, we must obviously take a different approach. As before, limit arguments are often invoked. Also as before, we shall avoid them.

Since the integral defining the Fourier Transform and that used to evaluate the inverse transform are quite similar (change of sign in the exponent differentiates them), one might guess that the transform of a constant is an impulse. That is, since an impulse transforms to a constant, a constant should transform to an impulse. In the hope that this reasoning is valid, let us find the inverse transform of an impulse,

$$\delta(f) \leftrightarrow \int_{-\infty}^{\infty} \delta(f)e^{j2\pi f t}\,df = 1 \qquad (1.89)$$

Our guess was correct! The inverse transform of $\delta(f)$ is a constant. Therefore, by a simple scaling factor, we have

$$A \leftrightarrow A\delta(f) \tag{1.90}$$

A straightforward application of the frequency shift theorem yields another useful transform pair,

$$Ae^{j2\pi f_0 t} \leftrightarrow A\delta(f-f_0) \tag{1.91}$$

The above "guess and check" technique deserves some comment. We stressed earlier that the uniqueness of the Fourier Transform is extremely significant. That is, given $S(f)$, $s(t)$ can be uniquely recovered. Therefore, the guess technique is perfectly rigorous to use in order to find the Fourier Transform of a time function. If we can somehow guess at an $S(f)$ which yields $s(t)$ when plugged into the inversion integral, we have found the one and only transform of $s(t)$. As in the above example, this technique is very useful if the transform integral cannot be easily evaluated, while the inverse transform integral evaluation is an obvious one.

Illustrative Example 1.14

Find the Fourier Transform of $s(t) = \cos 2\pi f_0 t$

Solution

We make use of Euler's identity to express the cosine function in the following form:

$$\cos 2\pi f_0 t = \tfrac{1}{2}e^{j2\pi f_0 t} + \tfrac{1}{2}e^{-j2\pi f_0 t}$$

By the linearity property of the Fourier Transform, we have

$$S(f) = \mathcal{F}[\tfrac{1}{2}e^{j2\pi f_0 t}] + \mathcal{F}[\tfrac{1}{2}e^{-j2\pi f_0 t}]$$
$$= \tfrac{1}{2}\delta(f-f_0) + \tfrac{1}{2}\delta(f+f_0) \tag{1.92}$$

This transform is sketched in Fig. 1.31.

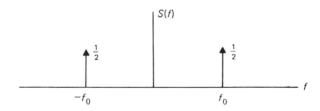

FIGURE 1.31   Transform of $\cos 2\pi f_0 t$ Illustrative Example 1.14.

We can now reveal something somewhat underhanded that has been attempted in this text. Although the definition of the Fourier Transform is a strictly mathematical definition, and "f" is just a dummy independent functional variable, we have slowly tried to brainwash the reader into thinking of it as a *frequency* variable. Indeed, the choice of "f" for the symbol of the independent variable brings the word "frequency" to mind. We know that the Fourier Transform of a real, even time function must be real and even. The transform can therefore not be identically zero for negative values of "f". Indeed, for any non-zero, real time function, the Fourier Transform cannot be identically zero for negative "f". Since negative values of frequency have no meaning, we could never be completely correct in calling "f" a frequency variable.

Let us view the positive f-axis only. For this region, the transform of $\cos 2\pi f_0 t$ is non-zero only at the point $f = f_0$. The only case for which we have previously experienced the definition of frequency is that in which the time function is a pure sinusoid. Since for the pure sinusoid, the positive f-axis appears to have meaning when interpreted as frequency, we shall consider ourselves justified in calling "f" a frequency variable. We shall continue to do this until such time that a contradiction arises (as it turns out, none will, since for other than a pure sinusoid, we will *define* frequency in terms of the Fourier Transform).

Another transform pair that will be needed in our later work is that of a unit step function and its transform. Here again, if one simply plugs this into the defining integral, the resulting integral does not converge. We could again attempt the guess technique, but due in part to the discontinuity of the step function, the technique is somewhat hopeless. The transform is relatively easy to evaluate once one realizes that

$$U(t) = \frac{1 + sgn(t)}{2} \qquad (1.93)$$

where the "sign" function is shown in Fig. 1.32 and is defined by

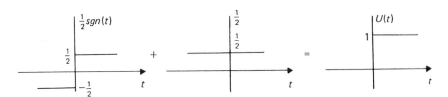

FIGURE 1.32   Representation of $U(t)$ as $\frac{1}{2}[1 + sgn(t)]$.

$$sgn(t) = \begin{cases} +1 & t > 0 \\ -1 & t < 0 \end{cases}$$

The transform of $\frac{1}{2}$ is $\frac{1}{2}\delta(f)$, and that of $sgn(t)$ can be evaluated by a simple limiting process (see Fig. 1.33).

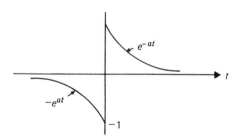

FIGURE 1.33 *sgn(t)* as a limit.

$$sgn(t) = \lim_{a \to 0} [e^{-a|t|} sgn(t)]$$

Assuming that the order of limiting and taking the transform can be interchanged (as engineers, we first do it, and if the result converges, we assume that it was a permissible thing to do), we have

$$\mathscr{F}[sgn(t)] = \lim_{a \to 0} \mathscr{F}[e^{-a|t|} sgn(t)]$$

$$= \lim_{a \to 0} \left[\frac{1}{j2\pi f + a} + \frac{1}{j2\pi f - a} \right]$$

$$\mathscr{F}[sgn(t)] = \lim_{a \to 0} \left[\frac{-4j\pi f}{(2\pi f)^2 + a^2} \right] = \frac{1}{j\pi f} \tag{1.94}$$

From Eqs. (1.93) and (1.94) we see that the transform of the unit step is given by

$$U(t) \leftrightarrow \frac{1}{j2\pi f} + \frac{1}{2}\delta(f) \tag{1.95}$$

The student may recall that the one-sided LaPlace Transform of the unit step is $1/s$. At first glance, it appears that the Fourier Transform of any function which is zero for negative "t" should be the same as the one-sided LaPlace Transform, with "s" replaced by "$2\pi jf$." However, we see that in the case of $f(t) = U(t)$, the two differ by a very important factor, $\frac{1}{2}\delta(f)$. The explanation of this apparent discrepancy requires a study of convergence in the complex s-plane. Since this is not critical to the applications of communication theory, we shall omit it in this text.

1.9   PERIODIC TIME FUNCTIONS

In Illustrative Example 1.14, we found the Fourier Transform of the cosine function. It was composed of two impulses occurring at the frequency of the cosine, and at the negative of this frequency. It will now be shown that the Fourier Transform of any periodic function of time is a discrete function of frequency. That is, the transform is non-zero only at discrete points along the f-axis. The proof follows from Fourier series expansions and the linearity of the Fourier Transform.

Suppose that we wish to find the Fourier Transform of a time function, $s(t)$, which happens to be periodic with period T. We choose to express the function in terms of the complex Fourier Series representation.

$$s(t) = \sum_{n=-\infty}^{\infty} c_n e^{jn2\pi f_0 t}$$

where

$$f_0 = \frac{1}{T}$$

Recall the transform pair, Eq. (1.91),

$$A e^{j2\pi f_0 t} \leftrightarrow A\delta(f - f_0)$$

From this transform pair, and the linearity of the transform, we have

$$\mathcal{F}[s(t)] = \sum_{n=-\infty}^{\infty} c_n \mathcal{F}[e^{jn2\pi f_0 t}]$$

$$= \sum_{n=-\infty}^{\infty} c_n \delta(f - nf_0) \qquad (1.96)$$

This transform is sketched in Fig. 1.34 for a representative $s(t)$. Thus, the Fourier

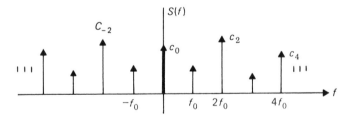

FIGURE 1.34   Transform of a periodic $s(t)$.

Transform of a periodic time function consists of a train of equally spaced delta functions, each of whose strength (area) is the corresponding c_n of the exponential Fourier Series representation of the time function. Note that the c_n may turn out to be complex numbers. Therefore, the area of the delta functions may also be complex.

This does not contradict anything which we said previously, so it should not bother anybody. The c_n will be real if the time function is real and even.[5]

Illustrative Example 1.15

Find the Fourier Transform of the periodic function made up of unit impulses, as shown in Fig. 1.35.

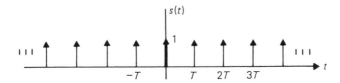

FIGURE 1.35   Periodic pulse train $s(t)$ for Illustrative Example 1.15.

$$s(t) = \sum_{n=-\infty}^{\infty} \delta(t - nT)$$

Solution

Using Eq. (1.96), we have

$$S(f) = \sum_{n=-\infty}^{\infty} c_n \delta(f - nf_0)$$

where

$$f_0 = \frac{1}{T}$$

and

$$c_n = \frac{1}{T} \int_{-T/2}^{T/2} s(t) e^{-jn2\pi f_0 t} \, dt$$

Within the range of integration, $-T/2$ to $+T/2$, the only contribution of $s(t)$ is that due to the impulse at the origin. Therefore,

$$c_n = \frac{1}{T} \int_{-T/2}^{T/2} \delta(t) e^{-jn2\pi f_0 t} \, dt = \frac{1}{T} \tag{1.97}$$

---

[5]From this point forth, unless otherwise stated, all time functions will be assumed to be real. This certainly conforms to real life, where one would never talk about a complex voltage or current.

This is an interesting Fourier Series expansion. All of the coefficients are equal. Each frequency component possesses the same amplitude as every other component. Is this analogous to the observation that the Fourier Transform of a single delta function is a constant?

Proceeding, we have

$$S(f) = \frac{1}{T} \sum_{n=-\infty}^{\infty} \delta(f - nf_0) \tag{1.98}$$

with

$$f_0 = \frac{1}{T}$$

This transform is sketched in Fig. 1.36.

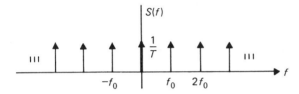

FIGURE 1.36   Transform of impulse train of Illustrative Example 1.15.

The Fourier Transform of a periodic train of impulses is itself a periodic train of impulses! This is the only case we will encounter in this text in which the Fourier Transform and the original time function resemble each other. This result will be used as a basis for an intuitive proof of the sampling theorem, a *remarkable* result which we shall now present.

1.10   THE SAMPLING THEOREM

The *sampling theorem* (sometimes called Shannon's theorem or Kotelnikov's theorem) states that, if the Fourier Transform of a time function is zero for $|f| > f_m$, and the values of the time function are known for $t = nT_s$ (all integer values of n), where $T_s < 1/2f_m$, then the time function is exactly known for ALL values of "t." Stated in a different way, $s(t)$ can be uniquely determined from its values at a sequence of equidistant points in time. Think about this. Knowing the values of the function at discrete points, we know its value *exactly* at all points between these discrete points. The upper limit of T_s, $1/2f_m$, is known as the *Nyquist sampling rate*.

The upper limit on T_s can be expressed in a more meaningful way by taking the reciprocal of T_s to obtain the sampling frequency denoted $f_s = 1/T_s$. The restriction then becomes

$$f_s > 2f_m \tag{1.99}$$

Thus, the sampling frequency must be at least twice the highest frequency of the signal being sampled. For example, a voice signal with 4-kHz maximum frequency must be sampled at least 8,000 times per second to comply with the conditions of the sampling theorem.

Before going further, we should observe that the spacing between the "sample points" is inversely related to the maximum frequency, f_m. This is intuitively satisfying, since the higher f_m is, the faster we would expect the function to vary. The faster the function varies, the closer together the sample points should be in order to permit reproduction of the function.

The sampling theorem is crucial to any form of pulse communication (we are getting way ahead of the game). If somebody wanted to transmit a time signal, $s(t)$, from one point to another, it would be sufficient to send a list of numbers which represented the values of $s(t)$ at the sampling instants. This is certainly easier than transmitting the value of $s(t)$ at every single instant of time (infinitely many values lie between any two points in time).

The theorem is so significant, we will present three different proofs of it. Each proof should provide a slightly different perspective and insight.

Proof 1

The first proof of the sampling theorem is a simple outgrowth of the Modulation Theorem discussed in Section 1.7. Let us first decide what a "sampler" would look like. Figure 1.37 shows a pulse train multiplying the original signal, $s(t)$. If the pulse

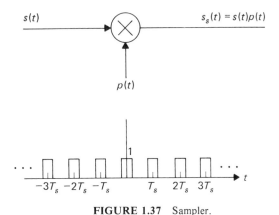

FIGURE 1.37 Sampler.

train consists of narrow pulses, you would probably say that the output of the multiplier is a sampled version of the original waveform. Figure 1.38 shows a representative $s(t)$ and the "sampled" waveform.

Multiplying $s(t)$ by a $p(t)$ of this type is really a form of time gating. It can be viewed as the opening and closing of a gate, or switch. The periodic multiplying function, $p(t)$, can consist of narrow pulses of any shape. In fact, the third proof we

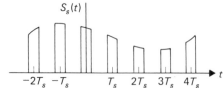

FIGURE 1.38 Sampled waveform.

present will use the same sampler but with ideal impulses for the multiplying function.

Our goal is to find the Fourier Transform of the sampled waveform, and to thereby show that the original $s(t)$ can be recovered. It is not possible to find the transform of the sampled waveform without assuming some shape for $S(f)$, the transform of $s(t)$. We don't wish to restrict the proof to a particular $s(t)$. However, one general property of $s(t)$ was assumed in the statement of the sampling theorem. The assumed conditions of the sampling theorem set an upper frequency of f_m for $S(f)$. We will sketch $S(f)$ as in Fig. 1.39. This is intended to be a general band limited $S(f)$, and we in no way mean to restrict $S(f)$ to this exact shape.

FIGURE 1.39 Fourier Transform of general band limited function.

Since the multiplying pulse train is assumed to be periodic, it can be expanded in a Fourier Series. We will make the simplifying assumption that $p(t)$ is even. If this is not the case, the Fourier Series would contain sine terms in addition to the cosine terms. This can be thought of as adding a phase shift to the cosine terms, and would not affect the magnitudes of the transforms. Thus, the proof would be essentially the same.

The output of the multiplier can therefore be expressed as follows:

$$s_s(t) = s(t)p(t)$$

$$= s(t)\left[a_0 + \sum_{n=1}^{\infty} a_n \cos 2\pi n f_s t\right]$$

$$= a_0 s(t) + \sum_{n=1}^{\infty} a_n s(t)\cos 2\pi n f_s t \qquad (1.100)$$

where $f_s = 1/T_s$. The goal is to separate out the first term, which is simply a constant times the original $s(t)$. We can then undo the effects of any constant multiplier with an amplifier or attenuator.

The Fourier Transform of $s_s(t)$ can be found by successive application of the modulation theorem. The first term in Eq. (1.100) is a scaled version of the original time function, and its Fourier Transform is the first section of $S_s(f)$ centered about zero frequency. All of the remaining terms represent $s(t)$ multiplied by a cosine, and the transform can be found from the modulation theorem. The transform of the sum is therefore a sum of shifted versions of $S(f)$, where the amount of shift is the frequency of the cosine, and the amplitude depends upon the associated Fourier Series coefficient, a_n. A representative $S_s(f)$ is sketched in Fig. 1.40. If the conditions of the sampling theorem obtain, that is, $f_s > 2f_m$, the blobs in Fig. 1.40 will not overlap. The question now arises whether $s(t)$ can be recovered from $s_s(t)$.

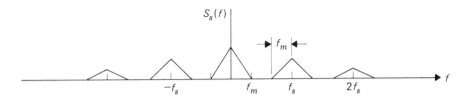

FIGURE 1.40   Fourier Transform of sampled waveform.

Signals and systems possess several basic properties. One that we will invoke many times is that it is trivial to separate signals if one of two (dual) conditions exist: The signals must be non-overlapping in time or non-overlapping in frequency. For example, in a conference telephone call, as long as no more than one person talks at any one time, the receiver can completely separate the signals. In a sense, the receiver is applying a type of "time gating" by disconnecting all but the desired source at any one time. There exists a complete duality between the time and frequency domain, and signals which do not overlap in frequency can be completely separated by using frequency gates (in Chapter 2, we will call them filters). Thus, as long as the condition of the sampling theorem obtains, the low frequency portion of $S_s(f)$ can be separated, thus recovering the original $s(t)$ and proving the theorem.

Proof 2

Since $S(f)$ is only non-zero along a finite portion of the f-axis (by assumption), we can expand it in a Fourier Series in the interval

$$-f_m < f \leq f_m$$

In expanding $S(f)$ in a Fourier Series, one must be a little careful not to let the change in notation confuse the issue. The "t" used in the Fourier Series previously presented is a dummy functional independent variable, and any other letter could have been substituted for it. Performing the Fourier Series expansion, we find

$$S(f) = \sum_{n=-\infty}^{\infty} c_n e^{jnt_0 f}$$

where

$$t_0 = \frac{\pi}{f_m}$$

The c_n in the above expansion are given by

$$c_n = \frac{1}{2f_m} \int_{-f_m}^{f_m} S(f) e^{-jnt_0 f} \, df \tag{1.101}$$

However, the Fourier inversion integral tells us that

$$s(t) = \int_{-\infty}^{\infty} S(f) e^{j2\pi ft} \, df = \int_{-f_m}^{f_m} S(f) e^{j2\pi ft} \, df \tag{1.102}$$

In Eq. (1.102), the limits were reduced since $S(f)$ is equal to zero outside of the interval between $-f_m$ and $+f_m$. Upon comparing Eq. (1.102) with Eq. (1.101), we can easily see that

$$c_n = \frac{1}{2f_m} s \left(-\frac{nt_0}{2\pi} \right) = \frac{1}{2f_m} s \left(-\frac{n}{2f_m} \right)$$

This says that the c_n are known once $s(t)$ is known at the points $t = n/2f_m$. Plugging these values of c_n into the series expansion for $S(f)$ we find

$$S(f) = \sum_{n=-\infty}^{\infty} c_n e^{jnt_0 f} = \frac{1}{2f_m} \sum_{n=-\infty}^{\infty} s \left(-\frac{n}{2f_m} \right) e^{jn\pi f/f_m} \tag{1.103}$$

Equation (1.103) is essentially a mathematical statement of the sampling theorem. It states that $S(f)$ is completely known and determined by the sample values, $s(-n/2f_m)$. These are the values of $s(t)$ at equally spaced points in time. We can plug this $S(f)$ given by Eq. (1.103) into the Fourier inversion integral to actually find $s(t)$ in terms of its samples. This will yield the following complex appearing result.

$$s(t) = \sum_{n=-\infty}^{\infty} \frac{1}{2f_m} \int_{-f_m}^{f_m} s \left(-\frac{n}{2f_m} \right) e^{jn\pi f/f_m} e^{j2\pi ft} \, df$$

$$= \sum_{n=-\infty}^{\infty} s \left(-\frac{n}{2f_m} \right) \left[\frac{\sin(2\pi f_m t + n\pi)}{2\pi f_m t + n\pi} \right] \tag{1.104}$$

Equation (1.104) tells us how to find $s(t)$ for any "t" from those values of $s(t)$ at the points $t = n/2f_m$. The student could never truly appreciate Eq. (1.104) unless a few terms of this sum are sketched for a representative $s(t)$.

As presented, this proof used the minimum sampling rate. That is, $T_s = 1/2f_m$. In order to use the same proof for smaller T_s, the period of the Fourier Series in Eq. (1.100) would have to be increased by adding some of the f-axis where $S(f) = 0$.

Proof 3

Given that $s(t)$ has a transform, $S(f)$, which is zero outside of the interval $-f_m < f < f_m$, we will again use the sketch of $S(f)$ as shown in Fig. 1.39 to represent a general $S(f)$ with this "bandlimited" property. We do not mean to restrict $S(f)$ to be of this exact shape. We now consider the product of $s(t)$ with a periodic train of delta functions, $s_\delta(t)$, which we shall call a sampling, or gating function.

$$s_\delta(t) = \sum_{n=-\infty}^{\infty} \delta(t - nT_s) \tag{1.105}$$

The sifting property of the impulse function tells us that

$$s(t)\delta(t - t_0) = s(t_0)\delta(t - t_0) \tag{1.106}$$

Using this property, we see that the sampled version of $s(t)$ (call it $s_s(t)$) is given by

$$s_s(t) = s(t)s_\delta(t) = \sum_{n=-\infty}^{\infty} s(t)\delta(t - nT_s)$$

$$= \sum_{n=-\infty}^{\infty} s(nT_s)\delta(t - nT_s) \tag{1.107}$$

A typical $s(t)$, a sampling function, and the sampled version of $s(t)$ are sketched in Fig. 1.41. $s_s(t)$ is truly a sampled version of $s(t)$ since it only depends upon the values

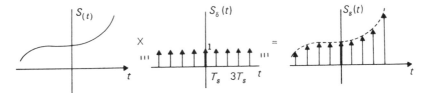

FIGURE 1.41   The sampling operation.

of $s(t)$ at the sample points in time. We wish to show that, under certain conditions, $s(t)$ can be recovered from $s_s(t)$. We do this by finding the transform of $s_s(t)$, $S_s(f)$. By the uniqueness of the Fourier Transform, if we can show that the transform of $s(t)$, $S(f)$, can be recovered from $S_s(f)$, then $s(t)$ can be recovered from $s_s(t)$. From the frequency convolution theorem, we have

$$s_s(t) \overset{\Delta}{=} s(t) s_\delta(t) \leftrightarrow S(f) * S_\delta(f) = S_s(f) \qquad (1.108)$$

We found $S_\delta(f)$ in Illustrative Example 1.15 (see Eq. (1.98))

$$S_\delta(f) = \frac{1}{T_s} \sum_{n=-\infty}^{\infty} \delta(f - nf_0)$$

where

$$f_0 = \frac{1}{T_s}$$

Recall that convolution with a delta function simply shifts the original function. Therefore,

$$S(f) * \delta(f - nf_0) = S(f - nf_0)$$

and

$$S_s(f) = S(f) * \frac{1}{T_s} \sum_{n=-\infty}^{\infty} \delta(f - nf_0)$$

$$= \frac{1}{T_s} \sum_{n=-\infty}^{\infty} S(f - nf_0) \qquad (1.109)$$

This $S_s(f)$ is sketched in Fig. 1.42, where we have let $1/T_s = k$. From the sketch of

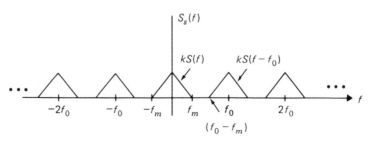

FIGURE 1.42 $S_S(f)$, the transform of the sampled wave.

$S_s(f)$, it is clear that $S(f)$ can be identified. In practice, we need only multiply $S_s(f)$ by a gating function, $H(f)$, as shown in Fig. 1.43. Devices which perform this multiplication in the frequency domain (frequency separation) are called filters, and are analyzed in detail in the next chapter.

One assumption which was made in the sketching of $S_s(f)$ will now be examined. The assumption is that the humps in the figure do not overlap. That is, the point labelled "$f_0 - f_m$" must fall to the right of that labelled "f_m." If this were not true, the center blob (centered about $f = 0$) would result from an interaction of several shifted versions of $S(f)$, and would therefore not be an accurate reproduction of $S(f)$.

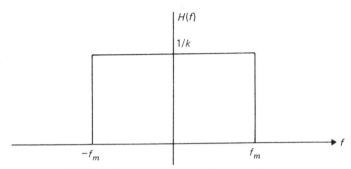

FIGURE 1.43 $H(f)$ to recover $S(f)$ from $S_s(f)$.

In this case, $s(t)$ could not be uniquely recovered. This restriction can be rewritten as follows:

$$f_0 - f_m > f_m \Rightarrow f_0 > 2f_m \qquad (1.110)$$

or in terms of the sampling period,

$$\frac{1}{f_0} = T_s < \frac{1}{2f_m} \qquad (1.111)$$

as was stated in the condition of the sampling theorem. This restriction says that the samples cannot be spaced more widely apart than $1/2f_m$. They can certainly be as close together, consistent with the $1/2f_m$ upper limit, as desired. We again note that as f_m increases, the samples must be closer together. This agrees with intuition since, as the frequency increases, one would expect the function to change value more rapidly.

The sampling function, $s_\delta(t)$, did not have to be a train of ideal impulses. Indeed, Proof 1 used a general periodic signal. We simply note at this time that the ideal train of impulses has the advantage that the sampled version of $s(t)$, $s_s(t)$, takes up a minimum amount of space on the time axis. (Actually, for ideal impulses, the time required for each sample is zero.) This will prove significant, and will be explored in detail when we study pulse modulation.

Illustrative Example 1.16

Consider the function, $s(t) = \sin 2\pi f_0 t$. For this function, $f_m = f_0$. That is, the Fourier Transform contains no components at frequencies above f_m. The sampling theorem therefore tells us that $s(t)$ can be completely recovered if we know its values at $t = nT_s$ where

$$T_s < \frac{1}{2f_m} = \frac{1}{2f_0}$$

Suppose that we pick the upper limit, that is $T_s = 1/2f_0$. Looking at $\sin 2\pi f_0 t$ we see that it is possible for each of the sample points to fall at a zero crossing of $s(t)$. (*See* Fig. 1.44) That is,

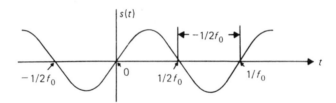

FIGURE 1.44 Sampling at zero points of sinusoid.

every one of the sample values may turn out to be zero. Obviously, $s(t)$ could not be recovered if this were the case.

Lesson To Be Learned from Illustrative Example 1.16: Looking back at the sketch of $S_s(f)$ (Fig. 1.42), we said that the humps must not overlap. However, if $T_s = 1/2f_m$, while they do not overlap, each pair does meet at one point, $f = nf_m$. While the value of $S(f)$ at one single point in frequency usually cannot affect the value of the time function, $s(t)$ [an integral operation is involved in going from $S(f)$ to $s(t)$], if $S(f)$ happens to be an impulse at this one point, $s(t)$ certainly can be changed. Since the transform of $\sin 2\pi f_0 t$ does indeed possess an impulse at $f = f_m$, this "point overlap" of the blobs cannot be tolerated. In the above example, the two impulses that fall at $f = f_m$ were cancelling each other out. To eliminate this possibility, the statement of the sampling theorem must carefully say

$$T_s < \frac{1}{2f_m}$$

where the inequality must mean "strictly less than" not "less than or equal to." In the above example, if T_s were a little bit less than $1/2f_0$, the sample values could not all equal zero, and $s(t)$ could be recovered from $s_s(t)$.

We can make another important observation from this same example. Assume that we sample at twice the minimum rate. That is, let the sampling period, $T_s = 1/4f_0$. We can recover $s(t)$ from the samples. Figure 1.45 shows the sampled

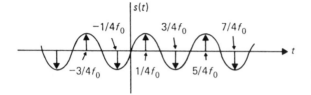

FIGURE 1.45 $\sin 2\pi f_0 t$ sampled at twice the minimum sampling rate.

version of $s(t) = \sin 2\pi f_0 t$. Note that every other sample falls at a zero crossing.

One may ask how we can be sure that the samples of Fig. 1.37 correspond to $s(t) = \sin 2\pi f_0 t$ and not, for example, the sawtooth type waveform of Fig. 1.46.

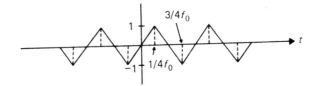

FIGURE 1.46 Sawtooth waveform with sample values of Fig. 1.45.

The answer lies in the observation that the waveform of Fig. 1.46 is not frequency bandlimited to f_m (in fact, it possesses infinite frequencies). Therefore, the conditions of the sampling theorem would have been violated.

In summation, we can uniquely recover a waveform from its sample values *provided* that we are sure the conditions of the sampling theorem ($T_s < 1/2f_m$) were not violated in producing the sampled waveform.

Errors in Sampling

The sampling theorem indicates that $s(t)$ can be perfectly recovered from its samples. If sampling is attempted in the real world, errors result from three major sources. Round-off errors occur when the various sample values are rounded off in the communication system. We will examine this type of error later when digital communication is studied and specific rules for rounding off are presented. Truncation errors occur if the sampling is done over a finite time. That is, the sampling theorem requires that samples be taken for all time in the infinite interval, and every sample is used to reconstruct the value of the original function at any particular time. In a real system, the signal is observed for a limited time. We can define an error function as the difference between the reconstructed time function and the original function, and upper bounds can be placed upon the magnitude of this error function. Such bounds involve sums of the rejected time sample values.

A third error results if the sampling rate is not high enough. This situation can be intentional or accidental. For example, if the original time signal has a Fourier Transform which asymptotically approaches zero with increasing frequency, a conscious decision is made to define a maximum frequency beyond which signal energy is negligible. In order to minimize this problem, the input signal is usually low pass filtered prior to sampling. On the other hand, we may design a system with a high enough sampling rate, yet an unanticipated high frequency signal appears at the input. In either case, the error caused by sampling too slowly is known as aliasing, a name which is derived from the fact that the higher frequencies disguise themselves in the form of lower frequencies. Analysis of aliasing is most easily performed in the frequency domain. Before doing that, we will illustrate the problem in the time domain. Figure 1.47 shows a sinusoid at a frequency of 3 Hz. Suppose we sample this sinusoid at 4 samples per second. The sampling theorem tells us that the minimum sampling rate for unique recovery is 6 samples per second, so 4 samples per second is not fast enough. The samples at the slower rate are indicated on the figure. But alas,

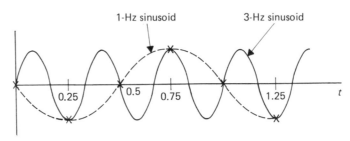

FIGURE 1.47   Demonstration of aliasing.

these are the same samples that would result from a sinusoid at 1 Hz as shown by the dashed curve. The 3-Hz signal is disguising itself (an alias) as a 1-Hz signal. The Fourier Transform of the sampled wave is found by periodically repeating the Fourier Transform of the original signal.

If the original signal has frequency components above one-half of the sampling rate, these components will fold back into the frequency band of interest. Thus in Fig. 1.47, the 3-Hz signal folded back to fall at 1 Hz. Figure 1.48 illustrates the case

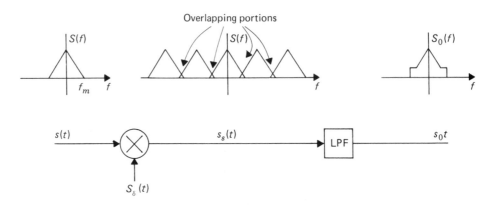

FIGURE 1.48   Sampling at less than the Nyquist rate.

where a representative signal is being sampled by an ideal train of impulses. Note that the transform at the output of the lowpass filter is no longer the same as the transform of the original signal. If we denote the filter output as $s_0(t)$, the error would be defined as

$$e(t) = s_0(t) - s(t) \tag{1.112}$$

Taking Fourier Transforms of both sides of Eq. (1.112), we see that

$$E(f) = S_0(f) - S(f) = S(f) + S(f - f_s) + S(f + f_s) - S(f)$$
$$= S(f - f_s) + S(f + f_s) \qquad |f| < f_m$$

Without assuming a specific form for $S(f)$, we cannot carry this example further. In general, various bounds can be placed upon the magnitude of the error function based upon properties of $S(f)$ for $|f| > f_s/2$.

Illustrative Example 1.17

Assume that the bandlimited function,

$$s(t) = \frac{\sin 20\pi t}{\pi t}$$

is sampled at 19 samples/second. The impulse sampled function forms the input to a lowpass filter with cutoff frequency of 10 Hz., as illustrated in Fig. 1.49. Find the output of the lowpass filter and compare this with the original $s(t)$.

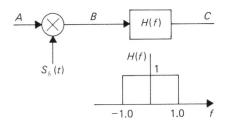

FIGURE 1.49   Sampled system for Illustrative Example 1.17.

Solution

The Fourier Transforms of the signals at points A, B, and C in Fig. 1.49 are shown in Fig. 1.50. The output time function is the inverse Fourier Transform of $S_0(f)$, and is given by

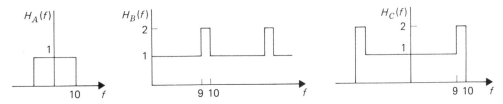

FIGURE 1.50   Fourier Transforms for Illustrative Example 1.17.

$$s_0(t) = \frac{\sin 20\pi t}{\pi t} + 2\frac{\sin \pi t}{\pi t}\cos 19\pi t \tag{1.113}$$

The second term in Eq. (1.113) represents the aliasing error. It has a maximum amplitude of 2 at $t = 0$, at which point the signal has amplitude of 20. Do not be tempted to calculate a percentage error by taking the ratio of the error amplitude to the desired signal amplitude. Since the first term in Eq. (1.113) goes to zero at periodic points, and the second term is not necessarily zero at these same points, the percentage error can approach infinity. Errors are often analyzed by looking at the energy of the time function representing the error. Energy is the area under the square of the function. We could therefore find the energy of the second term in Eq. (1.113). Once this energy is found, it is divided by the total energy of the desired signal (the first term) to get a percentage of error.

We conclude this section with the idea that the restriction on $S(f)$ imposed by the sampling theorem is not very severe in practice. All signals of interest in real life do possess Fourier Transforms which are zero (approximately) above some frequency. No physical device can transmit infinitely high frequencies since all wires contain inductance, and all circuits contain parasitic capacitances. The inductor opens and the capacitor shorts.

1.11   THE DISCRETE FOURIER TRANSFORM

We have seen that a continuous, bandlimited time signal can be represented by time samples. Practical system limitations often make it desirable to work with sample values rather than with continuous functions of time. As a trivial example, suppose a weather station were asked to monitor temperature. It could plot the temperature as a function of time, or it could report the temperature at periodic sampling times (e.g., every hour). The latter technique might be used if the weather station were responding to a request from a television network news bureau.

It is therefore often useful to be able to analyze sampled time functions. The advantages of using transform analysis were previously observed for continuous time functions. It would therefore seem reasonable to look toward transform techniques that would be applicable to sampled systems. We shall explore two types of transform analysis. The first is the *Discrete Fourier Transform (DFT)*, including a variation known as the *Fast Fourier Transform (FFT)*. The second is known as the *z-Transform*.

Suppose we have a series of N samples of a time function spaced apart by a sampling period of T sec. The total time record is therefore NT sec long. We wish to calculate the Fourier Transform of this time limited function. If the integral in the definition of the continuous Fourier Transform is replaced by a summation, the Discrete Fourier Transform results. Thus, instead of

$$S(f) = \int_0^{NT} s(t)e^{-j2\pi ft}\, dt$$

we have

$$S(f) = \sum_0^{N-1} s(nT)e^{-j2\pi fnT}T$$

Two additional modifications are now made. First, we sample the frequency at multiples of Ω, and second, we scale the transform by dividing by NT, the record length. (*Note:* We can cancel the effects of any constant scale factor when we get around to defining the equation for the inverse transform.) Doing this, we get

$$S(m\Omega) = \frac{1}{N} \sum_{0}^{N-1} f(nT)e^{-jmn2\pi\Omega T} \qquad 0 \le m \le N-1 \qquad (1.114)$$

To simplify Eq. (1.114) further, we now assume that the limited time record of $s(t)$ represents one period of a periodic function. If the period is NT sec, the fundamental frequency of the periodic function would be $1/NT$, and the Fourier Transform would have non-zero value only at multiples of this fundamental frequency. It would therefore make sense to choose Ω to be equal to this fundamental frequency. Using this value, ΩT becomes $1/N$, and Eq. (1.114) can be rewritten as

$$S(m\Omega) = \frac{1}{N} \sum_{0}^{N-1} s(nT)e^{-jmn2\pi/N} \qquad 0 \le m \le N-1 \qquad (1.115)$$

Finally, defining W as the Nth complex root of unity, that is, $W = e^{j2\pi/N}$, Eq. (1.115) becomes the simplified form

$$S(m\Omega) = \frac{1}{N} \sum_{0}^{N-1} s(nT)W^{-mn} \qquad 0 \le m \le N-1 \qquad (1.116)$$

This is the usual form of the Discrete Fourier Transform. The student should verify that this transform is periodic in m with period N. That is, $S[(N+m)\Omega] = S(m\Omega)$. For this reason, it is only necessary to calculate the transform for N frequency points.

Illustrative Example 1.18

Find the DFT of the following sampled time function:

t	0	T	$2T$	$3T$	$4T$	$5T$	$6T$	$7T$	$8T$
$s(t)$	1	1	1	1	1	1	0	0	0

This is a sampled pulse as shown in Fig. 1.51.

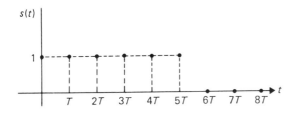

FIGURE 1.51   Sampled step of Example 1.15.

Solution

The number of sample points $N = 8$. Therefore,

$$S(m\Omega) = \tfrac{1}{8} \sum_{0}^{7} s(nT) W^{-mn}$$

$$= \tfrac{1}{8}(1 + W^{-m} + W^{-2m} + W^{-3m} + W^{-4m} + W^{-5m})$$

where

$$W = e^{j2\pi/8} = \frac{1 + j}{\sqrt{2}}$$

and

$$\Omega = \frac{1}{8T}$$

Therefore,

$$S(\Omega) = \frac{1}{8}(1 + e^{-j\pi/4} + e^{-2j\pi/4} + e^{-3j\pi/4} + e^{-4j\pi/4} + e^{-5j\pi/4})$$

$$= \frac{1}{8}\left(-j - \frac{j}{\sqrt{2}} - \frac{1}{\sqrt{2}}\right)$$

$$S(2\Omega) = \frac{1}{8}(1 - j)$$

$$S(3\Omega) = \frac{1}{8}\left(j + \frac{1}{\sqrt{2}} - \frac{j}{\sqrt{2}}\right)$$

$$S(4\Omega) = 0$$

$$S(5\Omega) = \frac{1}{8}\left(-j + \frac{1}{\sqrt{2}} + \frac{j}{\sqrt{2}}\right)$$

$$S(6\Omega) = \frac{1}{8}(1 + j)$$

$$S(7\Omega) = \frac{1}{8}\left(j + \frac{j}{\sqrt{2}} - \frac{1}{\sqrt{2}}\right)$$

Due to the periodicity of the DFT, there is no need to calculate any additional samples of the transform. We note from the above solution that the real part of the DFT is symmetric about the midpoint and that the imaginary part of the DFT is antisymmetric about the midpoint. This is true in general and is derived from the property of continuous Fourier Transforms; that is, $S(-\omega) = S^*(\omega)$. One can easily see from the DFT defining equation (Eq. 1.115)) that

$$S(-m\Omega) = S^*(m\Omega)$$

Since the DFT is periodic,

$$S[(m + N)\Omega] = S(m\Omega)$$

we have

$$S[(N - m)\Omega] = S^*(m\Omega) \tag{1.117}$$

which is a mathematical statement of the above observation. The DFT can be shown to have other properties similar to those of the continuous Fourier Transform. Some of these properties are explored in the problems at the end of this chapter.

The Discrete Fourier Transform is a convenient way to approximate the Fourier Transform when sampled waves are involved, and it lends itself readily to computation on a digital computer. Computation of the DFT as in Eq. (1.116) would involve summing N products (complex) for each of N frequency sample points. If N is large, this can become quite tedious.

The Fast Fourier Transform

The Fast Fourier Transform (FFT) is an algorithm which greatly simplifies and shortens the calculations which must be done to find the DFT. We begin by rewriting Eq. (1-116) in matrix form:

$$\overline{S(m\Omega)} = [W]\overline{s(nT)}$$

where $\overline{S(m\Omega)}$ is a column vector,

$$\overline{S(m\Omega)} = \begin{bmatrix} S(0) \\ S(1\Omega) \\ S(2\Omega) \\ S(3\Omega) \\ \cdot \\ \cdot \\ \cdot \\ S\{(N - 1)\Omega\} \end{bmatrix}$$

and $\overline{s(nT)}$ is a column vector,

$$\overline{s(nT)} = \begin{bmatrix} s(0) \\ s(1t) \\ s(2t) \\ s(3t) \\ \cdot \\ \cdot \\ \cdot \\ s\{(N - 1)t\} \end{bmatrix}$$

$[W]$ is a matrix of the form

$$[W] = \begin{bmatrix} W^0 & W^0 & W^0 & W^0 & \cdots & W^0 \\ W^0 & W^{-1} & W^{-2} & W^{-3} & \cdots & W^{-N} \\ W^0 & W^{-2} & W^{-4} & W^{-6} & \cdots & W^{-2N} \\ W^0 & W^{-3} & W^{-6} & W^{-8} & \cdots & W^{-3N} \\ \cdot & \cdot & \cdot & \cdot & \cdots & \cdot \\ W^0 & W^{-1N} & W^{-2N} & W^{-3N} & \cdots & W^{-N^2} \end{bmatrix}$$

The Fast Fourier Transform algorithm is derived by factoring the matrix [W] into a product of matrices. It is possible to do this in such a way that each of the new matrices contains many zeros, thereby simplifying the calculations.

Let us first take a more intuitive approach. We wish to calculate

$$S(m) = \sum_{n=1}^{N-1} s(n) W^{mn}$$

Let us now assume that N is even. We can then separate the sequence $s(n)$ into two sequences, each having $N/2$ points. Call these two sequences $s_e(q)$ and $s_o(r)$, where s_e is composed of the even points (every other point) of s, and s_o is composed of the odd points. This splitting is illustrated for a representative sequence in Fig. 1.52.

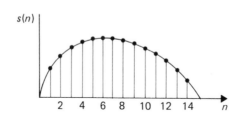

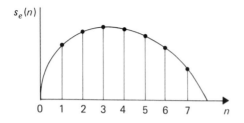 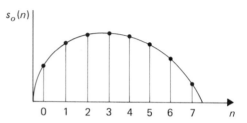

FIGURE 1.52   Splitting of samples.

Each of the two new sequences has a Discrete Fourier Transform formed by the addition of $N/2$ terms.

$$S_e(m) = \sum_{q=0}^{N/2-1} s_e(q) W^{2mq}$$

$$0 \le m \le N/2 \qquad (1.118)$$

$$S_o(m) = \sum_{r=0}^{N/2-1} s_o(r) W^{2mr}$$

The original Discrete Fourier Transform can be related to these separate DFT's by noting that $s_e(q) = s(2q)$ and $s_o(r) = s(2r+1)$. Therefore, the original DFT is given by

$$S(m) = \sum_{k=0}^{N/2-1} [s_e(k) W^{2mk} + s_o(k) W^{m(2k+1)}] \qquad 0 \le m \le N/2$$

$$= S_e(m) + W^m S_o(m)$$

This gives $S(m)$ for the first half of frequency points. To complete the picture, we note that $S_e(m)$ and $S_o(m)$ are periodic with period $N/2$. Therefore, to find $S(m)$ for $N/2 \le m < N$, we plug in $m = m + N/2$ and

$$S(m + N/2) = S_e(m + N/2) + W^{m+N/2} S_o(m + N/2)$$

$$S(m + N/2) = S_e(m) + W^m W^{N/2} S_o(m)$$

Since W is the Nth complex root of unity, $W^{N/2} = -1$, so

$$S(m + N/2) = S_e(m) - W^m S_o(m)$$

Finally putting this together, we have,

$$S(m) = S_e(m) + W^m S_o(m)$$

$$0 \le m \le N/2 - 1 \qquad (1.119)$$

$$S(m + N/2) = S_e(m) - W^m S_o(m)$$

Calculation of $S_e(m)$ requires summation of $N/2$ complex products for each of the $N/2$ frequency points. We thus do $N^2/4$ complex arithmetic operations to find $S_e(m)$. The same is true of $S_o(m)$. Therefore, finding $S(m)$ requires $N + (N^2/2)$ operations. This compares to N^2 operations if we had simply proceeded to find $S(m)$ using Eq. (1.116).

But why stop there? Let's take s_e and s_o and split each of them in the same manner. We can continue this process as long as the number of samples at each step is still divisible by 2.

If the original number of points in the discrete function is a power of 2, we can continue splitting in half until the problem has been reduced down to sequences of 1 sample each. We then note that the Discrete Fourier Transform of a single sample is the sample value itself. That is,

$$S(0) = \sum_{n=0}^{0} s(n) W^n = s(0)$$

Illustrative Example 1.19

Find the Discrete Fourier Transform of the function shown in Fig. 1.53.

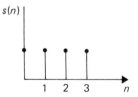

FIGURE 1.53 Function for Illustrative Example 1.19.

Solution

We start by successively splitting the samples. $s(n)$ is divided to form $s_1(n)$ and $s_2(n)$. Each of these is then divided one more time. The discrete functions are shown in Fig. 1.54. The DFT of

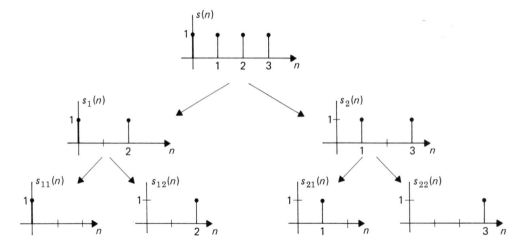

FIGURE 1.54 Splitting of samples for Illustrative Example 1.19.

$s(n)$ is then the sum of the DFT of s_1 with W^m times the DFT of s_2. But the DFT of s_1 can be expressed in terms of the DFT's of s_{11} and s_{12}, and so on. The required multiplications are most conveniently illustrated by the flow chart shown in Fig. 1.55.

We start at the left of the chart with the 4 reordered sample values. We then form 4 intermediate variables, denoted A_0 through A_3. Each of these is the sum of two preceding variables, one of which is weighted by a power of W. The weighting factor is shown adjacent to the arrow on the chart. If no number appears next to an arrow, multiplication by unity is assumed. Note that W is included to make the result more general, even though this could be replaced by 1.

After calculating the 4 values of the A variable, we proceed to calculate $S(0)$ through $S(3)$ in the same way.

Applying this flow chart to the example, we find

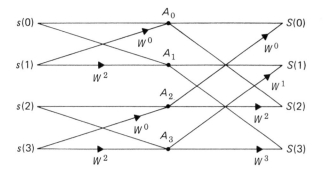

FIGURE 1.55   FFT flowchart for Illustrative Example 1.19.

$$A_0 = 2$$
$$A_1 = 1 + W^2$$
$$A_2 = 2$$
$$A_3 = 1 + W^2$$

and

$$S(0) = 4$$
$$S(1) = 1 + W^2 + W + W^3$$
$$S(2) = 2(1 + W^2)$$
$$S(3) = 1 + W^2 + W^3 + W^5$$

Now recall that W is the 4th root of unity, with powers as sketched in Fig. 1.56. If the various vectors in the above equations had been sketched, it would have become obvious that

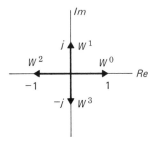

FIGURE 1.56   4th root of unity and its powers.

$S(0) = 4$ and all of the other transform values are zero. This is not terribly surprising when you realize we are taking the DFT of a sampled constant. That is, the result is a discrete impulse.

Examination of the previous example would show that the number of arithmetic operations required at each level was N, where N is the number of samples of the original function. The number of levels is $\log_2 N$. Therefore, the total number of arithmetic operations required to form the FFT is $N \log_2 N$. Figure 1.57 shows this function together with N, which was the number of operations required to find the DFT. The figure is included to illustrate the dramatic computational savings offered by the FFT. The savings are not very dramatic with only 4 samples as used in the example above. The real economy comes in when the number of samples is considerably greater than 4.

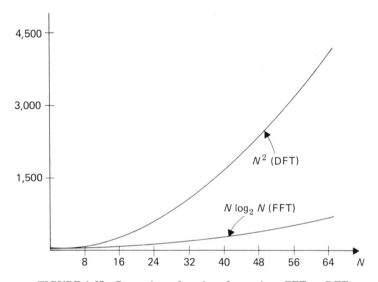

FIGURE 1.57 Comparison of number of operations: FFT vs. DFT.

1.12 THE z-TRANSFORM

As in the case of the DFT, the *z-Transform* is well suited to the analysis of discrete time signals and systems. The *z*-Transform is related to the Fourier Transform. It follows directly from the DFT. Calculation of the DFT often presents computational problems which make the use of a digital computer (and the FFT) almost necessary. This is not the case with the *z*-Transform. Finally, the time signal need not be time limited (i.e., periodic) to permit analysis using the *z*-Transform.

The *z*-Transform can be derived from Eq. (1.115) with the substitution

$$z = \exp(jm\,2\pi/N)$$

Given a sequence of numbers, $f(n)$, the *z*-Transform is defined by

$$F(z) = \sum_{n=0}^{\infty} f(n)z^{-n} \tag{1.120}$$

where z is a complex variable. $f(n)$ can be a sequence of numbers, or it can represent time samples of a continuous time function, $f(t)$.

Illustrative Example 1.20

Find the z-Transform of $f(n) = 1$ for all positive n. This is a unit step, often denoted by $u(n)$.

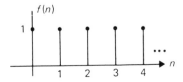

Solution

From Eq. (1.120), $F(z)$ is given by

$$F(z) = \sum_{n=0}^{\infty} 1 z^{-n}$$

$$= 1 + z^{-1} + z^{-2} + \cdots$$

$$= \frac{1}{1 - z^{-1}} = \frac{z}{z - 1}$$

This solution converges for all $|z| > 1$.

Illustrative Example 1.21

Find the z-Transform of

$$f(n) = \begin{cases} 1 & n = 0 \\ 0 & \text{otherwise} \end{cases}$$

This is a unit impulse, δ_n, sometimes known as the *Kronecker delta*.

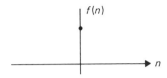

Solution

By direct substitution into Eq. (1.120), we find

$$F(z) = 1$$

Illustrative Example 1.22

Find the z-Transform of $f(n) = k^n$.

Solution

Again, we use the defining relationship, Eq. (1.120):

$$F(z) = \sum_{n=0}^{\infty} k^n z^{-n}$$

$$= \sum_{n=0}^{\infty} \left(\frac{z}{k} \right)^{-n}$$

$$= \frac{z}{z - k}$$

The z-Transform has properties which are similar to those of the Fourier Transform. Derivation of some of these properties is left to the exercises at the end of this chapter. One property which we will need here is the *time shift property*. Given that $F(z)$ is the z-Transform of $f(n)$, we wish to find the z-Transform of $f(n - m)$. This is a shifted sequence. If $f(n)$ were a sequence of time samples of a continuous signal, $f(n - m)$ would represent samples of the signal shifted by mT to the right, where T is the sampling period.

To derive the relationship between the transform of the original sequence and that of the shifted sequence, we proceed to calculate the z-Transform of $f(n - m)$, using the definition

$$\sum_{n=0}^{\infty} f(n - m)z^{-n} = \sum_{n=0}^{\infty} f(n - m)z^{-(n-m)}z^{-m}$$

$$= \sum_{k=-m}^{\infty} f(k)z^{-k}z^{-m}$$

$$= z^{-m}F(z)$$

In the last step of the above derivation, we assumed that $f(k)$ is equal to zero for k less than zero. Indeed, this is a necessary assumption since the z-Transform of $f(n)$ is independent of the values of $f(n)$ for negative argument. That is, using the z-Transform (as we defined it), negative arguments are ignored, and the transform is unique only if we make some uniform assumption about these values (e.g., assume all sequences are zero for negative argument).

Now that we know what the z-Transform is and have found some representative z-Transforms, we arrive at the next logical question, Who cares? Why was the z-Transform invented in the first place?

Just as the *convolution* operation justified the existence of the Fourier Transform, the same is true of the z-Transform. The only difference is that we must now consider a discrete form of the convolution operator. We shall find this discrete convolution operation to be critical to the analysis of discrete linear systems, as presented in Chapter 2.

Given two sequences, $x(n)$ and $h(n)$, the discrete convolution of these two sequences results in a third sequence, $y(n)$, according to

$$y(n) = x(n)*h(n)$$

$$= \sum_{m=0}^{\infty} x(m)h(n-m) \qquad (1.121)$$

The summation starts at zero in order to be consistent with our previous assumption about all sequences—that they are zero for negative argument.

Equation (1.21) requires an infinite summation for every value of n. This is clearly a tedious computation to perform. However, just as we found in the case of continuous convolutions, the transform will greatly simplify the problem. In the case of continuous convolution, we found that the Fourier Transform of the result is equal to the product of the transforms of each of the input functions. The exact same relationship is true of discrete convolution and the z-Transform. That is, the z-Transform of the convolution of two sequences is the product of the z-Transforms of each of the two sequences. To prove this relationship, we evaluate the z-Transform of $y(n)$ in Eq. (1.121):

$$\sum_{n=0}^{\infty} y(n)z^{-n} = \sum_{n=0}^{\infty} \sum_{m=0}^{\infty} x(m)h(n-m)z^{-n} \qquad (1.122)$$

We now use the time shift property derived previously to find

$$\sum_{n=0}^{\infty} h(n-m)z^{-n} = z^{-m}H(z) \qquad (1.123)$$

Substituting Eq. (1.123) into Eq. (1.122) yields

$$\sum_{n=0}^{\infty} y(n)z^{-n} = \sum_{m=0}^{\infty} z^{-m}x(m)H(z)$$

$$= X(z)H(z) \qquad (1.124)$$

Equation (1.124) is the desired result. That is, the z-Transform of the convolution of two sequences is the product of the z-Transforms of each of the sequences.

PROBLEMS

1.1. Find the x and y components of a unit vector which is orthogonal to $\bar{x}$, where

$$\bar{x} = 2\bar{a}_x + 3\bar{a}_y$$

Note that there are two possible answers to this problem.

1.2. Find the squared error that exists when we approximate the vector

$$\bar{x} = 2\bar{a}_x + 3\bar{a}_y + 4\bar{a}_z$$

by

$$\hat{\bar{x}} = k_1\bar{a}_x + k_2\bar{a}_y$$

where k_1 and k_2 are chosen to minimize this error.

1.3. Show that the two functions

$$\cos t \quad \text{and} \quad \cos 2t$$

are *not* orthogonal over the integral $-\pi/2 < t \leq \pi/2$.

1.4. Evaluate the Fourier Series expansion of each of the following periodic functions. Use either the complex exponential or the trigonometric form of the series.

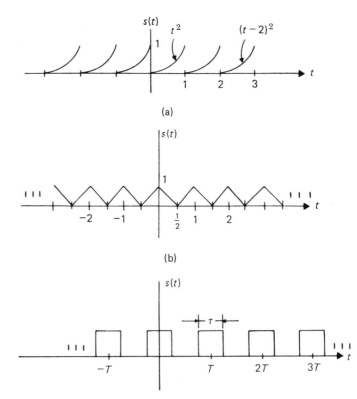

(a)

(b)

(c)

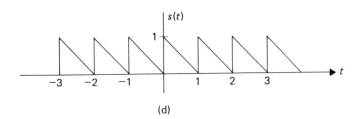

(d)

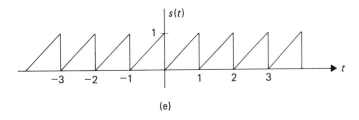

(e)

1.5. Evaluate the Fourier Series representation of the periodic function, $s(t)$.

$$s(t) = 2 \sin t + 3 \sin 2t$$

1.6. The periodic function of Problem 1.4(c) is expressed in a trigonometric Fourier Series. Find the error if only 3 terms of the Fourier Series are used. Repeat for 4 terms and 5 terms.

1.7. Find a Fourier Series expansion of $s(t)$ which applies for $-\tau/2 < t < \tau/2$.

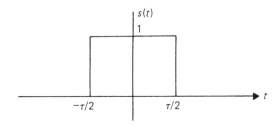

Does your answer surprise you? Does this seem to contradict the answer to Problem 1.4(c)? (Don't just answer "yes" or 'no.")

1.8. Find the complex Fourier Series representation $s(t) = t^2$ which applies in the interval $0 < t < 1$. How does this compare with your answer to 1.4(a)?

1.9. Find a trigonometric Fourier Series representation of the function $s(t) = \cos t$ in the interval $0 < t < 2\pi$.

1.10. Which of the following could not be the Fourier Series expansion of a periodic signal? Explain your answer.

(a) $s(t) = 2 \cos t + 3 \cos 3t$
(b) $s(t) = 2 \cos \frac{1}{2}t + 3 \cos 3.5t$
(c) $s(t) = 2 \cos \frac{1}{4}t + 3 \cos 0.00054t$
(d) $s(t) = 2 \cos \pi t + 3 \cos 2t$
(e) $s(t) = 2 \cos \pi t + 3 \cos 2\pi t$

1.11. Find the error made in approximating $|\cos t|$ by a constant, $2/\pi$, as in Illustrative Example 1.3. Then include the first term in the Fourier Series expansion and recalculate the error.

1.12. Differentiate Eq. (1.34) with respect to C and show that the derivative goes to zero at $C = 2/\pi$.

1.13. Evaluate the Fourier Transform of each of the following time functions.

(a) $s(t) = \dfrac{\sin at}{\pi t}$

(b)
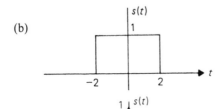

(c)
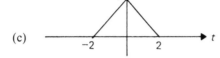

(d)
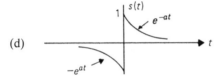

1.14. Evaluate the Fourier Transform of the following functions.
(a) $s(t) = e^{-at} U(t)$
(b) $s(t) = \cos 2t\ U(t)$
(c) $s(t) = t e^{-at} u(t)$

1.15. Convolve $e^{at} U(-t)$ with $e^{-at} U(t)$. These two functions are sketched below.

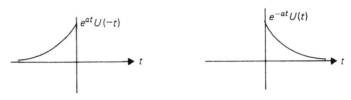

1.16. Convolve the following two functions together using the convolution integral. Repeat the calculation using graphical convolution.

1.17. The Fourier Transform of a time function, $s(t)$, is $S(f)$ as shown. Find the time function, $s(t)$.

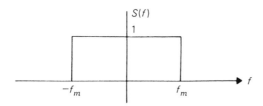

1.18. Find $\int_{-\infty}^{\infty} e^{-2t} U(t) \, dt$ by using Parseval's Theorem. (Hint: Use the fact that $e^{-2t} U(t) = |e^{-t} U(t)|^2$.)

1.19. Evaluate the following integral using Parseval's theorem.

$$\int_{-\infty}^{\infty} \frac{\sin 2t}{t} \cos 1,000t \frac{\sin t}{t} \cos 2,000t \, dt$$

1.20. Evaluate the Fourier Transform of $s(t) = \cos 5t$ starting with the Fourier Transform of $\cos t$ and using the time scaling property.

1.21. Given that the Fourier Transform of $s(t)$ is $S(f)$:

(a) What is the Fourier Transform of ds/dt in terms of $S(f)$?

(b) What is the Fourier Transform of $\int_{-\infty}^{t} s(\sigma) \, d\sigma$ in terms of $S(f)$?

(Hint: Write the Fourier inversion integral and differentiate or integrate both sides of the expression.)

1.22. Use the time shift theorem to find the Fourier Transform of

$$\frac{s(t + T) - s(t)}{T}$$

From this result find the Fourier Transform of ds/dt and compare this with your answer to 1.21(a).

1.23. Show that if $S(f) = 0$ for $|f| > \sigma$, then

$$s(t) * \frac{\sin at}{\pi t} = s(t)$$

provided $a/2\pi > \sigma$. Note that the Fourier Transform of $(\sin at)/\pi t$ is as shown in the accompanying diagram.

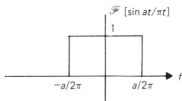

1.24. (a) Write the convolution of $s(t)$ with $U(t)$, $s(t) * U(t)$, in integral form. See if you can identify this as the integral (anti-derivative) of $s(t)$.

(b) Using the convolution theorem, what is the transform of $s(t)*U(t)$?

(c) Compare your answer to (a) and (b) with that which you found in 1.21(b).

1.25. Prove the relationship expressed by Eq. (1.55), the frequency convolution theorem.

1.26. Any arbitrary function can be expressed as the sum of an even and an odd function.

$$s(t) = s_e(t) + s_0(t)$$

where

$$s_e(t) = \tfrac{1}{2}[s(t) + s(-t)]$$
$$s_0(t) = \tfrac{1}{2}[s(t) - s(-t)]$$

(a) Show that $s_e(t)$ is indeed even and that $s_0(t)$ is odd.
   (Hint: Use the definition of an even and an odd function.)

(b) Show that $s(t) = s_e(t) + s_0(t)$.

(c) Find $s_e(t)$ and $s_0(t)$ for $s(t) = U(t)$, the unit step.

(d) Find $s_e(t)$ and $s_0(t)$ for $s(t) = \cos 10t$.

1.27. Given a function $s(t)$ which is zero for negative t [i.e., $s(t) = s(t)U(t)$], find a relationship between $s_e(t)$ and $s_0(t)$. Can this relationship be used to find a relationship between the real and imaginary parts of the Fourier Transform of $s(t)$?

1.28. A time signal, $s(t)$, is put through a gate and truncated in time. The gate is closed for $1 < t < 2$. Therefore

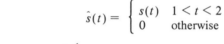

$$\hat{s}(t) = \begin{cases} s(t) & 1 < t < 2 \\ 0 & \text{otherwise} \end{cases}$$

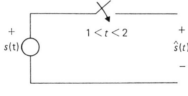 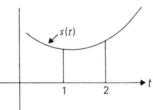

Find the Fourier Transform of $\hat{s}(t)$ in terms of $\mathscr{F}[s(t)]$.

(Hint: Can you express $\hat{s}(t)$ as the product of two time functions and use the frequency convolution theorem?

1.29. (a) Find the time derivative of the function shown.

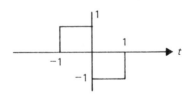

(b) Find the Fourier Transform of this derivative.

(c) Using the formula for the transform of the integral of a time function [Problems 1.21(b) or 1.24(a)] find the transform of the time function in part (a) using the answer to part (b).

1.30. Find the Fourier Transform of $s(t) = \sin 2\pi f_0 t$ from the Fourier Transform of $\cos 2\pi f_0 t$ and the time shift theorem.

1.31. Find the Fourier Transform of $\cos^2 2\pi f_0 t$ using the frequency convolution theorem and the transform of $\cos 2\pi f_0 t$. Check your answer by expanding $\cos^2 2\pi f_0 t$ by trigonometric identities.

1.32. Evaluate the following integrals:

(a) $\displaystyle\int_{-\infty}^{\infty} \frac{\sin 3\tau}{\tau} \, \delta(t - \tau) \, d\tau$

(b) $\displaystyle\int_{-\infty}^{\infty} \frac{\sin 3(\tau - 3)}{(\tau - 3)} \frac{\sin 5(t - \tau)}{(t - \tau)} \, d\tau$

(c) $\displaystyle\int_{-\infty}^{\infty} \delta(1 - t)(t^3 + 4) \, dt$

1.33. In Eq. (1.104) (the final statement of the sampling theorem) verify that $s(t)$ given in this form coincides with the sample values at the sample points ($t = n/2f_m$). Sketch the first few terms in the summation for a typical $s(t)$.

1.34. The $s(t)$ of Problem 1.4c with $T = 1$ and $\tau = 0.5$ multiplies a time function $g(t)$, with $G(f)$ as shown below.

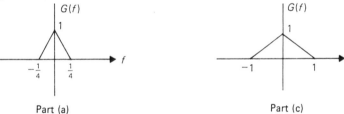

Part (a)   Part (c)

(a) Sketch the Fourier Transform of $g_s(t) = g(t)s(t)$.
(b) Can $g(t)$ be recovered from $g_s(t)$?
(c) If $G(f)$ is as sketched in Part (c), can $g(t)$ be recovered from $g_s(t)$? Explain your answer.

1.35. You are given a function of time,

$$s(t) = \frac{\sin t}{t}$$

This function is sampled by being multiplied by $s_\delta(t)$ as shown below.

$$s_s(t) = s(t)s_\delta(t)$$

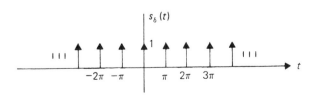

(a) What is the Fourier Transform of the sampled function, $S_s(f)$?

(b) From your answer to part (a), what is the actual form of $s_s(t)$?

(c) Should your answer to part (b) have been a train of impulses? Did it turn out that way? Explain any apparent discrepancies.

(d) Describe a method of recovering $s(t)$ from $s_s(t)$ and show that it works in this specific case.

1.36.   A signal, $s(t)$, with $S(f)$ as shown, is sampled by two different sampling functions, $s_{\delta 1}(t)$ and $s_{\delta 2}(t)$, where

$$s_{\delta 2}(t) = s_{\delta 1}\left(t - \frac{T}{2}\right) \quad \text{and} \quad T = \frac{1}{2f_m}$$

Find the Fourier Transform of the sampled waveforms,

$$s_{s1}(t) = s_{\delta 1}(t)s(t)$$

$$s_{s2}(t) = s_{\delta 2}(t)s(t)$$

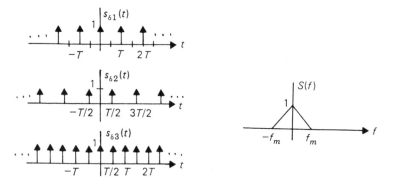

Now consider $s(t)$ sampled by $s_{\delta 3}(t)$, a train of impulses spaced $T/2$ apart. Note that this new sampling function is equal to $s_{\delta 1}(t) + s_{\delta 2}(t)$. Show that the transform of $s_{s3}(t)$ is equal to the sum of the transforms of $s_{s2}(t)$ and $s_{s1}(t)$. (Hint: Don't forget the phase of $s_{\delta 2}(t)$. You should find that when $S_{\delta 1}(f)$ is added to $S_{\delta 2}(f)$, every second impulse is cancelled out.)

1.37.   The function, $s(t) = \cos 2\pi t$, is sampled every $\frac{3}{4}$ second. Evaluate the aliasing error.

1.38.   Find the DFT of the following sampled function.

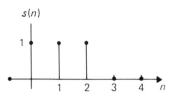

1.39.   Show that (a) the DFT of an even function is real and (b) the DFT of an odd function is imaginary.

1.40.   What is the effect upon the DFT of a function, $s(t)$, if the function is delayed by n sampling periods? That is, find the DFT of $s(t - nT_s)$ in terms of the DFT of $s(t)$.

1.41.   Find the DFT of the convolution $s(t)*g(t)$ in terms of the DFT of $s(t)$ and the DFT of $g(t)$.

1.42.   Find the DFT of the following function.

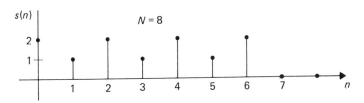

1.43.   Find the DFT of the following function.

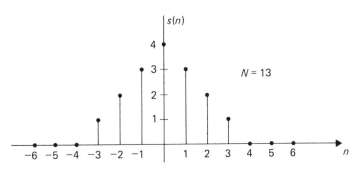

1.44.   Find the FFT of the signals of Problems 1.38, 1.42, and 1.43, and compare the answers with those found earlier.

1.45.   Find the z-Transform of the signal of Problem 1.43. Compare this to the Fourier Transform of a triangular pulse.

1.46.   Find the z-Transform of $f(n) \cos 10n$ in terms of the z-Transform of $f(n)$.

1.47.   You are given a time function, $s(t)$, with z-Transform, $S(z)$. That is, the z-Transform of a sampled version of $s(t)$ is $S(z)$. $s(t)$ is now multiplied by e^{-at}. What is the effect of this multiplication upon the z-Transform?

Chapter 2

Linear Systems

In Chapter 1 of this text, the basic mathematical tools required for waveform analysis were developed. We shall now apply these techniques to the study of linear systems in order to determine system capabilities and features. We will then be in a position to pick those particular characteristics of linear systems which are desirable in communication systems.

It is first necessary to define some common terms. A *system* is defined as a set of rules that associates an "output" time function to every "input" time function. This is sometimes shown in block diagram form as Fig. 2.1.

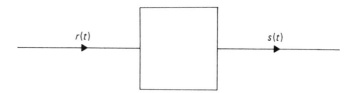

FIGURE 2.1 Block diagram of a system.

The input, or source signal, is $r(t)$; and $s(t)$ is the output, or response signal due to this input. The actual physical structure of the system determines the exact relationship between $r(t)$ and $s(t)$.

For example, if the system under study is an electric circuit, $r(t)$ can be a voltage or current signal, and $s(t)$ can be a voltage or current anywhere in the circuit. Don't be disturbed by the fact that two wires are needed to describe a voltage. The lines in Fig. 2.1 are not restricted to representing wires.

In the special case of a two terminal electrical network, $r(t)$ can be a sinusoidal voltage across the two terminals, and $s(t)$ can be the current flowing into one of the terminals and out of the other due to this impressed voltage. In this case, the relationship between $r(t)$ and $s(t)$ is the familiar complex impedance between the two terminals of the network.

Any system can be described by specifying the response, $s(t)$, associated with every possible input, $r(t)$. This is an obviously exhaustive process. We would certainly hope to find a much simpler way of describing the system.

Before exploring alternate techniques of characterizing systems, some additional basic definitions are needed.

A system is said to obey *superposition* if the output due to a sum of inputs is equal to the sum of the corresponding individual outputs. That is, given that the response (output) due to an excitation (input) of $r_1(t)$ is $s_1(t)$, and that the response due to $r_2(t)$ is $s_2(t)$, then the output due to $r_1(t) + r_2(t)$ is $s_1(t) + s_2(t)$.

A single ended arrow is often used as a shorthand method of relating an input to its resulting output. That is,

$$r(t) \rightarrow s(t) \tag{2.1}$$

and is read, "an input, $r(t)$, causes an output, $s(t)$."

In terms of this notation, the definition of superposition is as follows. If $r_1(t) \rightarrow s_1(t)$ and $r_2(t) \rightarrow s_2(t)$, then

$$r_1(t) + r_2(t) \rightarrow s_1(t) + s_2(t) \tag{2.2}$$

Some thought will convince you that in order for a system to obey superposition, the source free, or transient response (response due to initial conditions) must be zero. In practice, one often replaces a circuit having non-zero initial conditions with one containing zero initial conditions. Additional sources are added to replace the initial condition contributions.

A concept closely related to superposition is *linearity*. Assume again that $r_1(t) \rightarrow s_1(t)$ and $r_2(t) \rightarrow s_2(t)$. The system is said to be linear if the following relationship holds true for all values of the constants a and b:

$$ar_1(t) + br_2(t) \rightarrow as_1(t) + bs_2(t)$$

In the remainder of this text, we will use the words *linearity* and *superposition* interchangeably.

A system is said to be *time invariant* if the response due to an input is not dependent upon the actual time of occurrence of the input. That is, a time shift in input signal causes an equal time shift in the output waveform. In symbolic form, if $r(t) \rightarrow s(t)$, then $r(t - t_0) \rightarrow s(t - t_0)$ for all real t_0.

It should be clear that a sufficient condition for an electrical network to be time invariant is that its component element values do not change with time (assuming constant initial conditions). That is, resistances, capacitances, and inductances remain constant.

2.1   THE SYSTEM FUNCTION

Returning to the task of characterizing a system, we shall see that for a time invariant system which obeys superposition, a very simple description is possible. That is, instead of requiring that we know the response due to *every* possible input, it will turn out that we need only know it for one "test" input. The derivation of this highly significant result follows.

Recall that the convolution of any function with an impulse yields the original function. That is, we can write any general input, $r(t)$, in the following form,

$$r(t) = r(t)*\delta(t),$$

$$= \int_{-\infty}^{\infty} r(\tau)\delta(t - \tau)\, d\tau \tag{2.3}$$

This integral can be considered as a limiting case of a weighted sum of delayed delta functions. That is,

$$r(t) = \lim_{\Delta\tau \to 0} \sum_{-\infty}^{\infty} r(n\Delta\tau)\delta(t - n\Delta\tau)\Delta\tau \tag{2.4}$$

Suppose that we know the system response due to $\delta(t)$. Suppose also that the system is time invariant and obeys superposition. If we call this impulse response, $h(t)$, then the response due to

$$r(t) = a\delta(t - t_1) + b\delta(t - t_2)$$

will be

$$s(t) = ah(t - t_1) + bh(t - t_2)$$

That is, if

$$\delta(t) \to h(t)$$

then

$$a\delta(t - t_1) + b\delta(t - t_2) \to ah(t - t_1) + bh(t - t_2) \tag{2.5}$$

Generalizing this to the infinite sum (integral) input, we see that the output due to $r(t)$ in Eq. (2.4), call it $s(t)$, is given by

$$r(t) \to s(t) = \lim_{\Delta\tau \to 0} \sum_{-\infty}^{\infty} r(n\Delta\tau)h(t - n\Delta\tau)\, \Delta\tau$$

$$= \int_{-\infty}^{\infty} r(\tau)h(t - \tau)\, d\tau$$

$$= r(t)*h(t) \tag{2.6}$$

In words, the output due to any general input is found by convolving that input with the system's response to an impulse. That is, if the impulse response is known, then the response to any other input is automatically also known. Equation (2.6) is sometimes called the *superposition integral equation*. It is probably the most significant result of system theory.

Recall from Chapter 1 that the Fourier Transform of $\delta(t)$ is unity. Therefore, in some sense, $\delta(t)$ contains all frequencies to an equal degree. Perhaps this observation hints at the delta function's suitability as a test function for system behavior.

Taking the Fourier Transform of both sides of Eq. (2.6), we have the corresponding equation in the frequency domain,

$$S(f) = R(f)H(f) \tag{2.7}$$

The Fourier Transform of the output due to $r(t)$ as an input to a linear (we are being sloppy about linearity as promised) system is equal to the product of the Fourier Transform of the input with the Fourier Transform of the impulse response. $H(f)$ is sometimes given the name *system function* since it truly characterizes the linear system. It is also sometimes given the name, *transfer function* since it is the ratio of the output Fourier Transform to the input Fourier Transform.

2.2   COMPLEX TRANSFER FUNCTION

In the sinusoidal steady state analysis of an electrical network, a complex transfer function is defined. It represents the ratio of the output phasor (complex number giving amplitude and relative phase of a sinusoid) to the input phasor. This ratio is a complex function of frequency. In the special case in which the input is a current flowing between two terminals and the output is the voltage due to this current, the transfer function becomes the familiar complex impedance between these two terminals.

For example, in the circuit of Fig. 2.2, if $i_1(t)$ is the input, and $v(t)$ is the output, the transfer function is

$$H(f) = \frac{4j\pi f}{1 + 4j\pi f} \tag{2.8}$$

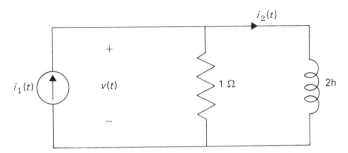

FIGURE 2.2   Circuit with Eqs. (2.8) and (2.9) as two possible transfer functions.

Alternatively, if $i_2(t)$ is regarded as the output and $i_1(t)$ as input, then the transfer function is

$$H(f) = \frac{1}{1 + 4j\pi f} \tag{2.9}$$

It is the intention of this section to show that this $H(f)$, derived for the simple sinusoidal steady state case, is actually the system function, $H(f)$ (i.e., the transform of the impulse response).

For example, in the previous circuit the complex transfer function between $i_1(t)$ and $i_2(t)$ was found as

$$H(f) = \frac{1}{1 + 4j\pi f}$$

We now claim that the transform of the impulse response is

$$H(f) = \frac{1}{1 + 4j\pi f}$$

That is, if $i_1(t) = \delta(t)$, then the transform of $i_2(t)$ is

$$I_2(f) = \frac{1}{1 + 4j\pi f}$$

If this is indeed the case, the output, $i_2(t)$, due to $i_1(t) = \delta(t)$ would be given by the Fourier inversion integral,

$$i_2(t) = \int_{-\infty}^{\infty} \frac{1}{1 + 4j\pi f} e^{j2\pi ft} \, df \tag{2.10}$$

The general proof of this result follows.

Suppose that the input to a circuit is
$$r(t) = e^{j2\pi f_0 t}$$

Then the output due to this input is found by sinusoidal steady state techniques to be

$$s(t) = H(f)e^{j2\pi f_0 t} \tag{2.11}$$

The complex transfer function evaluated at f_0 is $H(f_0)$. This gives us the input-output pair,

$$e^{j2\pi f_0 t} \rightarrow H(f_0)e^{j2\pi f_0 t}$$

Taking the transform of the input and of the corresponding output (noting that $H(f_0)$ is a constant) yields

$$e^{j2\pi f_0 t} \leftrightarrow \delta(f - f_0) = R(f)$$
$$H(f_0)e^{j2\pi f_0 t} \leftrightarrow H(f_0)\delta(f - f_0) = S(f) \tag{2.12}$$

Recalling that, for a linear system, $R(f)H(f) = S(f)$ (Eq. 2.7), where $H(f)$ is the *system* function, we have

$$\delta(f - f_0)H(f) = H(f_0)\delta(f - f_0) \qquad (2.13)$$

If we invoke the "sifting" property of the impulse function,

$$r(t)\delta(t - t_0) = r(t_0)\delta(t - t_0)$$

Eq. (2.13) becomes

$$\delta(f - f_0)H(f_0) = H(f_0)\delta(f - f_0)$$

and

$$H(f_0) = H(f_0) \qquad (2.14)$$

Equation (2.14) must be viewed cautiously. The left side is the Fourier Transform of the impulse response, $h(t)$, while the right side is the sinusoidal steady state transfer function. The equation holds for any value of f_0, and thus, our proof is complete.

Illustrative Example 2.1

In the circuit of Fig. 2.3, the capacitor is initially uncharged. If $v(t) = \delta(t)$, find $i(t)$.

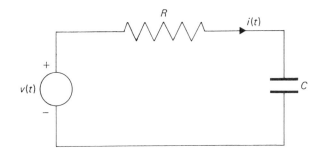

FIGURE 2.3   Circuit for Illustrative Example 2.1.

Solution

The complex transfer function relating $i(t)$ to $v(t)$ is

$$H(f) = \frac{1}{R + 1/j2\pi fC} = \frac{j2\pi fC}{1 + j2\pi fRC} \qquad (2.15)$$

The problem asks for $i(t)$ due to $v(t) = \delta(t)$. This is simply the impulse response, whose transform is the system function, $H(f)$. Since $H(f)$ is the Fourier Transform of $h(t)$, we have

$$i(t) = \int_{-\infty}^{\infty} \frac{j2\pi fCe^{j2\pi ft}}{1 + j2\pi fCR} \, df \qquad (2.16)$$

At this point, it should be emphasized that the evaluation of integrals cannot be considered as a significant part of communication theory. One can look up the answer in a table or as a last resort approximate the evaluation on a computer. Therefore, in this text we will only evaluate integrals in those cases when the result is of instructional significance. Otherwise, the final solutions will be left in integral form.

In this particular example, we need not evaluate the integral at all since the inversion can be performed by inspection. Those familiar with the LaPlace Transform will notice that this is analogous to partial fraction expansions. We note that $H(f)$ can be rewritten as follows,

$$H(f) = \frac{j2\pi fC}{1 + j2\pi fCR} = \frac{1}{R} - \frac{1/R}{1 + j2\pi fCR}$$

$$i(t) = \frac{1}{R}\,\delta(t) - \frac{1}{R^2 C}\,c^{-t/RC}U(t) \tag{2.17}$$

This current waveform is sketched as Fig. 2.4.

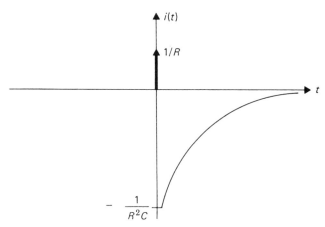

FIGURE 2.4 $i(t)$ for Illustrative Example 2.1 (circuit of Fig. 2.3).

Note that the impulse in $i(t)$ has appeared automatically. This is noteworthy since so-called classical (traditional) techniques handle impulses with a great deal of difficulty or by resorting to physical reasoning.

It should be realized that the above solution is only correct for zero initial charge on the capacitor. Otherwise, superposition is violated (prove it) and the output is not the convolution of $h(t)$ with the input.

This apparent shortcoming of the Fourier Transform analysis of systems with non-zero initial conditions is easily circumvented by treating initial conditions as sources. However, the consideration of initial conditions is not critical to communication theory as we will use it, and we will therefore always assume zero initial conditions.

2.3   FILTERS

The word "filter" refers to the removal of the undesired parts of something. In linear system theory, it was probably originally applied to systems which eliminate undesired frequency components from a time waveform. It has been modified to include systems which simply weigh the various frequency components of a function.

In other words, a linear filter is any linear system which was intended to have the characteristics that it does have. Not as dramatic as you had hoped, perhaps, but extremely useful.

We shall be using the term "ideal distortionless filter" many times in the work to follow. It would therefore be appropriate to define *distortion* at this time.

We define a distorted time signal to be one whose basic shape has been altered. That is, $r(t)$ can be multiplied by a constant and shifted in time without distorting the actual waveform.

In mathematical terms, we consider $Ar(t - t_0)$ to be an undistorted version of $r(t)$, where A and t_0 are any real constants (and, of course, $A \neq 0$). The transform of $Ar(t - t_0)$ can be found from the time shift theorem.

$$Ar(t - t_0) \leftrightarrow Ae^{-j2\pi f t_0}R(f) \qquad (2.18)$$

where

$$r(t) \leftrightarrow R(f)$$

We can consider this as the output of a linear system with input $r(t)$ and system function, $H(f) = Ae^{-j2\pi f t_0}$ as shown in Fig. 2.5. Here, since $H(f)$ is complex, both

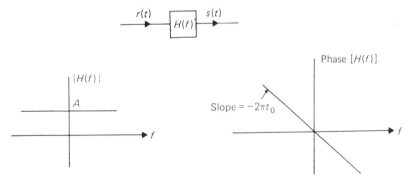

FIGURE 2.5.   System function of a distortionless system.

the magnitude and phase of $H(f)$ have been plotted. The real and imaginary parts would have also sufficed, but would not have been as instructive.

When we view Fig. 2.5, it seems reasonable that the magnitude function turned out to be a constant. This indicates that all frequencies of $r(t)$ are multiplied by the same factor. Why did the phase turn out to be a linear function of frequency? Why aren't all frequencies shifted by the same "amount"? The answer is clear from a simple example. Suppose that we wished to shift the function, $r(t) = \cos 2\pi t$ by $\frac{1}{4}$ second in time. This would represent a shift of $\pi/2$ radians, or $90°$ in the argument of the cosine. If we now wished to shift a function of twice the frequency, $r(t) = \cos 4\pi t$ by the same $\frac{1}{4}$ second, it represents a shift of π radians, or $180°$. The extension to a continuous representation of frequencies should now be obvious.

Ideal Lowpass Filter

An ideal lowpass filter is a linear system which acts like an ideal distortionless system, provided that the input signal contains no frequency components above the "cutoff" frequency of the filter. Frequency components above this cutoff are completely blocked, and do not appear in the output.

Denoting the maximum output frequency (cutoff of the filter) as f_m, (the "m" denotes maximum), we see that the system function is given by

$$H(f) = \begin{cases} Ae^{-j2\pi ft_0} & |f| < f_m \\ 0 & |f| > f_m \end{cases} \qquad (2.19)$$

This transfer function is sketched in Fig. 2.6.

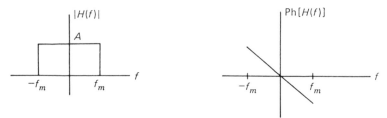

FIGURE 2.6   Ideal lowpass filter characteristic.

Illustrative Example 2.2

A sampler is followed by an ideal lowpass filter, as shown in Fig. 2.7. $p(t)$ is a periodic signal (sometimes a pulse train) of fundamental frequency, f_s. The lowpass filter passes

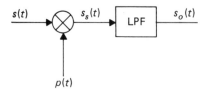

FIGURE 2.7   System for Illustrative Example 2.2.

frequencies up to the maximum frequency of $s(t)$, f_m. We assume $f_s > 2f_m$. Find the output of the lowpass filter.

Solution

The input to the filter was found in Section 1.10 to be

$$s_s(t) = a_0 s(t) + \sum_n a_n s(t) \cos n2\pi f_s t$$

with Fourier Transform as illustrated in Fig. 1.40. If this Fourier Transform is multiplied by the ideal lowpass filter characteristic, the result is a transform which is the center section of the transform of Fig. 1.40. This low frequency portion of the transform of $s_s(t)$ is simply a_0 times the transform of the original $s(t)$. Therefore, the filter output is $a_0 s(t)$. This example illustrates that the lowpass filter can be used to recover $s(t)$ from the sampled waveform.

The impulse response of the ideal lowpass filter is found by computing the inverse Fourier Transform of $H(f)$.

$$h(t) = \int_{-\infty}^{\infty} H(f) e^{j2\pi ft} \, df$$

$$= \int_{-f_m}^{f_m} A e^{-j2\pi ft_0} e^{j2\pi ft} \, df$$

$$= \frac{A \sin 2\pi f_m(t - t_0)}{\pi(t - t_0)} \tag{2.20}$$

This $h(t)$ is shown in Fig. 2.8.

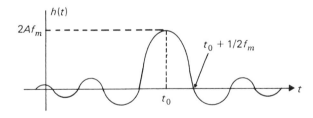

FIGURE 2.8   Impulse response of ideal lowpass filter.

Ideal Bandpass Filter

Rather than transmit (pass) frequencies between zero and f_m as the lowpass filter does, one might desire a system which passes frequencies between two other limits, say f_1 and f_2.

An ideal bandpass filter is such a linear system. It acts like an ideal distortionless system provided that the input signal contains no frequency components outside of the filter's "pass band."

The system function of an ideal bandpass filter with pass band $f_1 < |f| < f_2$ is given by

$$H(f) = \begin{cases} Ae^{-j2\pi f t_0} & f_1 < |f| < f_2 \\ 0 & \text{otherwise} \end{cases} \tag{2.21}$$

This $H(f)$ is shown in Fig. 2.9.

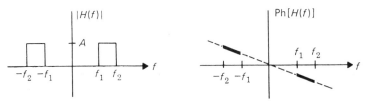

FIGURE 2.9   Ideal bandpass filter characteristic.

Illustrative Example 2.3

A sampler is followed by an ideal bandpass filter, as shown in Fig. 2.10. Find the output of the bandpass filter.

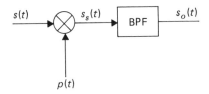

FIGURE 2.10   System for Illustrative Example 2.3.

Solution

The input to the filter was found in Section 1.10 to be

$$s_s(t) = a_0 s(t) + \sum_n a_n s(t) \cos n2\pi f_s t$$

with Fourier Transform as illustrated in Fig. 1.40. If this Fourier Transform is multiplied by the ideal bandpass filter characteristics, the resulting transform is as shown in Fig. 2.11. This can be recognized as the transform of the a_1 term in the Fourier Series expansion of $s_s(t)$.

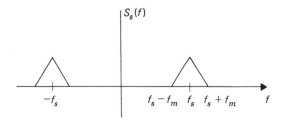

FIGURE 2.11   Output Fourier transform for Illustrative Example 2.3.

Therefore, the output of the bandpass filter is $a_1 s(t) \cos 2\pi f_s t$. The bandpass filter could have been centered around the frequency of any one of the terms in the summation, and therefore could have been used to separate out whatever term is desired.

The impulse response of the bandpass filter, $h(t)$, can be found from the result for the lowpass filter and the frequency shifting theorem. That is, given $H_{1p}(f)$ as shown in Fig. 2.12, $H(f)$ is given by

$$H(f) = H_{lp}\left[f - \frac{f_1 + f_2}{2}\right] + H_{lp}\left[f + \frac{f_1 + f_2}{2}\right]. \tag{2.22}$$

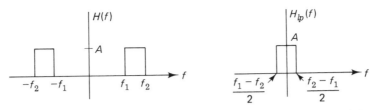

FIGURE 2.12 Bandpass characteristic and corresponding lowpass characteristic.

Note that for convenience, we have chosen $t_0 = 0$. This makes the phase equal to zero which simplifies the derivation. We can always reinsert a time shift into the final answer for $h(t)$. Therefore, from Eq. (2.22) we have

$$h(t) = h_{lp}(t)e^{j2\pi f_{av}t} + h_{lp}(t)e^{-j2\pi f_{av}t} \tag{2.23}$$

where, for notational convenience, f_{av}, the average frequency is defined as

$$f_{av} \triangleq \tfrac{1}{2}(f_1 + f_2)$$

The result from Eq. (2.23) follows from the frequency shift theorem. Continuing, we have

$$h(t) = h_{lp}(t)[e^{j2\pi f_{av}t} + e^{-j2\pi f_{av}t}]$$
$$= 2h_{lp}(t)\cos 2\pi f_{av}t$$
$$= 2h_{lp}(t)\cos \pi(f_1 + f_2)t \tag{2.24}$$

$h_{lp}(t)$ is the impulse response of the lowpass filter with $t_0 = 0$. From Eq. (2.20), we have

$$h_{lp}(t) = \frac{A}{\pi t} \sin \pi(f_2 - f_1)t$$

Finally, for the bandpass filter, the impulse response is given by

$$h(t) = \frac{2A}{\pi t} \sin[\pi(f_2 - f_1)t]\cos[\pi(f_1 + f_2)t] \tag{2.25}$$

If the filter phase factor is included, it can be simply accounted for by a time shift. For the ideal bandpass filter, we therefore have

$$h(t) = \frac{2A}{\pi(t - t_0)} \sin[\pi(f_2 - f_1)(t - t_0)] \cos[\pi(f_1 + f_2)(t - t_0)]$$

This impulse response is sketched in Fig. 2.13. The outline of this sketch resembles

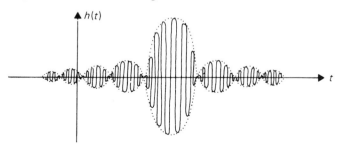

FIGURE 2.13   Impulse response of ideal bandpass filter.

the impulse response of the lowpass filter. We note that as the two limiting frequencies become high compared to the difference between them, the impulse response starts resembling a shaded-in version of the lowpass impulse response (and its mirror image). In more precise terms, this happens when the center frequency of the bandpass filter becomes large compared to the width of its pass band. This observation will prove highly significant in our later studies of amplitude modulation.

Distortion

Distortion is defined as anything that a filter does to a signal other than pure delay or constant multiplication. We shall assume that the filters we will encounter are linear and therefore cannot create frequencies that are not present in the input.

Linear distortion can cause time spreading of the input signal. Assume that the filter transfer function is of the form:

$$H(f) = A(f)e^{-j\theta(f)} \tag{2.26}$$

$A(f)$ is the amplitude factor and $\theta(f)$ is the phase factor, so all we have done is changed these factors to functions of frequency from the constant (or linear) values of Eq. (2.19).

Distortion arises from the two frequency-dependent terms in Eq. (2.26). If $A(f)$ is not a constant in f, we have what is known as amplitude distortion, and if $\theta(f)$ is not linear in f, we have phase distortion. Each of these two types of distortion will be analyzed separately.

Amplitude Distortion

Let us first assume that $\theta(f)$ is linear with frequency. That is, we assume no phase distortion is present. The filter transfer function is therefore of the form

$$H(f) = A(f)e^{-j2\pi f t_0}$$

As in Eq. (2.19), t_0 represents the delay between input and output of the filter.

One way to analyze this for all types of amplitude variation is to expand $A(f)$ in a series. As an example, it may be possible to expand $A(f)$ in a Fourier Series (we are usually dealing with a limited band of frequencies, so $A(f)$ can be repeated as a periodic function). Doing so yields

$$H(f) = \sum_{n=0}^{N} H_n(f)$$

where the terms in the summation are of the form

$$H_n(f) = a_n \cos \frac{n\pi f}{f_m} e^{-j2\pi f t_0} \qquad |f| < f_m \qquad (2.27)$$

Each term in Eq. (2.27) represents a cosine filter whose amplitude characteristic follows a cosine wave in the pass band. This is illustrated in Fig. 2.14 for $N = 1$. The characteristic of Fig. 2.14 represents a non-ideal lowpass filter. The system function can be expressed as follows in the pass band of the filter.

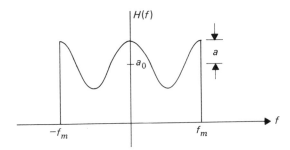

FIGURE 2.14   Cosine filter.

$$H(f) = \left(A + a \cos \frac{2\pi f}{f_m} \right) e^{-j2\pi f t_0}$$

$$= Ae^{-j2\pi f t_0} + \tfrac{1}{2}a \left(\exp \left[j2\pi f \left(\frac{1}{f_m} - t_0 \right) \right] \right.$$

$$\left. + \exp \left[j2\pi f \left(-\frac{1}{f_m} - t_0 \right) \right] \right)$$

This system function can be viewed as the sum of three distortionless system functions with three different delays. Suppose that the input to this filter, $r(t)$, has a Fourier Transform with no component above f_m. Then the output of the cosine filter would be

$$s(t) = Ar(t - t_0) + r\left(t + \frac{1}{f_m} - t_0\right) + r\left(t - \frac{1}{f_m} - t_0\right)$$

The extra two terms can be thought of as echoes, one advanced and one delayed by $1/f_m$ seconds from the undistorted signal.

Returning to the general filter case, we see that the output due to amplitude distortion is a sum of shifted inputs. Thus with

$$H(f) = \sum_{n=0}^{\infty} a_n \cos \frac{n\pi f}{f_m} e^{-j2\pi f t_0}$$

$$= \sum_{n=0}^{\infty} (\tfrac{1}{2}a_n e^{jfn\pi/f_m} + a_n e^{-jfn\pi/f_m}) e^{-j2\pi f t_0}$$

the output, $s(t)$, due to an input, $r(t)$, is given by

$$s(t) = \sum_{n=0}^{\infty} \tfrac{1}{2}a_n \left[r\left(t + \frac{n}{2f_m} - t_0\right) + r\left(t - \frac{n}{2f_m} - t_0\right) \right]$$

$$= a_0 r(t - t_0) + \sum_{n=1}^{\infty} \tfrac{1}{2}a_n \left[r\left(t + \frac{n}{2f_m} - t_0\right) + r\left(t - \frac{n}{2f_m} - t_0\right) \right]$$

$$\tag{2.28}$$

This approach to general amplitude distortion is only practical if the Fourier Series contains relatively few significant terms. In these cases, the distortion is evident as a series of echoes.

Illustrative Example 2.4

Consider the triangular filter characteristic shown in Fig. 2.15. Assume that the phase

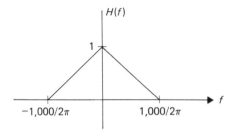

FIGURE 2.15 Filter characteristic for Illustrative Example 2.4.

characteristic is linear, with slope $-2\pi t_0$. This is not an ideal lowpass filter, since the amplitude characteristic is not a constant. Find the output of this filter when the input is

$$r(t) = \sin \frac{200t}{\pi t}$$

Solution

Although the output can be found by simply multiplying $R(f)$ by $H(f)$ and taking the inverse Fourier Transform, we shall use the series technique to illustrate the echoes in the output function.

It is first necessary to expand $H(f)$ in a Fourier Series. The period is $1000/\pi$, so the fundamental frequency (actually has units of time) is $t_0 = \pi/1000$. The series becomes

$$H(f) = \tfrac{1}{2} + \frac{4}{\pi^2} \cos \frac{2\pi^2 f}{1000} + \frac{4}{9\pi^2} \cos \frac{6\pi^2 f}{1000} + \dots$$

The output can be found directly from Eq. (2-28). Retaining the first three non-zero terms yields

$$s(t) = \tfrac{1}{2} r(t - t_0) + \frac{2}{\pi^2} r \left(t - \frac{\pi}{1000} - t_0 \right) + r \left(t + \frac{\pi}{1000} - t_0 \right)$$

$$+ \frac{2}{9\pi^2} r \left(t - \frac{3\pi}{1000} - t_0 \right) + r \left(t + \frac{3\pi}{1000} - t_0 \right) \qquad (2.29)$$

The first term in Eq. (2.29) is the undistorted output. The error, or distortion, can be analyzed by comparing the function in Eq. (2.29) with an undistorted version of $r(t)$ and finding the average (or, preferably, average squared) difference. Since the consequences of the error, and therefore acceptable distortion levels, are a function of the application, we will not carry this example further. Problems at the end of this chapter investigate errors in more detail.

Phase Distortion

Phase distortion is caused by the phase characteristic varying from the distortionless (linear phase) case. It can be characterized by variations in the slope of the phase characteristic and in the slope of a line from the origin to a point on the curve. Two characteristics of the phase plot can be given physical meaning. We define *group delay* (also known as *envelope delay*) and *phase delay* as follows:

$$t_{ph}(f) = \frac{\theta(f)}{2\pi f}$$

$$t_{gr}(f) = \frac{d\theta(f)}{2\pi df}$$

Figure 2.16 illustrates these definitions for a representative phase characteristic. For an ideal distortionless channel, the phase characteristic is linear and the group and phase delays are both constant for all f. In the ideal case, both of these delays are equal to the time delay of the input signal, t_o.

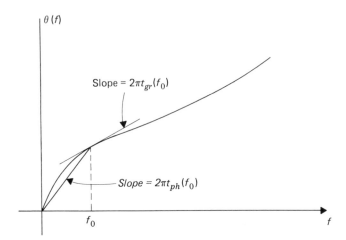

FIGURE 2.16 Definition of group and phase delay.

Suppose the filter input occupied a relatively small frequency band around f_0. We could approximate $\theta(f)$ as a straight line in this small region. The equation of the straight line is

$$\theta(f) = \theta(f_0) + \frac{d\theta(f_0)}{df}(f - f_0)$$

or using the definition of group and envelope delay, this becomes

$$\theta(f) = t_{ph}(f_0)2\pi f_0 + 2\pi(f - f_0)t_{gr}(f_0)$$

Let us now examine the case of a filter which has phase distortion but no amplitude distortion. That is, the amplitude characteristic is a constant. We assume that the filter has an input which is bandlimited to some range of frequencies. The easiest way to form such a signal is to start with a low frequency time signal and multiply it by a cosine waveform. By the modulation theorem, the Fourier Transform of this waveform is

$$\tfrac{1}{2}[R(f - f_0) + R(f + f_0)]$$

The transform of the filter output, $S(f)$, is given by Eq. (2.30) for positive values of f, and for negative f, it is the conjugate of this expression.

$$S(f) = \tfrac{1}{2}[R(f - f_0)]A \exp(jt_{ph}(f_0)2\pi f_0)\exp(j2\pi(f - f_0)t_{gr}(f_0)) \qquad (2.30)$$

Equation (2.30) can be simplified (see Exercise 2.15 at the end of this chapter) to yield the output time function,

$$s(t) = Ar(t - t_{gr}(f_0))\cos 2\pi f_0(t - t_{ph}(f_0))$$

This is the product of $f(t - t_{gr}(f_0))$ with a delayed cosine. Comparing this with the given function, we see that the low frequency signal has been delayed by $t_{gr}(f_0)$, and the multiplying cosine has been delayed by $t_{ph}(f_0)$. Although time shift is not a form of distortion, note that a shift of the function would require that $t_{gr}(f_0) = t_{ph}(f_0)$.

2.4   CAUSALITY

Things are going much too well. We must now darken the picture by showing that it is impossible to build ideal filters. We do this by introducing the idea of *causality*. The concept of causality refers to the "cause and effect" relationship. In loose terms, it states that the effect, or response, due to a cause, or input, cannot anticipate the input in a causal system. That is, a causal system's output at any particular time depends only upon the input prior to that time, and not upon any future values of the input.

$$s(t_0) \text{ depends only upon } r(t) \qquad (t \le t_0)$$

In order for a linear system to be causal, it is necessary and sufficient that the impulse response, $h(t)$, be zero for $t < 0$. It follows that the response due to a general $r(t)$ input depends only upon past values of $r(t)$. To prove this, we write the expression for the output of a linear system in terms of its input and impulse response.

$$s(t) = \int_{-\infty}^{\infty} h(\tau) r(t - \tau) \, d\tau$$

If $h(t) = 0$ for $t < 0$, this becomes

$$s(t) = \int_{0}^{\infty} h(\tau) r(t - \tau) \, d\tau \qquad (2.31)$$

If we now change variables, letting $t - \tau = k$, this becomes

$$s(t) = \int_{-\infty}^{t} r(k) h(t - k) \, dk \qquad (2.32)$$

which, as stated, shows that $s(t)$ is known in terms of the values of $r(k)$ for $k \le t$, that is all past values of the input. This proves sufficiency. In order to prove necessity, we note that $\delta(t) = 0$ for all $t < 0$. Therefore, in a causal system, the inputs $r(t) = 0$ and $r(t) = \delta(t)$ must yield the same outputs at least until time $t = 0$. That is, if the output cannot anticipate the input, it has no way of telling the difference between "0" and $\delta(t)$ prior to time $t = 0$. But in a linear system, an input which is identically equal to zero yields an output which is also identically equal to zero. Therefore, $h(t)$ must equal zero for $t < 0$, and the necessity part of the theorem is proven.

The study of causality is important since, in general, causal systems are physically realizable, and non-causal systems are not (much to the dismay of astrologers, fortune tellers, earthquake predictors, and gamblers).

The above criterion is easy to apply if $h(t)$ is explicitly known. That is, one need simply examine $h(t)$ and see if it is zero for negative "t." Since the system function, $H(f)$, is often known, and the integral evaluation of $h(t)$ from $H(f)$ is sometimes difficult to perform, it would be nice to translate the "$h(t) = 0$ for $t < 0$" requirement into a corresponding restriction on $H(f)$. Then, if causality were the only thing we were looking for, it would not be necessary to find the time function corresponding to $H(f)$. The *Paley-Wiener criterion* does just this. If

$$\int_{-\infty}^{\infty} \frac{|\ln H(f)|}{1 + (2\pi f)^2} \, df < \infty$$

and

$$\int_{-\infty}^{\infty} |H(f)|^2 df < \infty \qquad (2.33)$$

then, for an appropriate choice of phase function for $H(f)$, $h(t) = 0$ for $t < 0$.

We note that Eq. (2.33) does not take the phase of $H(f)$ into account at all. The actual form of $h(t)$ certainly does depend upon this phase. Indeed, if some $H(f)$ corresponds to a causal system, we can form a non-causal system by shifting the original impulse response to the left in time. This time shift corresponds to a linear change in the phase of $H(f)$, which does not affect Eq. (2.33). Thus, the Paley-Wiener criterion really tells us whether it is possible for $H(f)$ to be the transform of a causal time function (i.e., a necessary but not sufficient condition). Assuming that Eq. (2.33) were satisfied, the phase of $H(f)$ would have to be examined before final determination about the causality of the system could be made. Because of all these complications, we will only use the Paley-Wiener condition to make one simple, but significant observation.

Since $\ln (0) = -\infty$, one can observe that if $H(f) = 0$ for any range along the f-axis, the first integral of Eq. (2.33) does not converge, and the system cannot be physically realizable. In particular, all of our ideal lowpass and bandpass filters must be non-causal. We already knew this from the actual observation of their impulse responses, $h(t)$, which could be easily evaluated for the ideal filters. For this reason, practical filters must be, at best, approximations to their ideal versions.

The previous observation about $H(f)$ is a special case of a much more general theorem. Suppose $h(t)$ is such that

$$\int_{-\infty}^{\infty} |h(t)|^2 dt < \infty \qquad (2.34)$$

With this restriction, if $h(t)$ is identically zero for any finite range along the t-axis, then its Fourier Transform cannot be zero for any finite range along the f-axis, and vice versa. This is sometimes stated in loose terms as "a function which is time limited cannot be bandlimited."

This leads to an unfortunate fact of life. Any time function which does not exist for all time cannot be bandlimited. Nothing in real life does exist for all time since no sources were turned on at the dawn of creation (even if they were, that would not be early enough). Alas, it is time to approximate again.

The next section briefly discusses some of the most common approximations that are used in place of the ideal filters. We emphasize that the remainder of this text will be presented as if the results using ideal filters were exact representations of what happens in real life. The student must bear in mind the fact that, while the mathematics is exact, approximations must be involved in applying it to real life situations. Some of the problems at the back of this chapter deal with this concept.

The next section will show that, provided enough electrical elements are available, the ideal characteristics can be approached arbitrarily closely. We prematurely emphasize this fact now since, the student who already believes it, can skip the next section without any loss of continuity. Once one possesses a vague awareness of practical limitations, the design of filters can be divorced from the discipline of communication theory. Communication theorists must leave some work for the filter designers to do.

2.5   PRACTICAL FILTERS

We now present actual circuits which approximate the ideal bandpass and lowpass filters. Throughout this section, we will assume that the closer $H(f)$ approaches the system function of an ideal filter, the more the filter will behave in an ideal manner in our applications. This fact is not at all obvious. A very "small" change in $H(f)$ can lead to large changes in $h(t)$. One can examine the consequences and categorize the effects of deviation from the constant amplitude characteristic or from the linear phase characteristic of the ideal system functions. We will not concern ourselves with this aspect of the analysis.

Before any of the known filter synthesis techniques can be applied to filter design, a mathematical expression must be found for $H(f)$. It is not sufficient to work from a sketch of $H(f)$.

Let us begin by analyzing the lowpass filter.

Lowpass Filter

The amplitude characteristic of the ideal lowpass filter can be approximated by the function

$$|H_n(f)| = \frac{1}{\sqrt{1 + (2\pi f)^{2n}}} \tag{2.35}$$

This is sketched for several values of "n" in Fig. 2.17. We have only illustrated the positive half of the f-axis, since the function is an even function of "f," as it must be in

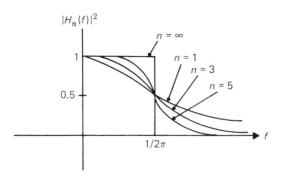

FIGURE 2.17 Approximation to ideal lowpass filter characteristic (Eq. (2.35)).

order to represent the magnitude of the transform of a real impulse response. We have chosen $f_m = 2\pi$ for simplicity, but this does not restrict our approach to the synthesis.

Note that as "n" gets larger, the amplitude characteristic approaches that of the ideal filter.

Filters which have amplitude characteristics as in Eq. (2.35) are known as *Butterworth filters*. They are used very often since they represent the "best" approximation on the basis of "maximal flatness" in the filter pass band. The concept of maximal flatness relates to the number of derivatives of the magnitude function which are zero at $f = 0$. The more derivatives which are zero, the more slowly one would expect the function to turn away from its value at $f = 0$. (Think of a Taylor series expansion of the magnitude function.) One can verify that the first $2n - 1$ derivatives of the amplitude function of Eq. (2.35) are zero at $f = 0$.

There are other useful approximations to the ideal lowpass filter characteristic, most of which optimize with respect to a criterion other than maximal flatness. We will only examine the Butterworth filter since our only intention is to demonstrate realizability of one good approximation to the ideal filters.

Before actually designing the Butterworth filter, we must say something about the phase of $H(f)$. The result of Problem 1.27 indicates that, for a causal time function, the odd part of the Fourier Transform can be found from the even part. In a similar way, the phase is determined from the amplitude. Thus, for a causal filter, we are not free to choose the amplitude and phase of $H(f)$ independently. Equation (2.35) therefore gives us all of the information that we need to design the filter.

Recall that, for any real time function, $h(t)$,

$$H(f) = H^*(-f) \tag{2.36}$$

where the asterisk (*) indicates the operation of taking the complex conjugate. We use this fact in factoring $|H(f)|^2$.

$$|H(f)|^2 = H(f)H(-f) \tag{2.37}$$

In splitting $|H(f)|^2$ into the two factors, we must make sure that $H(f)$ is the Fourier

Transform of a causal time function. We illustrate this factoring procedure by an example.

Illustrative Example 2.5

In this example, we shall design a third order ($n = 3$) Butterworth filter with $f_m = (2\pi)^{-1}$. From Eq. (2.35),

$$|H(f)|^2 = \frac{1}{1 + (2\pi f)^6} \tag{2.38}$$

If we make a change of variables, $s = j2\pi f$, we can take advantage of the properties of pole-zero diagrams in the s-plane.

$$|H(s)|^2 = H(s)H(-s) = \frac{1}{1 - s^6} \tag{2.39}$$

The poles of $|H(s)|^2$ are the six roots of unity as sketched in Fig. 2.18. Three of the poles are

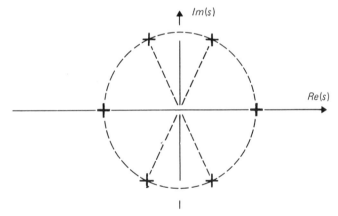

FIGURE 2.18   Six roots of unity.

associated with $H(s)$ and the other three with $H(-s)$. We associate the three poles in the left half plane with $H(s)$, since this will lead to an $h(t)$ which is causal. We are assuming that the reader is familiar with LaPlace Transforms and basic circuit theory. Otherwise this result may simply be accepted.

Finally, $H(s)$ is found from its poles,

$$H(s) = \frac{1}{(s - p_1)(s - p_2)(s - p_3)} = \frac{1}{s^3 + 2s^2 + 2s + 1} \tag{2.40}$$

There are well known techniques for synthesizing a circuit given $H(s)$. If $v(t)$ is the response and $i(t)$, the source, the above system function corresponds to the circuit of Fig. 2.19. Details

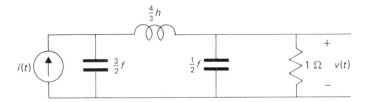

FIGURE 2.19 Third order Butterworth filter.

can be found in the references. The references also contain formulae to use in the design of Butterworth filters of any order.

Higher order filters (larger n) would lead to other ladder networks with additional elements. That is, additional series inductor, parallel capacitor combinations would appear.

We state without elaboration that, as n increases, the filter approaches an ideal filter in performance. However, as n approaches infinity, the delay between input and output (i.e., slope of the phase characteristic) also approaches infinity.

Bandpass Filter

One can derive a bandpass filter corresponding to every lowpass filter by simply substituting element combinations for each element in the lowpass filter circuit. This has the effect of shifting the frequency term in the filter transfer function. Thus the Butterworth bandpass filter has poles lying on a circle centered at the center frequency of the filter. This contrasts to the poles lying on a circle centered at the origin for the lowpass filter. A typical pole-zero diagram for a bandpass filter is sketched as Fig. 2.20.

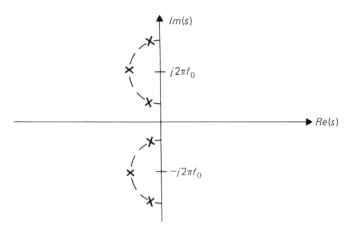

FIGURE 2.20 Poles of Butterworth bandpass filter.

The bandpass version of the Butterworth filter is what is actually used in household radios as we shall see in Chapter 4.

The synthesis of filters becomes routine with a digital computer. Since computer processing is becoming more and more popular, synthesis of actual discrete element filter circuitry is becoming less common. Entire systems are being constructed without the use of a single capacitor, inductor, or resistor. The received signals are modified and fed into a computer. The computer is programmed to simulate the operations of the various electronic devices. The output can then be printed on paper, converted to an analog electrical signal, or even displayed on an oscilloscope screen. The versatility of the computerized system is quite impressive. One can change the system function of a filter by simply changing one parameter in the program. Indeed, if the input is known for all time, one can even construct a non-causal system in the computer.

2.6   DIGITAL FILTERS

A *digital filter* is a system which converts one sequence of numbers (the input) into another sequence of numbers (the output). Each output number is a linear combination of past input numbers. This leads naturally to the convolution relationship of

$$y(n) = \sum_{m=0}^{n} x(m)h(n-m) \qquad (2.41)$$

This type of filter is known as *nonrecursive*. Digital filter outputs may also depend upon past output values in addition to past input values. That is,

$$y(n) = \sum_{m=0}^{n} x(m)h(n-m) + \sum_{m=0}^{n-1} y(m)k(n-m) \qquad (2.42)$$

The latter type of filter is known as *recursive* since the outputs must be calculated in sequence. That is, for example, before $y(3)$ can be found, $y(0)$, $y(1)$, and $y(2)$ must first be evaluated.

Realization of Digital Filters

We present herein only one form of practical realization of each type of filter.

Illustrative Example 2.6

Derive a block diagram of a nonrecursive digital filter with $h(n) = 1, 1, 2, 4$ for $n = 0, 1, 2$, and 3, respectively. $h(n) = 0$ for n greater than 3.

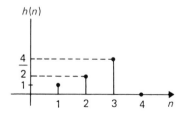

Solution

We expand the summation of Eq. (2.41) to get

$$y(n) = 4x(n - 3) + 2x(n - 2) + x(n - 1) + x(n)$$

This can be realized with the block diagram in Fig. 2.21.

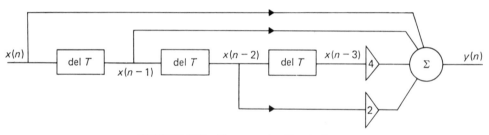

FIGURE 2.21   Non-recursive filter for Example 2.5.

The "delay T" block is often denoted as follows:

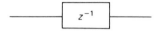

This notation is used since z^{-1} is the z-Transform of a system which delays the input by one period.

Illustrative Example 2.7

Derive the block diagram of a recursive filter with $h(n)$ and $k(n)$ as sketched below.

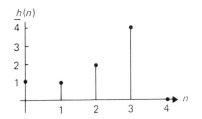

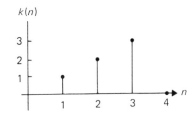

Solution

Expanding the summation in Eq. (2.42), we have

$$y(n) = 4x(n-3) + 2x(n-2) + x(n-1) + x(n) + y(n-1)$$
$$+ 2y(n-2) + 3y(n-3)$$

This equation can be implemented with the block diagram in Fig. 2.22.

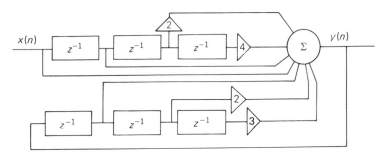

FIGURE 2.22   Recursive filter for Example 2.5.

The block diagrams of Fig. 2.21 and 2.22 can be implemented using *tapped delay lines* in analog form or by using logic circuits (e.g., registers or, indeed, computers) in the digital form.

2.7   ACTIVE FILTERS

Section 2.5 presented some realizations of filters using inductors, capacitors, and resistors. Such filters are called *passive* since all component parts either absorb or store energy.

A filter is called *active* if there are devices within it which give out energy. Active filters have the advantage of not absorbing part of the desired signal energy. They are versatile and simple to design, and arbitrary causal transfer functions can be realized. For some applications, such as audio filtering, the passive filter would require impractically large inductors and capacitors.

The basic building block of an active filter is the *operational amplifier* (*OP AMP*). The operational amplifier has characteristics which approach those of an ideal, infinite gain, amplifier. That is, it approximately possesses infinite input resistance, zero output resistance, and infinite voltage gain. Such amplifiers can be purchased as integrated circuits (ICs), and several operational amplifiers are usually included within one IC package. While an ideal OP AMP has infinite bandwidth, practical devices often have highly limited bandwidth.

The OP AMP is almost always used with a large amount of negative feedback. This is true since, as a basic amplifier, the open loop device would have very high gain, but this gain would vary widely with, among other parameters, input frequency.

Assume that an OP AMP is connected as shown in Fig. 2.23. The OP AMP of Fig. 2.23 is being operated as a differential amplifier. It amplifies the difference in voltage between the terminal marked "+" and that marked "−."

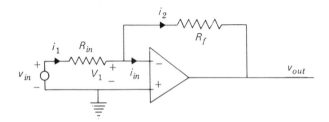

FIGURE 2.23   OP AMP with feedback.

Since the amplifier has infinite input resistance, we can assume that $i_{in} = 0$. Therefore, $i_1 = i_2$. In a practical system, v_{out} is limited by the supply voltages within the system. Since the amplification is assumed to be infinite, the assumption of finite v_{out} leads to the simplifying conclusion that $v_1 = 0$. Therefore,

$$i_1 = \frac{v_{in}}{R_{in}}$$

and

$$i_2 = -\frac{v_{out}}{R_f}$$

Since $i_1 = i_2$, we have

$$\frac{v_{in}}{R_{in}} = -\frac{v_{out}}{R_f}$$

and

$$\frac{v_{out}}{v_{in}} = -\frac{R_f}{R_{in}} \tag{2.43}$$

Thus, the amplification of the system is a function only of the elements attached to the operational amplifier.

Suppose now that R_{in} is replaced by a capacitor, as in Fig. 2.24. Equation (2.43) now becomes

$$Cv'_{in} = -\frac{v_{out}}{R_f}$$

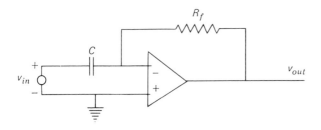

$$v_{out} = -R_f C v'_{in}$$

FIGURE 2.24 OP AMP with capacitor input.

and

$$v_{out} = -R_f C v'_{in}$$

where v'_{in} is the time derivative, dv_{in}/dt. Thus, in the configuration of Fig. 2.24, the OP AMP acts as a differentiator. The student should derive relationships for the case in which the capacitor replaces R_f instead of R_{in}. In addition, the effects of using an inductor instead of a capacitor should be easily derivable.

If more complex frequency sensitive circuits replace R_{in} and/or R_f, various filter characteristics result. As one example, a first order lowpass filter is shown in Fig. 2.25. Since neither input to the OP AMP is grounded, we start with the assumption

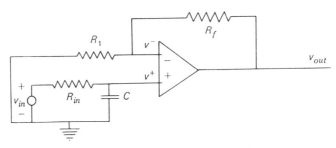

FIGURE 2.25   First order lowpass filter.

that $v^+ - v^- = 0$. Therefore, $v^+ = v^-$, and from the previous analysis,

$$v_{out} = -\left(\frac{R_f}{R_1}\right)v^- = -\left(\frac{R_f}{R_1}\right)v^+$$

Therefore, the entire system operates simply as the circuitry attached to the v^+ input. In this case, v_{out} is proportional to the voltage across the capacitor, and the transfer function is basically that of the RC circuit.

The advantage of the active system over the simple RC circuit is that v_{out} is isolated from the RC circuit by the OP AMP. In a passive RC circuit, if the voltage across the capacitor feeds further circuitry, the current drain can seriously affect the relationship between capacitor voltage and input voltage.

The simple first order lowpass filter is presented as an example and is not intended to imply that the filter of Fig. 2.25 is often used. The reader is referred to the references for additional information.

2.8   RISE TIME AND PULSE WIDTH

In designing a communication system, an important consideration is that of the bandwidth of the system. The bandwidth is the range of frequencies which the system must be capable of handling.

Bandwidth clearly relates to the Fourier Transform of a time function. It is not directly definable in terms of the time function. That is, one cannot look at a time function and say, "Aha; the bandwidth of that time function is . . . ." The Fourier Transform must first be calculated.

The Fourier Transform does *not* exist anywhere in real life. One cannot directly display it on any device in the laboratory. It is simply a mathematical definition which proves useful in the analysis of real life systems. We would therefore like to relate the concept of bandwidth to properties of signals which *do* exist in real life.

Both the minimum width of a time pulse and the minimum time in which the output of a system can jump from one level to another are truly physical terms. We wish to show that both the minimum width of a time pulse and the minimum transition time between two levels are intimately related to the system bandwidth. We shall start with a specific example, and then attempt to generalize the results.

Returning to the ideal lowpass filter response, recall that the impulse response was found to be (we repeat Eq. 2.20)

$$h(t) = \frac{\sin 2\pi f_m(t - t_0)}{\pi(t - t_0)} \tag{2.44}$$

Both this $h(t)$ and the corresponding $H(f)$ are shown in Fig. 2.26.

Note that $h(t)$ is symmetrical about $t = t_0$, and has its maximum, at this point, $t = t_0$.

A general result for time invariant linear systems is that the output due to the integral of a particular signal is the integral of the output due to the signal itself. That is, if $r(t) \rightarrow s(t)$, then

$$\int_{-\infty}^{t} r(\tau)d\tau \rightarrow \int_{-\infty}^{t} s(\tau)\, d\tau \tag{2.45}$$

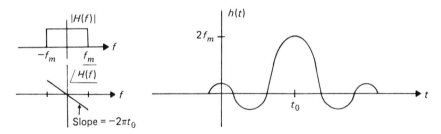

FIGURE 2.26 $H(f)$ and $h(t)$ for ideal lowpass filter.

Although we will not need it here, it is also true that

$$\frac{dr}{dt} \to \frac{ds}{dt} \tag{2.46}$$

This derivative relationship is easily proven since, for a time invariant linear system,

$$\frac{1}{\Delta t}[r(t + \Delta t) - r(t)] \to \frac{1}{\Delta t}[s(t + \Delta t) - s(t)]$$

Letting Δt approach zero, we get

$$\frac{dr}{dt} \to \frac{ds}{dt}$$

The integral property can be derived from the derivative relationship by a change of variables (do it!).

Using this result, we can find the response of a lowpass filter when its input is a unit step function. Since the unit step is the integral of an impulse function, the step response, sometimes denoted as $a(t)$, is simply the integral of $h(t)$.

$$U(t) \to a(t) = \int_{-\infty}^{t} h(\tau)\, d\tau \tag{2.47}$$

The step response of a lowpass filter is sketched in Fig. 2.27.

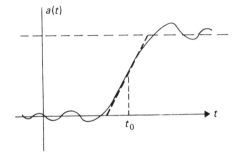

FIGURE 2.27 Step response of ideal lowpass filter.

Since the instantaneous slope of $a(t)$ is equal to $h(t)$, the maximum slope, $2f_m$, occurs at the point $t = t_0$.

There are several common definitions of rise time. They all try to mathematically define the length of time it takes the output to respond to a change, or jump, in the input. In practice, it is difficult to actually define the point in time when the output has finished responding to the input jump. We present one particular definition which happens to work well when applied to the ideal lowpass filter example.

The rise time may be defined as the time required for a signal to go from the initial to the final value (steady state value) along a ramp with constant slope equal to the maximum slope of the actual function. That is,

$$\text{Rise time} = t_r \overset{\Delta}{=} \frac{f_{\text{final}} - f_{\text{initial}}}{df/dt \text{ maximum}} \qquad (2.48)$$

For the lowpass filter example, the rise time is therefore given by

$$t_r = \frac{1}{2f_m}$$

The ramp with slope equal to the maximum derivative is shown as a dashed line in Fig. 2.27. We note that since the lowpass filter in question passes frequencies between zero and f_m, its bandwidth is f_m. Therefore the preceding result becomes

$$t_r = \frac{1}{2BW}$$

From this result, we make the bold generalization that, for any bandlimited linear system (i.e., $H(f) = 0$ outside of some interval in the f-axis), the rise time is inversely proportional to the bandwidth of the system. The bandwidth is defined as the width of the interval occupied by $H(f)$ on the positive f-axis. The actual constant of proportionality depends upon the shape of $H(f)$, upon the definition of rise time, and upon the definition of bandwidth. The bandwidth definition is required in cases where $H(f)$ does not equal exactly zero outside the interval, but, for example, may asymptotically approach zero.

In a slightly more general way, we would now like to show that an inverse relationship also exists between the width of a pulse in the time domain and the bandwidth of its Fourier Transform (i.e., the width of the corresponding pulse in the frequency domain). The similarity between pulse width and rise time should be somewhat obvious. The student should intuitively justify the statement that the minimum pulse width is related to twice the rise time.

Since the time limited pulse cannot be bandlimited (*see* "Paley-Wiener condition"), a new definition of bandwidth is required. Actually our definition to follow will require neither the time function nor its transform to be strictly limited.

We assume a positive pulse, $s(t)$, symmetrical about $t = 0$ (even). We shall choose a definition of pulse width which yields a simple result, even though it is not analogous to our previous definition of rise time. The pulse width T, will be defined as the width of a rectangular pulse of height $s(0)$, and area the same as that of the original pulse (*see* Fig. 2.28). Thus,

$$T \overset{\Delta}{=} \frac{\int_{-\infty}^{\infty} s(t)\, dt}{s(0)} \tag{2.49}$$

Since $s(t)$ is assumed to be real and even, its transform $S(f)$, will be a real and even function of f. We define the bandwidth of $S(f)$ in exactly the same way in which we defined the pulse width of $s(t)$. The bandwidth, W, is then given by (*see* Fig. 2.28)

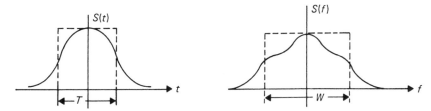

FIGURE 2.28   Pictorial definition of pulse width and bandwidth.

$$W \overset{\Delta}{=} \frac{\int_{-\infty}^{\infty} S(f)\, df}{S(0)} \tag{2.50}$$

Note that the definition in Eq. (2.50) does not restrict the function to resemble the sketch of Fig. 2.28. That is, the transform of a time pulse does not necessarily resemble a pulse in frequency. Indeed, it doesn't necessarily have a maximum at $f = 0$. (Can you think of an example in which it doesn't?)

We would now like to find a relationship between T and W. From the definition of the Fourier Transform,

$$S(f) \overset{\Delta}{=} \int_{-\infty}^{\infty} s(t) e^{-j2\pi ft}\, dt \tag{2.51}$$

we see that

$$S(0) = \int_{-\infty}^{\infty} s(t)\, dt \tag{2.52}$$

From the inverse Fourier Transform theorem,

$$s(t) = \int_{-\infty}^{\infty} S(f) e^{j2\pi ft}\, df \tag{2.53}$$

we see that

$$s(0) = \int_{-\infty}^{\infty} S(f)\,df \qquad (2.54)$$

Therefore, Eq. (2.49) becomes

$$T = \frac{S(0)}{s(0)}$$

and Eq. (2.50) becomes

$$W = \frac{s(0)}{S(0)}$$

Multiplying W by T, we get the desired result,

$$WT = 1$$

or

$$W = \frac{1}{T} \qquad (2.55)$$

The bandwidth and pulse width are inversely related, the constant of proportionality again depending upon the actual definition of width.

Therefore, the faster we desire a signal to change from one level to another, the more space on the frequency axis we must allow for it. This fact of life will prove significant in the digital communication schemes to be discussed later. It also has a more general interpretation, this being the "uncertainty principle." Given a duality such as that existing between time and frequency, the more accurately one wishes to measure one parameter, the worse will be the accuracy of the other measurement.

2.9   POWER AND ENERGY

A primary goal of many communication systems is the enhancement of the signal and simultaneous suppression of the noise (use an intuitive definition of noise as the "unwanted stuff" since we have not yet formally defined it). More specifically, we will want to decrease the noise power at the output of our processing system without decreasing the signal power. The system will (ideally) increase the so-called *signal to noise ratio*. Before systems can be evaluated with respect to this property, facility must be gained in calculating the power content of a signal and in determining what effect a linear system has on this power content.

We shall associate a number, E_r, with any function of time, $r(t)$.

$$E_r \triangleq \int_{-\infty}^{\infty} |r^2(t)|\,dt \qquad (2.56)$$

E_r is called the *energy* of $r(t)$. This definition makes intuitive sense since, if $r(t)$ were the voltage across, or current through, a 1-ohm resistor, E_r would be the total energy dissipated as heat in the resistor. This number, E_r, is defined by the integral of Eq. (2.56) for any time function, $r(t)$, even if the function does not represent an electrical voltage or current. This distinction between the mathematical definition of energy and the electrical definition is emphasized in Problem 2.27 at the end of this chapter.

E_r is often infinite. We therefore are forced to also define the *average power* of $r(t)$ as the average time derivative of the energy. It is certainly possible for a function to be infinite, while its derivative remains finite (consider a ramp function). Defining average power in this way, we have,

$$P_r \overset{\Delta}{=} \text{average power} = \overline{r^2(t)}$$

$$= \lim_{T \to \infty} \frac{1}{2T} \int_{-T}^{T} |r(t)|^2 \, dt \qquad (2.57)$$

The student should become convinced that if E_r is finite, P_r must equal zero, and that if P_r is non-zero, E_r must be infinite.

We will use the average power of $r(t)$, P_r, to classify time functions into one of three groups.

Group I: $P_r = 0$ This group includes all finite energy signals.

Group II: $0 < P_r < \infty$ This group includes, among other things, periodic functions of time.

Group III: $P_r = \infty$ These are messy, unbounded signals. While several common *theoretical* waveforms (i.e., an impulse train) are included in this group, we shall ignore these signals as far as power and energy studies are concerned since we cannot handle them mathematically (this approach is used more often than you think). Fortunately, this class does not occur in the real world.

We shall now say as much as possible about each of the first two groups. It is unfortunate that, as in the development of any new concept, several definitions will first have to be presented. An attempt will be made to supply some physical insight into the definitions. This will be meant only as an aid in the memorization since justification of a definition is never necessary (by definition of a definition).

Signals with Finite Energy (Group I)

By invoking Parseval's Theorem, the total energy in a time function, $r(t)$, can be rewritten as

$$E_r = \int_{-\infty}^{\infty} |r^2(t)| \, dt = \int_{-\infty}^{\infty} |R(f)|^2 \, df \qquad (2.58)$$

We choose to define the integrand of the right hand term in Eq. (2.58) as a new variable, using the Greek letter, psi (ψ).

$$\psi_r(f) \overset{\Delta}{=} |R(f)|^2 \tag{2.59}$$

Since the magnitude of the Fourier Transform of a real time function is an even function of "f" $\psi_r(f)$ will necessarily be even and Eq. (2.58) can be rewritten as

$$E_r = \int_{-\infty}^{\infty} \psi_r(f)\, df = 2 \int_{0}^{\infty} \psi_r(f)\, df \tag{2.60}$$

Does the question, "What part of the total energy of $r(t)$ lies between frequencies $f = f_1$ and $f = f_2$?" have any reasonable interpretation?

Suppose that $r(t)$ is the input to an ideal bandpass filter which only passes (transmits) input signals with frequencies between f_1 and f_2. (*See* Fig. 2.29.)

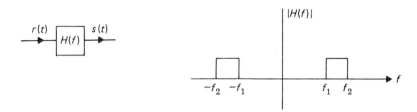

FIGURE 2.29   Ideal bandpass filter.

If $r(t)$ is a finite energy signal (Group I), it is easy to intuitively convince oneself that $s(t)$ is also a finite energy signal. If this were not true, we would have succeeded in solving a significant ecological problem by inventing a simple linear system which could replace all of the atomic, fossil fuel, and hydro-electric generating plants in the world.

Thus, the total energy of $s(t)$ is well defined. Since $s(t)$ is, in some sense, that part of $r(t)$ with frequencies between f_1 and f_2, we claim that the total energy of $s(t)$ can be thought of as that part of the energy of $r(t)$ which lies between f_1 and f_2. This total energy of $s(t)$ is given by

$$E_s = \int_{-\infty}^{\infty} |s^2(t)|\, dt$$
$$= 2 \int_{0}^{\infty} |S(f)|^2\, df \tag{2.61}$$

From basic linear system theory, we know that

$$S(f) = R(f)H(f)$$

Therefore,

$$E_s = 2 \int_0^\infty |H(f)R(f)|^2 df = 2 \int_0^\infty |H(f)|^2 |R(f)|^2 \, df$$

$$E_s = 2 \int_0^\infty |H(f)|^2 \psi_r(f) \, df = 2 \int_{f_1}^{f_2} \psi_r(f) \, df \qquad (2.62)$$

Thus, the energy of $r(t)$ between frequencies f_1 and f_2 is found by integrating $\psi_r(f)$ between these two frequency limits and multiplying the result by 2. Because of this property, $\psi_f(\omega)$ is called the *two-sided energy spectral density* (or simply energy density) of $f(t)$. We can avoid the multiplying operation by defining a "one-sided" density which is twice that found above.

The energy spectral density of the output of a linear system is determined in terms of the input density function and the system function, $H(f)$, as follows,

$$\psi_s(f) = |S(f)|^2 = |R(f)H(f)|^2 = \psi_r(f)|H(f)|^2 \qquad (2.63)$$

This is a very useful relationship. It shows us how to find the energy density, and therefore the total energy, of the output of a linear system without first finding the actual time function, (t).

In practice, given $r(t)$, $\psi_r(f)$ is sometimes hard to find using Eq. (2.59). Is it possible to find $\psi_r(f)$ from $r(t)$ in a more direct way? The answer is yes. However, another definition is first required. The only reason that we continue at this point is that this new definition will prove very useful in later work.

We define the *autocorrelation* of $r(t)$ as the inverse Fourier Transform of $\psi_r(f)$. Thus, if autocorrelation is denoted by $\varphi_r(t)$, we have

$$\varphi_r(t) \overset{\Delta}{=} \mathscr{F}^{-1}[\psi_r(f)] = \mathscr{F}^{-1}[|R(f)|^2]$$

$$\varphi_r(t) = \mathscr{F}^{-1}[R(f)R^*(f)] \qquad (2.64)$$

We have used the relationship

$$|R(f)|^2 = R(f)R^*(f)$$

where the asterisk (*) indicates complex conjugate.

By the convolution theorem, this last equality becomes

$$\varphi_r(t) = \mathscr{F}^{-1}[R(f)] * \mathscr{F}^{-1}[R^*(f)] \qquad (2.65)$$

We need a simple relationship before simplifying this further. For real $r(t)$, if

$$r(t) \leftrightarrow R(f)$$

then

$$r(-t) \leftrightarrow R^*(f)$$

This can be seen from the Fourier Transform integral.

$$\mathcal{F}[r(-t)] = \int_{-\infty}^{\infty} r(-t)e^{-j2\pi ft}\, dt$$

Changing variables, we let $\tau = -t$.

$$\mathcal{F}[r(-t)] = \int_{-\infty}^{\infty} r(\tau)e^{+j2\pi f\tau}\, d\tau$$

$$= \left[\int_{-\infty}^{\infty} r(t)e^{-j2\pi ft}\, dt\right]^{*} = R^{*}(f) \qquad (2.66)$$

We have assumed that $r(t)$ is real.

Finally, using this relationship, Eq. (2.65) can be rewritten as

$$\varphi_r(t) = r(t)*r(-t) \qquad (2.67)$$

Since $r(t)*s(t) = s(t)*r(t)$ (i.e., convolution is commutative),

$$\varphi_r(-t) = r(-t)*r(t) = \varphi_r(t)$$

and the autocorrelation is an even function of time. This could have been predicted since its Fourier Transform, the energy spectral density, is real and even.

Using the definition of convolution in order to write Eq. (2.68) in its expanded integral form,

$$\varphi_r(t) = \int_{-\infty}^{\infty} r(\tau)r(\tau - t)\, d\tau = \int_{-\infty}^{\infty} r(\tau)r(\tau + t)\, d\tau \qquad (2.68)$$

The integral of Eq. (2.68) takes $r(\tau)$, shifts it by "t," and compares (correlates) it with the unshifted version. As stated previously, it is often easier to use this integral form to find $\varphi_r(t)$ from the time function, $r(t)$, and then find the energy spectral density by taking the Fourier Transform. Indeed, at this point the *only* apparent use of the autocorrelation is as an intermediate step in finding the energy spectral density. Other applications will be developed later in the study of communications (Chapter 3). In fact, the autocorrelation will form a major tie between this elementary theory and the study of random noise.

Illustrative Example 2.8

A signal, $r(t) = e^{-2t}U(t)$ is passed through an ideal lowpass filter with cutoff frequency, $f_m = 1/2\pi$ Hz. Find the ratio of output energy to input energy.

Solution

It is first necessary to determine whether $r(t)$ is indeed a finite energy signal. To do this, we check the integral,

$$E_r = \int_{-\infty}^{\infty} |r^2(t)| \, dt = \int_0^{\infty} e^{-4t} \, dt = \tfrac{1}{4} < \infty \qquad (2.69)$$

The signal is therefore one with bounded energy. We must now find the energy spectral density of the output signal. Calling this output signal $s(t)$, we have

$$\psi_s(f) = \psi_r(f) |H(f)|^2$$

where

$$\psi_r(f) = |R(f)|^2 = \left| \frac{1}{j2\pi f + 2} \right|^2 = \frac{1}{4\pi^2 f^2 + 4}$$

$$|H(f)|^2 = \begin{cases} 1 & |f| < f_m = 1/2\pi \\ 0 & \text{otherwise} \end{cases} \qquad (2.70)$$

Therefore,

$$\psi_s(f) = \begin{cases} \dfrac{1}{4\pi^2 f^2 + 4} & |f| < 1/2\pi \\ 0 & \text{otherwise} \end{cases} \qquad (2.71)$$

Finally, the total energy contained in $s(t)$ is given by

$$2 \int_0^{\infty} \psi_s(f) \, df = 2 \int_0^{1/2\pi} \frac{1}{4\pi^2 f^2 + 4} \, df$$

$$= \frac{1}{2\pi} \tan^{-1} \left(\frac{1}{2} \right) \qquad (2.72)$$

Therefore, the ratio of output to input energy is given by

$$\frac{(1/2\pi)\tan^{-1}(\tfrac{1}{2})}{\tfrac{1}{4}} = \frac{2}{\pi} \tan^{-1} \left(\frac{1}{2} \right) = 0.29$$

Thus, 71% of the input energy is absorbed by the lowpass filter itself. Only 29% reaches the output (i.e., lies in the pass band of the filter).

Signals with Finite Power (Group II)

We have worked so hard in order to derive the previous results for finite energy signals, it would only be just that they be applicable to finite power signals. This is, of course, impossible since the energy of a finite power signal is infinite. We will, however, use the previous results as an intermediate step in the derivation of analogous power formulae.

We start with a finite power signal, $r(t)$, and modify it in such a way that it becomes a finite energy signal. One way of doing this is to truncate $r(t)$ in time. That is, define a new signal, $r_T(t)$ as follows,

$$r_T(t) \triangleq \begin{cases} r(t) & |t| < \dfrac{T}{2} \\ 0 & \text{otherwise} \end{cases} \tag{2.73}$$

Figure 2.30 shows an example of an $r(t)$ and its associated truncated version, $r_T(t)$. If

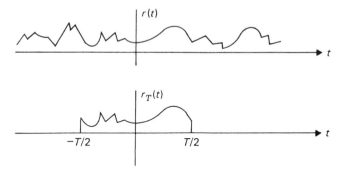

FIGURE 2.30   A representative $r(t)$ and $r_\tau(t)$.

$r(t)$ is well behaved, $r_T(t)$ will be a bounded energy signal for finite T. That is,

$$\int_{-\infty}^{\infty} |r_T(t)|^2 \, dt = \int_{-T/2}^{T/2} |r(t)|^2 \, dt < \infty$$

We can therefore define an energy spectral density of $r_T(t)$.

$$\psi_{r_T}(f) = |R_T(f)|^2$$

where

$$R_T(f) = \int_{-\infty}^{\infty} r_T(t) e^{-j2\pi ft} \, dt = \int_{-T/2}^{T/2} r(t) e^{-j2\pi ft} \, dt \tag{2.74}$$

Now we proceed to some handwaving. This energy spectral density, $\psi_{r_T}(f)$, is actually the density of the energy contained in $r(t)$ in the interval $-T/2 < t < T/2$. Therefore, the density of the average power of $r(t)$ in this interval is given by this energy density divided by the length of the interval.

$$\frac{\psi_{r_T}(f)}{T} = \text{Density of average power in } -\frac{T}{2} < t < \frac{T}{2}$$

We will call this the average power spectral density of $r(t)$ in the interval $-T/2 < t < T/2$. This is given the symbol, $G_T(f)$. Therefore,

$$G_T(f) = \frac{\psi_{r_T}(f)}{T} = \frac{1}{T} |R_T(f)|^2 \tag{2.75}$$

Since we are really interested in the average power for all time rather than in this finite interval, we let T approach infinity to get the *average power spectral density*, $G(f)$, of the finite power signal, $r(t)$.

$$G(f) = \lim_{T \to \infty} G_T(f) = \lim_{T \to \infty} \frac{1}{T} |R_T(f)|^2 \qquad (2.76)$$

where

$$R_T(f) = \int_{-T/2}^{T/2} r(t)e^{-j2\pi ft} dt$$

Since $G(f)$ is a density function, the average power of $r(t)$ in the frequency band between f_1 and f_2 is given by

$$2 \int_{f_1}^{f_2} G(f) \, df \qquad (2.77)$$

just as in the case of the energy spectral density. From Eq. (2.76) we can see that the power spectral density of the output of a linear system, $G_s(f)$, is related to that of the input, $G_r(f)$, in the same way that the energy spectral densities were related.

$$G_s(f) = G_r(f) |H(f)|^2 \qquad (2.78)$$

In the power case, we define an autocorrelation, $R_r(t)$, as the inverse Fourier Transform of $G_r(f)$.

$$R_r(t) \overset{\Delta}{=} \mathscr{F}^{-1}[G_r(f)] \qquad (2.79)$$

It is unfortunate that both $R(t)$ and $\varphi(t)$ are called the autocorrelation since their definitions are completely different. However, given an $r(t)$ it will become obvious which of the two definitions will apply. If the wrong one is accidentally used, an autocorrelation of either zero or infinity will result.

As before, $R_r(t)$, the autocorrelation, was invented as an intermediate step in finding the power spectral density. It would therefore be nice to be able to find $R_r(t)$ directly from the time function, $r(t)$. The derivation is analogous to that used for finite energy signals. Defining $R_T(t)$ as the inverse transform of $S_T(f)$, we have

$$R_T(t) \leftrightarrow G_T(f) = \frac{1}{T} |R_T(f)|^2$$

$$R_T(t) = \frac{1}{T} \mathscr{F}^{-1}[R_T(f)R_T^*(f)] \qquad (2.80)$$

where the asterisk (*) denotes complex conjugate.

From the basic properties of the Fourier Transform, and the convolution theorem,

$$R_T(t) = \frac{1}{T}[r_T(t) * r_T(-t)]$$

$$= \frac{1}{T} \int_{-\infty}^{\infty} r_T(\tau) r_T(\tau - t) \, d\tau \tag{2.81}$$

Taking the definition of $r_T(t)$ into account and carefully changing the limits of integration,

$$R_T(t) = \frac{1}{T} \int_{-(T/2)+t}^{T/2} r(\tau) r(\tau - t) \, d\tau$$

Finally, letting T approach infinity, we get, for finite t,

$$R(t) = \lim_{T \to \infty} \frac{1}{T} \int_{-T/2}^{T/2} r(\tau) r(\tau - t) \, d\tau \tag{2.82}$$

Finite Power Signals	Finite Energy Signals
Power spectral density	Energy spectral density
$G_r(f) = \lim\limits_{T \to \infty} \dfrac{1}{T} \lvert R_T(f) \rvert^2$	$\psi_r(f) = \lvert R(f) \rvert^2$
where	where
$R_T(f) = \displaystyle\int_{-T/2}^{T/2} r(t) e^{-j2\pi ft} \, dt$	$R(f) = \displaystyle\int_{-\infty}^{\infty} r(t) e^{-j2\pi ft} \, dt$
Power between f_1 and f_2	Energy between f_1 and f_2
$2 \displaystyle\int_{f_1}^{f_2} G_r(f) \, df$	$2 \displaystyle\int_{f_1}^{f_2} \psi_r(f) \, df$
Total average power	Total energy
$2 \displaystyle\int_{0}^{\infty} G_r(f) \, df$	$2 \displaystyle\int_{0}^{\infty} \psi_r(f) \, df$
Output density-linear system	Output density-linear system
$G_s(f) = G_r(f) \lvert H(f) \rvert^2$	$\psi_s(f) = \psi_r(f) \lvert H(f) \rvert^2$
Autocorrelation	Autocorrelation
$R_r(t) = \mathcal{F}^{-1}[G_r(f)]$	$\varphi_r(t) = \mathcal{F}^{-1}[\psi_r(f)]$
$R_r(t) = \lim\limits_{T \to \infty} \dfrac{1}{T} \displaystyle\int_{-T/2}^{T/2} r(\tau) r(\tau \pm t) \, d\tau$	$\varphi_r(t) = \displaystyle\int_{-\infty}^{\infty} r(\tau) r(\tau \pm t) \, d\tau$

FIGURE 2.31   Table of results on power and energy.

Figure 2.31 summarizes the results of this section.

Illustrative Example 2.9

Find the total average power of

$$r(t) = A \cos 2\pi f_0 t$$

Solution

If one attempts to apply the definition,

$$G_T(f) = \lim_{T \to \infty} \frac{1}{T} |R_T(f)|^2$$

in order to find the power spectral density of $r(t)$, *it will be found that the limit is difficult to evaluate for $f = f_0$* (try it!). We therefore attempt the autocorrelation route.

$$R(t) = \lim_{T \to \infty} \frac{1}{T} \int_{-T/2}^{T/2} r(\tau) r(\tau + t) \, d\tau$$

$$= \lim_{T \to \infty} \frac{1}{T} \int_{-T/2}^{T/2} A \cos 2\pi f_0 \tau A \cos 2\pi f_0 (\tau + t) \, d\tau \qquad (2.83)$$

Using trigonometric identities, this becomes

$$R(t) = \lim_{T \to \infty} \frac{A^2}{2T} \int_{-T/2}^{T/2} [\cos 2\pi f_0 t \cos \omega_0 (2\tau + t)] \, d\tau \qquad (2.84)$$

Since $\cos 2\pi f_0 t$ is not a function of the variable of integration,

$$R(t) = \lim_{T \to \infty} \frac{A^2}{2T} \left[\cos 2\pi f_0 t \int_{-T/2}^{T/2} d\tau + \frac{\sin 2\pi f_0 (2\tau + t)}{4\pi f_0} \Bigg|_{-T/2}^{T/2} \right]$$

$$R(t) = \lim_{T \to \infty} \frac{A^2}{2} \left[\frac{T \cos 2\pi f_0 t}{T} + \frac{\sin 2\pi f_0 (T + t)}{4\pi f_0 T} + \frac{\sin 2\pi f_0 (t - T)}{4\pi f_0 T} \right] \qquad (2.85)$$

In the limit as T approaches infinity, $(\sin x)/T$ must go to zero. This is true since, for any value of x, the absolute value of the numerator is bounded by 1, and the denominator is increasing without limit. Therefore,

$$R(t) = \frac{A^2}{2} \cos 2\pi f_0 t \qquad (2.86)$$

Taking the transform of this, we find $G(f)$, the power spectral density of $r(t)$.

$$G(f) = \mathscr{F}[R(t)] = \frac{A^2}{4} [\delta(f - f_0) + \delta(f + f_0)] \qquad (2.87)$$

The total average power of $r(t)$ is therefore given by

$$2 \int_0^\infty G(f)\, df = 2 \int_0^\infty \frac{A^2}{4} [\delta(f-f_0) + \delta(f+f_0)]\, df$$

$$\text{Average power} = P_{\text{avg}} = \frac{A^2}{2} \int_0^\infty \delta(f-f_0)\, df$$

since $\delta(f+f_0)$ doesn't fall within the range of integration. Finally,

$$P_{\text{avg}} = \frac{A^2}{2} \int_0^\infty \delta(f-f_0)\, df = \frac{A^2}{2} \tag{2.88}$$

Illustrative Example 2.10

A signal with power spectral density,

$$G(f) = \tfrac{1}{2}[\delta(f - 1/2\pi) + \delta(f + 1/2\pi)]$$

is the input to the linear system shown in Fig. 2.32. Find the average power of the output voltage, $e_{\text{out}}(t)$.

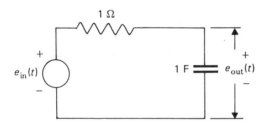

FIGURE 2.32 Circuit for Illustrative Example 2.9.

Solution

We must first find the power spectral density of $e_{\text{out}}(t)$.

$$G_{e_{\text{out}}}(f) = G_{e_{\text{in}}}(f)\,|H(f)|^2$$

where

$$H(f) = \frac{1/j2\pi f}{1 + 1/j2\pi f} = \frac{1}{1 + j2\pi f}$$

Therefore,

$$|H(f)|^2 = \frac{1}{1 + 4\pi^2 f^2}$$

and

$$G_{e_{\text{out}}}(f) = \frac{1/2}{1 + 4\pi^2 f^2}\,[\delta(f - 1/2\pi) + \delta(f + 1/2\pi)] \tag{2.89}$$

The average power at the output may now be found.

$$P_{avg} = 2 \int_0^\infty G_{e_{out}}(f)\, df$$

$$= \int_0^\infty \frac{1}{1 + 4\pi^2 f^2} [\delta(f - 1/2\pi) + \delta(f + 1/2\pi)]\, df \qquad (2.90)$$

$$= \frac{1}{1 + 1^2} = \frac{1}{2}$$

The student should verify that the average input power is equal to unity in this case.

2.10   SPECTRUM ANALYSIS

The Fourier Transform does not exist in real life. It is a mathematical tool to aid in the analysis of systems. The Fast Fourier Transform is one technique of approximating the Fourier Transform of a continuous time function in real life. The time signal can be sampled, A/D-converted, and then FFT'ed.

There are severe limitations when one attempts to find the continuous Fourier Transform of a time signal by using an analog system. First, since any system we build must be causal, the best one can hope to do is find the Fourier Transform based upon past input values, that is,

$$\int_{-\infty}^t r(t) e^{-j2\pi ft}\, dt$$

Suppose now that $r(t)$ forms the input to an ideal (non-causal) narrow bandpass filter with $H(f)$ as shown in Fig. 2.33. The output transform is given by

FIGURE 2.33   Narrowband pass filter.

$$S(f) = \begin{cases} R(f) & \text{for } f_1 < |f| < f_2 \\ 0 & \text{otherwise} \end{cases}$$

$s(t)$ is given by the inverse transform of $S(f)$,

$$s(t) = \int_{f_1}^{f_2} R(f) e^{j2\pi ft}\, df + \int_{-f_2}^{-f_1} R(f) e^{j2\pi ft}\, df \qquad (2.91)$$

If f_2 is very close to f_1, we can assume that the integrand is approximately constant over the entire range of integration. Thus, Eq. (2.91) reduces to

$$s(t) = (f_2 - f_1)(R(f_1)e^{j2\pi f_1 t} + R(-f_1)e^{-j2\pi f_1 t}) \tag{2.92}$$

Now since $R(-f_1) = R^*(f_1)$, we have

$$s(t) = (f_2 - f_1)(Re[R(f_1)]\cos 2\pi f_1 t + Im[R(f_1)]\sin 2\pi f_1 t)$$
$$= (f_2 - f_1)|R(f_1)|\cos\left(2\pi f_1 t + \angle R(f_1)\right) \tag{2.93}$$

The magnitude of the output is proportional to the magnitude of the input transform evaluated at f_1, and the phase is shifted by the phase of $R(f_1)$.

In many practical spectrum analyzers, the bandpass filter is swept across a range of frequencies (we are oversimplifying), and the magnitude of the output varies approximately with $|R(f)|$. There are three primary sources of error. First, while the filter is narrow, the bandwidth is not zero. Second, the filter is causal and non-ideal. Finally, when the center frequency of the filter is varying with time, the output will not necessarily reach its steady state value. Caution must therefore be observed in choosing a bandwidth for the filter and a rate of sweep.

PROBLEMS

2.1. Prove that the circuit shown below, where $v_{in}(t)$ is the input and $v_{out}(t)$ is the output, obeys superposition.

2.2. Does the circuit shown below, with $v_{in}(t)$ as input and $v_{out}(t)$ as output, obey superposition?

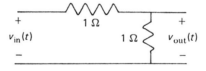

2.3. Which of the following systems are time invariant?

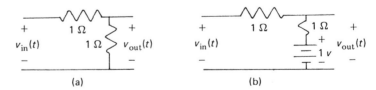

(a)   (b)

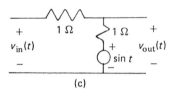

(c)

2.4. You are given a system with input $r(t)$ and output $s(t)$. You are further told that when $r(t) = 0$, $s(t)$ is not equal to zero. Show that this system cannot possibly obey superposition.

2.5. Find $H(f)$ and $h(t)$ for the following system, where $v_{in}(t)$ is the input and $v_{out}(t)$ is the output.

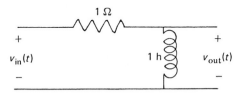

2.6. The initial charge on the capacitor in the circuit of Fig. 2.3 is not equal to zero. Show that under this condition, the circuit does not obey superposition.

2.7. Find $v_{out}(t)$ in the following circuit when $v_{in}(t)$ is equal to
(a) $\delta(t)$.
(b) $U(t)$.
(c) $e^{-2t}U(t)$.

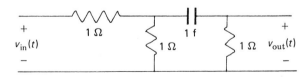

2.8. Consider the ideal lowpass filter with system function as shown below. Show that the response of this filter to an input $(\pi/K)\delta(t)$ is the same as its response to $\sin Kt/Kt$.

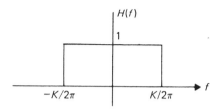

2.9. You are given the ideal lowpass filter with input as shown below. An error function is defined as the difference between input and output.

$$e(t) = x(t) - y(t)$$

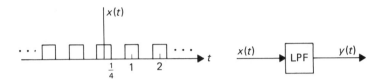

Find the error function if the filter cutoff frequency is given by

(a) $f_m = 2.5$
(b) $f_m = 3.5$
(c) $f_m = 4.5$

2.10. You are given the ideal lowpass filter with input as shown below.

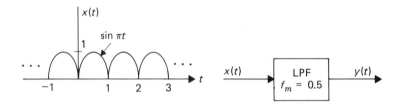

The mean square error is defined by

$$mse = \frac{1}{T} \int_0^T [y(t) - x(t)]^2 \, dt$$

(a) Show that $y(t)$ is the average value of $x(t)$.
(b) Find the mean square error.

2.11. A filter has a sinusoidal amplitude response as shown below, and a linear phase response with slope $-2\pi t_0$.
(a) Find the system response due to an input of $\cos t$.
(b) Find the system response due to an input of $(\sin t)/t$.
(c) Find the system response due to an input of $(\sin 10t)/t$.

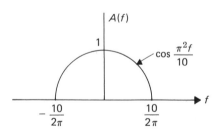

2.12. Repeat Problem 2.11 for the amplitude response shown below.

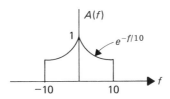

2.13.   A system is as shown below.

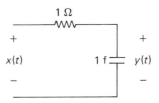

Assume that $\theta(f) = -2\pi ft_0$. Expand $A(f)$ in a series and use this to find the response due to a unit-amplitude unit-width square pulse.

2.14.   You are given the system shown below.

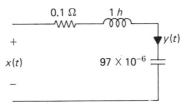

The output of the system is $i(t)$. Find the phase distortion when the input is given by

$$v_{in}(t) = \frac{\sin t}{t} \cos 200t$$

2.15.   Starting with Eq. (2.30), show that the time function corresponding to this Fourier Transform is

$$s(t) = Ar(t - t_{gr}(f_0))\cos 2\pi f_0(t - t_{ph}(f_0))$$

Hint: You may find it useful to prove the following relationships first.

$$r(t - t_0)\cos 2\pi f_0 t \leftrightarrow \tfrac{1}{2}[R(f - f_0)e^{-j2\pi(f-f_0)t_0} + R(f + f_0)e^{-j2\pi(f+f_0)t_0}]$$

$$r(t - t_1)\cos 2\pi f_0(t - t_1) \leftrightarrow \tfrac{1}{2}[R(f - f_0) + R(f + f_0)]e^{-j2\pi ft_1}$$

$$r(t - t_0)\cos 2\pi f_0(t - t_1) \leftrightarrow \tfrac{1}{2}[R(f - f_0)e^{-j2\pi(f-f_0)(t_0 - t_1)}$$
$$+ R(f + f_0)e^{-j2\pi(f+f_0)(t_0 - t_1)}]e^{-j2\pi ft_1}$$

2.16.   You wish to construct an ideal lowpass filter with

$$h_0(t) = \frac{\sin(t - 10)}{t - 10}$$

Because of causality, you actually build a filter with

$$h(t) = \begin{cases} h_0(t) & t > 0 \\ 0 & t < 0 \end{cases}$$

(a) Find $H(f)$ and compare this to the system function of the ideal filter.
(b) Find the output when the input is

$$r(t) = \frac{\sin t}{t}$$

(c) Find the error (difference between output and input) for the input of part (b).

2.17. Find and plot the phase, $\angle H(f)$, of the third order Butterworth filter shown in Fig. 2.13. Compare this to the phase of the ideal lowpass filter.

2.18. You are told that if $r(t) \rightarrow s(t)$ for a time invariant linear system, then $dr/dt \rightarrow ds/dt$. Using this fact, show that

$$\int_{-\infty}^{t} r(\sigma)\, d\sigma \rightarrow \int_{-\infty}^{t} s(\sigma)\, d\sigma$$

2.19. (a) Find the system function, $H(f)$, of the linear system shown below. $v_{in}(t)$ is the input and $v_{out}(t)$ is the output.
(b) Find the response of this system to an impulse, $v_{in}(t) = \delta(t)$. Using this impulse response the the convolution property of linear systems, find the output due to $v_{in}(t) = e^{-t}U(t)$.

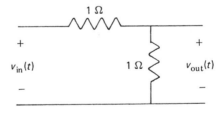

2.20. You are given a linear system with system function

$$H(f) = \delta(f - f_c) + \delta(f + f_c)$$

You are told that the output due to a certain input is given by

$$s(t) = e^{-3t}U(t)$$

Can you find $r(t)$, the input that caused this output? If not, why?

2.21. Evaluate the transfer function of the active filter shown below. What function does this filter perform?

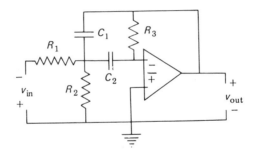

2.22. Convolve the following two sequences together and sketch the resulting sequence:

$$x(m) = \begin{cases} 1 & 0 \le m \le 5 \\ 0 & \text{otherwise} \end{cases}$$

$$h(n) = \begin{cases} 1 & 0 \le n \le 2 \\ 0 & \text{otherwise} \end{cases}$$

2.23. Convolve the following two sequences together and sketch the resulting sequence:

$$x(m) = \begin{cases} 1 & 0 \le m \le 4 \\ 0 & \text{otherwise} \end{cases}$$

$$h(n) = \begin{cases} n & 0 \le n \le 4 \\ 0 & \text{otherwise} \end{cases}$$

2.24. Draw a block diagram of a nonrecursive digital filter with $h(n)$ as given in Problem 2.22. Repeat for $h(n)$ as given in Problem 2.23.

2.25. A time function, $x(t) = \cos \pi t$, is sampled once each second to form $x(n)$.
   (a) This $x(n)$ forms the input to a digital filter with $h(n)$ as given in Problem 2.22. Find the output of the filter.
   (b) $x(n)$ forms the input to a filter with $h(0) = h(1) = 2$, and $h(n) = 0$ for all other values of n. Find the output of this filter, and compare with your answer to part (a). Describe the action of this filter.

2.26. Find the impulse response for an ideal bandpass filter. Show that the filter bandwidth is inversely proportional to the width of this impulse response.

2.27. In the circuit shown below the input power spectral density is given by $G_{v_i}(f)$ as shown. Find the power spectral density of the voltages across the resistors, $G_{v_1}(f)$ and $G_{v_2}(f)$. Find the total

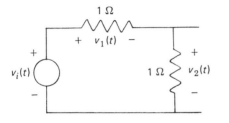 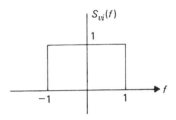

input power and the total power of $v_1(t)$ and $v_2(t)$. Does conservation of power seem to be in trouble? (Hint: Carefully reread the definition of energy.)

2.28. Find the average power, $\overline{v_{out}^2(t)}$ at the output of the circuit shown if $v_{in}(t)$ is a finite power signal with power density spectrum

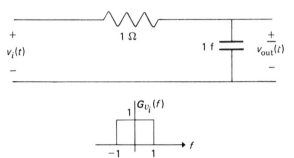

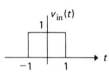

(a) $G_{v_i}(f) = K$.
(b) $G_{v_i}(f) = \delta(f+1) + \delta(f-1)$.
(c) $G_{v_i}(f)$ as shown:
(d) The time signal, $v_{in}(t)$, as shown:

2.29. If $r(t) = A$, is $r(t)$ a finite power or finite energy signal? Find its autocorrelation and power (or energy) density spectrum.

2.30. If $r(t) = AU(t)$, is $r(t)$ a finite power or finite energy signal? Find its autocorrelation and power (or energy) density spectrum.

2.31. The impulse response of a linear system is

$$h(t) = e^{-2t}U(t)$$

The autocorrelation of its input is $R_f(t) = \delta(t)$.
(a) Find the input power spectral density.
(b) Find the output power spectral density, $G_g(f)$.
(c) Find the total output power.
(d) Find the output autocorrelation (use table of transforms or leave in integral form).

2.32. A signal $r(t) = A$ is the input to an ideal bandpass filter with system function as shown below. Find the total output power or energy.

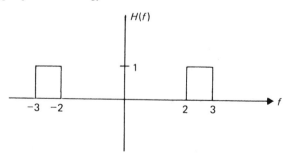

2.33. The signal, $r(t)$, goes through a linear system with transfer function, $H(f) = 2\pi jf$. This is a differentiator. If $S_f(f)$ is the power spectrum of the input, what is the power spectrum of the output? What is the autocorrelation of the output in terms of $R_f(t)$? (Use the time differentiation property of the Fourier Transform.)

2.34. Find the autocorrelation of $r(t)$, where

$$r(t) = 2\cos t + 3\cos 3t + 4\sin 4t$$

Find the average power of this signal. From this result you can reason what the average power of $r(t)$ would be if $r(t)$ were given by

$$r(t) = \sum_{n=1}^{\infty} a_n \cos n\, 2\pi f_0 t$$

Chapter 3
Probability and Random Analysis

3.1 INTRODUCTION

Studying communication systems without taking noise into consideration is analogous to learning how to drive a car by practicing in a huge abandoned parking lot. It is a valuable first step, though it is not very realistic. Were it not for what is known as noise, the communication of signals would be trivial indeed. One could simply build a circuit with a battery and a microphone and hang the wire outside the window. The intended receiver (perhaps on the other side of the globe) would simply hold a wire outside the window, receive a very weak signal, and use lots and lots of amplification.

In real life, the person at the receiver would receive a complete mess made up of an infinitesimal portion of the desired signal mixed with static (automotive ignition, people dialing telephones and turning lights on and off, distant electrical storms, etc.) and also mixed with the signals of all other people wishing to communicate at the same time. The real challenge of communication is therefore that of separating the desired signal from the undesired junk.

Anything other than the desired signal is called *noise*. For example, a radar system attempting to track a particular object will consider the signals returned from other objects to be noise. The point is that the definition of noise goes far beyond the usual assumption of static. If you tune channel 4 on your television and receive a background image of channel 3, that background image is noise.

The most common annoying types of noise do fall into the "static" category and are made up of a combination of numerous sources. In fact, those sources are so numerous that one cannot hope to describe the noise in a deterministic way (i.e., by giving a formula). We must therefore resort to discussing noise in terms of *averages*.

The study of averages requires a knowledge of basic probability theory, which justifies this chapter.

3.2   BASIC ELEMENTS OF PROBABILITY THEORY

Probability theory can be approached either on a strictly mathematical level or in an empirical setting. The mathematical approach embeds probability theory into the realm of abstract set theory, and is itself somewhat abstract. The empirical approach satisfies one's intuition. Fortunately, this latter appraoch will be sufficient for our projected applications.

We use the so-called "relative frequency" definition of probability. Before we present this definition, some other related definitions will be required.

An *experiment* is a set of rules governing an operation which is performed.

An *outcome* is a result realized after performing the experiment once.

An *event* is a combination of outcomes.

For example, consider the experiment defined by flipping a single die. There are six possible outcomes, these being any one of the six surfaces of the die facing upward after the performance of the experiment. There are many possible events (64 to be precise). For example, one event would be that of "an even number of dots showing." This event is a combination of the three outcomes; two dots, four dots, and six dots. Another possible event would be "one dot." This latter event is called an *elementary event* since it is actually equal to one of the outcomes.

We can now define what is meant by the probability of an event. Suppose that the experiment is performed N times, where N is very large. Suppose also that in "n" of these N experiments, the outcome belongs to a given event. If N is large enough, the probability of this event is given by the ratio n/N. That is, it is the fraction of times that the event occurred. Formally, we define the probability of an event, A, as

$$Pr\{A\} \triangleq \lim_{N \to \infty} \left[\frac{n_A}{N} \right] \tag{3.1}$$

where n_A is the number of times that the event A occurs in N performances of the experiment. This definition is intuitively satisfying. For example, if a coin were flipped many times, the ratio of the number of heads to the total number of flips would approach $\frac{1}{2}$. We would therefore define the probability of a head to be $\frac{1}{2}$.

Suppose that we now consider two different events, A and B, with probabilities

$$Pr\{A\} = \lim_{N \to \infty} \left[\frac{n_A}{N} \right] \text{ and } Pr\{B\} = \lim_{N \to \infty} \left[\frac{n_B}{N} \right]$$

If A and B could not possibly occur at the same time, we call them _disjoint_. The events "an even number of dots" and "two dots" are not disjoint in the die-throwing example, while the events "an even number of dots" and "an odd number of dots" are disjoint.

The probability of event A or event B would be the ratio of the number of times A or B occurs divided by N. If A and B are disjoint, this is seen to be

$$Pr\{A \text{ or } B\} = \lim_{N \to \infty} \frac{n_A + n_B}{N} = Pr\{A\} + Pr\{B\} \qquad (3.2)$$

Equation (3.2) expresses the "additivity" concept. That is, if two events are disjoint, the probability of their "sum" is the sum of their probabilities.

Since each of the elementary events (outcomes) is disjoint from every other outcome, and each event is a sum of outcomes, we see that it would be sufficient to assign probabilities to the elementary events only. We could derive the probability of any other event from these. For example, in the die-flipping experiment, the probability of an even outcome is the sum of the probabilities of a "2 dots," "4 dots" and "6 dots" outcome.

Illustrative Example 3.1

Consider the experiment of flipping a coin twice. List the outcomes, events, and their respective probabilities.

Solution

The outcomes of this experiment are (letting H denote heads and T, tails)

$$HH, HT, TH, \text{ and } TT$$

We shall assume that somebody has used some intuitive reasoning or has performed this experiment enough times to establish that the probability of each of the four outcomes is $\frac{1}{4}$. There are sixteen events, or combinations of these outcomes. These are

$\{HH\}, \{HT\}, \{TH\}, \{TT\}$
$\{HH, HT\}, \{HH, TH\}, \{HH, TT\}, \{HT, TH\}, \{HT, TT\}, \{TH, TT\}$
$\{HH, HT, TH\}, \{HH, HT, TT\}, \{HH, TH, TT\}, \{HT, TH, TT\},$
$\{HH, HT, TH, TT\}, \text{ and } \{\emptyset\}$

Note that the comma within the curly brackets is read as "or." Thus, the events $\{HH, HT\}$ and $\{HT, HH\}$ are identical. For completeness we have included the zero event, denoted $\{\emptyset\}$. This is the event made up of none of the outcomes.

By the additivity rule, the probability of each of these events would be the sum of the probabilities of the outcomes comprising each event. Therefore,

$Pr\{HH\} = Pr\{HT\} = Pr\{TH\} = Pr\{TT\} = \frac{1}{4}$
$Pr\{HH, HT\} = Pr\{HH, TH\} = Pr\{HH, TT\}$
$\qquad\qquad\qquad = Pr\{HT, TH\} = Pr\{HT, TT\} = Pr\{TH, TT\} = \frac{1}{2}$
$Pr\{HH, HT, TH\} = Pr\{HH, HT, TT\} = Pr\{HH, TH, TT\} = Pr\{HT, TH, TT\} = \frac{3}{4}$
$Pr\{HH, HT, TH, TT\} = 1$
$Pr\{\emptyset\} = 0$

The last two probabilities indicate that the event made up of all four outcomes is the "certain" event. It has probability "1" of occurring since each time the experiment is performed, the outcome must belong to this event. Similarly, the zero event has probability zero of occurring since each time the experiment is performed, the outcome does not belong to the zero event (sometimes called the "null set").

Random Variables

We would like to perform several forms of analysis upon these probabilities. As such, it is not too satisfying to have symbols such as "heads," "tails," and "two dots" floating around. It would be much preferable to work with numbers. We therefore will associate a real number with each possible outcome of an experiment. Thus, in the single flip of the coin experiment, we could associate the number "0" with "tails" and "1" with "heads." Similarly, we could just as well associate "π" with heads and "2" with tails.

The mapping (function) which assigns a number to each outcome is called a *random variable.*

With a random variable so defined, many things can now be done that could not have been done before. For example, we could plot the various outcome probabilities as a function of the random variable. In order to make such a plot, we first define something called the *distribution function, $F(x)$.* If the random varible is denoted by[1] "X," then the distribution function, $F(x)$, is defined by

$$F(x_0) = Pr\{X \leq x_0\} \qquad (3.3)$$

We note that the set $\{X \leq x_0\}$ defined an event, a combination of outcomes.

Illustrative Example 3.2

Assign two different random variables to the "one flip of a die" experiment, and plot the two distribution functions.

Solution

The first assignment which we shall use is the one that is naturally suggested by this particular experiment. That is, we will assign the number "1" to the outcome described by the face with one dot ending up in the top position. We will assign "2" to "two dots," "3" to "three dots," etc. We therefore see that the event $\{X \leq x_0\}$ includes the one dot outcome if x_0 is between 1 and 2. If x_0 is between 2 and 3, the event includes the one dot and the two dot outcomes. Thus the distribution function is easily found, and is shown in Fig. 3.1 (a).

If we now assign the random variable as follows,

Outcome	Random Variable
one dot	1
two dots	π
three dots	2
four dots	$\sqrt{2}$
five dots	11
six dots	5

[1]We shall use capital letters for random variables and lower case letters for the values that they take on. Thus $X = x_0$ means that the random variable X is equal to the number x_0.

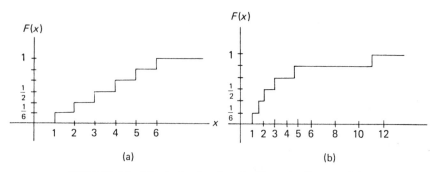

(a)   (b)

FIGURE 3.1   Distribution functions for Illustrative Example 3.2.

we find, for example, that the event $\{X \leq 3\}$ is the event made up of three outcomes; one dot, three dots, and four dots. The distribution function is easily found and is shown in Fig. 3.1(b).

Note that a distribution function can never decrease with increasing argument. This is true since an increase in argument can only add outcomes to the event. We also easily verify that

$$F(-\infty) = 0 \quad \text{and} \quad F(+\infty) = 1 \tag{3.4}$$

Density function

Whenever a function exists, somebody is going to come along and differentiate it. In the particular case of $F(x)$, the derivative has great significance. We therefore define

$$p_X(x) \triangleq \frac{dF(x)}{dx} \tag{3.5}$$

and call $p_X(x)$ the *density function* of the random variable, X. By definition of the derivative,

$$F(x_0) = \int_{-\infty}^{x_0} p_X(x)\, dx \tag{3.6}$$

The random variable can be used to define any event. For example, $\{x_1 < X \leq x_2\}$ defines an event. Since the events $\{X \leq x_1\}$ and $\{x_1 < X \leq x_2\}$ are disjoint, the additivity principle can be used to prove that

$$Pr\{X \leq x_1\} + Pr\{x_1 < X \leq x_2\} = Pr\{X \leq x_2\}$$

or

$$Pr\{x_1 < X \leq x_2\} = Pr\{X \leq x_2\} - Pr\{X \leq x_1\} \tag{3.7}$$

Therefore in terms of the density,

$$Pr\{x_1 < X \le x_2\} = \int_{-\infty}^{x_2} p_X(x)\, dx - \int_{-\infty}^{x_1} p_X(x)\, dx$$

$$= \int_{x_1}^{x_1} p_X(x)\, dx \tag{3.8}$$

We can now see why $p_X(x)$ was called a density function. The probability that X is between any two limits is given by the integral of the density function between these two limits.

The properties of the distribution function indicate that the density function can never be negative and that the integral of the density function over infinite limits must be unity.

The examples given previously (die and coin) would result in densities which contain impulse functions. A more common class of experiments gives rise to random variables whose density functions are continuous. This class is logically called the class of *continuous random variables*. Three very common density functions are presented below.

The first, and most common, density encountered in the real world is called the *Gaussian density function*. It arises whenever a large number of factors (with some broad restrictions) contribute to an end result, as in the case of static previously discussed. It is defined by

$$p_X(x) = \frac{1}{\sqrt{2\pi}\,\sigma} \exp\left[\frac{-(x-m)^2}{2\sigma^2}\right] \tag{3.9}$$

where "m" and "σ" are given constants. m is known as the *mean*, and σ as the *standard deviation*. This density function is sketched in Fig. 3.2. From the figure one

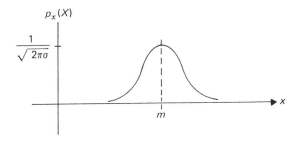

FIGURE 3.2   The Gaussian density function.

can see that the parameter "m" dictates the center position and "σ", the spread of the density function. In order to find probabilities of various events, the density must be integrated. The integration of the Gaussian density function cannot be performed in closed form. It is so important that the integral has been tabulated under the name *error function*. Since the Gaussian density is completely specified given its mean and variance, we would need a separate table of integrals for each of the infinite combinations of these two parameters. Instead, we present one table of the *normalized* integral, since any other integral can be expressed in this form with a change of variables.

The error function, as presented in Appendix IV, is defined as

$$\text{erf}(x) = \frac{2}{\sqrt{\pi}} \int_0^x e^{-u^2}\, du \tag{3.10}$$

It is clear from the table (and Eq. (3.10)) that $\text{erf}(0) = 0$ and $\text{erf}(x)$ approaches 1 as x approaches infinity. The error function is illustrated in Fig. 3.3.

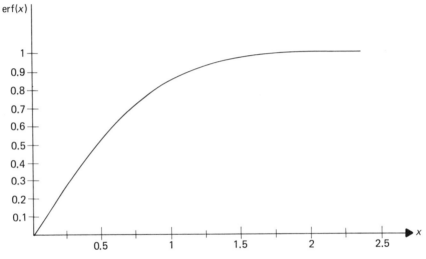

FIGURE 3.3   The error function.

Since we have

$$\text{erf}(\infty) = \frac{2}{\sqrt{\pi}} \int_0^\infty e^{-u^2} du = 1$$

$$\frac{2}{\sqrt{\pi}} \int_0^x e^{-u^2} du + \frac{2}{\sqrt{\pi}} \int_x^\infty e^{-u^2} du = 1$$

and

$$\frac{2}{\sqrt{\pi}} \int_x^\infty e^{-u^2} du = 1 - \text{erf}(x)$$

This is defined as the *complementary error function*, erfc (x).

Illustrative Example 3.3

A random variable, X, is Gaussian distributed with a mean of 5 and a variance of 4. Find the probability that X is

variance $= \sigma^2$
standard deviation $= \sigma$

(a) greater than 10;
(b) less than 2;
(c) greater than 0.

Solution

(a) The probability that X is greater than 10 is given by the integral of the density between 10 and infinity:

$$Pr(X > 10) = \int_{10}^{\infty} \frac{1}{2\sqrt{2\pi}} e^{-(x-5)^2/8} \, dx$$

Making a change of variables, we let

$$u = \frac{(x-5)}{2\sqrt{2}}$$

The probability then becomes

$$Pr(x > 10) = \frac{1}{\sqrt{\pi}} \int_{5/2\sqrt{2}}^{\infty} e^{-u^2} du$$

$$= \frac{\text{erfc}(5/2\sqrt{2})}{2}$$

$$= \frac{\text{erfc}(1.768)}{2} = 0.006$$

(b) The probability that X is less than 2 is the same as the probability that X is greater than 8. This can be seen from the symmetry of the Gaussian distribution. If we were to write this integral, it would be of the same form as the integral of part (a), and the same change of variables would reduce it to the error function. Thus, jumping to the last step, we have

$$Pr(x < 2) = Pr(x > 8) = \frac{\text{erfc}(3/2\sqrt{2})}{2}$$

$$= \frac{\text{erfc}(1.06)}{2}$$

$$= 0.066$$

(c) The probability that X is greater than zero poses a slightly different problem. If we proceed as above, we would end up with error functions with negative arguments. A simple observation gets us out of this predicament. Since X must be either greater than or less than zero, the sum of these two probabilities must be unity. Thus, the probability that x is greater than zero can be written as

$$Pr(X > 0) = 1 - Pr(X < 0)$$

But the probability that X is less than zero is the same as the probability that X is greater than 10. In part (a) we found that to be 0.006. Thus,

$$Pr(X > 0) = 1 - 0.006$$

$$= 0.994$$

Illustrative Example 3.4

You are given a Gaussian random variable with mean m and variance $v = \sigma^2$. Find the probability that the variable is

(a) within σ of its mean;
(b) within 2σ of its mean;
(c) within 3σ of its mean.

Solution

The probability that the random variable, X, is within any positive number, a, of its mean is given by the integral

$$Pr(m - a < X < m + a) = \frac{1}{\sqrt{2\pi}\sigma} \int_{m-a}^{m+a} \exp\left(\frac{(x-m)^2}{2\sigma^2}\right) dx$$

With the following change of variables,

$$u = \frac{x - m}{\sigma\sqrt{2}}$$

this becomes

$$Pr(m - a < X < m + a) = \frac{1}{\sqrt{\pi}} \int_{a/b\sqrt{2}}^{a/b\sqrt{2}} e^{-u^2} du$$

$$= \frac{2}{\sqrt{\pi}} \int_{0}^{a/b\sqrt{2}} e^{-u^2} du$$

$$= \operatorname{erf}(a/\sigma\sqrt{2})$$

From the table of error functions, we then have

$$Pr(m - \sigma < X < m + \sigma) = \operatorname{erf}(1/\sqrt{2}) = 0.723$$

$$Pr(m - 2\sigma < X < m + 2\sigma) = \operatorname{erf}(2/\sqrt{2}) = 0.954$$

$$Pr(m - 3\sigma < X < m + 3\sigma) = \operatorname{erf}(3/\sqrt{2}) = 0.997$$

The results of Illustrative Example 3.4 have practical significance. We see that the probability that a Gaussian variable is within σ of its mean is 0.723, within 2σ of its mean is 0.954, and within 3σ of the mean is 0.997. These are sometimes known as confidence limits. Thus, if we know the mean and standard deviation of a Gaussian variable, we can say that a measurement of this variable will be no more than 3 times the standard deviation away from the mean 99.7% of the time we perform the experiment.

Illustrative Example 3.5

A *binary* communication system is one which sends only two possible messages. The simplest form of binary system is one in which either *zero* or *one* volt is sent. Consider such a system in which the transmitted voltage is corrupted by additive atmospheric noise. If the receiver receives anything above $\frac{1}{2}$ volt, it assumes that a *one* was sent. If it receives anything below $\frac{1}{2}$ volt, it assumes that a *zero* was sent. Measurements have shown that if one volt is transmitted, the received signal level is random and has a Gaussian density with unit mean and $\sigma = \frac{1}{2}$. Find the probability that a transmitted *one* will be interpreted as a *zero* at the receiver.

Solution

The received signal level when a "one" is transmitted has density (assigning the random variable V as equal to the voltage level)

$$p_V(v) = \frac{1}{\sqrt{2\pi}[\frac{1}{2}]} \exp \left[\frac{-(v-1)^2}{2(\frac{1}{2})^2} \right]$$

Assuming that anything received above a level of $\frac{1}{2}$ is called "1" (and anything below $\frac{1}{2}$ is called "0"), the probability that a transmitted "1" will be interpreted as a "0" at the receiver is simply the probability that the random variable V is less than $\frac{1}{2}$. This is given by

$$\int_{-\infty}^{1/2} p_V(v)\, dv = \frac{1}{\sqrt{2\pi}[\frac{1}{2}]} \int_{-\infty}^{1/2} \exp[-2(v-1)^2]\, dv$$

Now with a change of variables, if we let

$$u = \sqrt{2}\,(v-1)$$

the integral becomes

$$\frac{1}{\sqrt{\pi}} \int_{-\infty}^{-\sqrt{2}/2} e^{-u^2}\, du = \frac{1}{2} \left(1 - \text{erf} \left(\frac{\sqrt{2}}{2} \right) \right)$$

Reference to a table of error functions (*see* Appendix IV) indicates that this integral is approximately equal to 0.16. Thus, on the average, one would expect 16 out of every 100 transmitted 1's to be misinterpreted as 0's at the receiver.

The second commonly encountered density function is the so-called *uniform density function*. This is sketched in Fig. 3.4, where x_0 is a given parameter. As an example of a situation in which this density arises, suppose you were asked to turn on a sinusoidal generator. The output of the generator would be of the form

$$v(t) = A \cos(2\pi f_0 t + \theta)$$

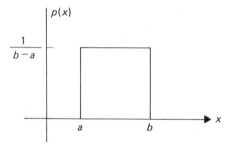

FIGURE 3.4 The uniform density function.

Since the absolute time at which you turn on the generator is random, it would be reasonable to expect that θ is uniformly distributed between 0 and 2π. It would therefore have the density shown in Fig. 3.4 with $a = 0$, $b = 2\pi$.

A third commonly encountered density function is the *Rayleigh* density function, described by

$$p_X(x) = \begin{cases} \dfrac{x}{\sigma^2} \exp\left\{ \dfrac{-x^2}{2\sigma^2} \right\} & x > 0 \\ 0 & x < 0 \end{cases} \qquad (3.11)$$

where σ is a given constant. Figure 3.5 shows two representative sketches of this density function.

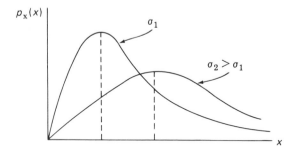

FIGURE 3.5 The Rayleigh density function.

The Rayleigh density is related to the Gaussian density function. In fact, the square root of the sum of the squares of two Gaussian distributed random variables is, itself, Rayleigh. Thus, if we transform from rectangular to polar coordinates, the radius is given by

$$r = \sqrt{x^2 + y^2}$$

Therefore, if x and y are Gaussian with the same mean and variance, in most cases (the restriction is one of *independence*, a term we have not yet defined) r will be Rayleigh. As an ideal example, suppose you were throwing darts at a target on a dart board and that the horizontal and vertical components of your error were Gaussian distributed (neglecting gravity effects). The distance from the center of the target (radial error) would then be Rayleigh distributed.

If we were to consider r^2 instead of r, we would find that variable to be *chi-square* distributed. That is, a chi-square distribution results from summing the squares of Gaussian variables. If we sum two such variables, the result is chi-square with 2 *degrees of freedom*. In general, if

$$z = x_1^2 + x_2^2 + x_3^2 + \cdots + x_n^2$$

and $x_1, x_2, \ldots, x_n$ are Gaussian with $\sigma = 1$, z will be chi-square with n degrees of freedom. The form of the density is given by

$$p(z) = \begin{cases} \dfrac{(z)^{n/2-1}}{2^{n/2}(n/2-1)!} \exp\left(\dfrac{-z}{2}\right) & z > 0 \\ 0 & z < 0 \end{cases} \quad (3.12)$$

In Equation (3.12), "!" is the *factorial*.

What if more than one parameter were required to describe the outcome of an experiment? For example in the die-tossing experiment, consider the output as the face showing *and* the time at which the die hits the floor. We would therefore have to define two random variables. The density function would have to be two dimensional, and would be defined by

$$Pr\{x_1 < X \le x_2 \quad \text{and} \quad y_1 < Y \le y_2\} = \int_{y_1}^{y_2} \int_{x_1}^{x_2} p(x, y)\, dx\, dy \quad (3.13)$$

The question now arises as to whether or not we can tell if one random variable has any effect upon the other. In the die experiment, if we knew the time at which the die hit the floor would it tell us anything about which face was showing? This question leads naturally into a discussion of *conditional* probabilities.

Let us abandon the density functions for a moment and speak of two events, A and B. The probability of event A given that event B has occurred is defined by

$$Pr[A/B] \overset{\Delta}{=} \frac{Pr\{A \text{ and } B\}}{Pr\{B\}} \quad (3.14)$$

Thus, for example, if A is the event of "two dots appearing in the die experiment" and B is the event of "an even number of dots," the probability of A given B would (intuitively) be $\frac{1}{3}$. That is, given that either 2, 4, or 6 dots appeared, the probability of 2 dots is $\frac{1}{3}$. Now using Eq. (3.14) the probability of A and B is the probability of getting two *and* an even number of dots simultaneously. This is simply the probability of 2 dots, that is, $\frac{1}{6}$. The probability of B is the probability of 2, 4, or 6 dots, which is $\frac{1}{2}$. The ratio is $\frac{1}{3}$ as expected.

Similarly we could have defined event A as "an even number of dots" and event B as "an odd number of dots." The event "A *and* B" would therefore be the zero event, and $Pr[A/B]$ would be zero. This is reasonable since the probability of an even outcome given that an odd outcome occurred is clearly zero.

It is now logical to define independence as follows: Two events, A and B, are said to be *independent* if

$$Pr[A/B] = Pr\{A\} \tag{3.15}$$

The probability of A given that B occurred is simply the probability of A. Knowing that B has occurred tells nothing about A. Plugging this in Eq. (3.14) shows that independence implies,

$$Pr\{A \text{ and } B\} = Pr\{A\}Pr\{B\} \tag{3.16}$$

Now going back to the density functions, this indicates that two variables, X and Y, are independent if

$$p(x, y) = p_X(x)p_Y(y) \tag{3.17}$$

In words, the joint density of two independent random variables is the product of their individual densities.

Illustrative Example 3.6

A coin is flipped twice. Four different events are defined.

A is the event of getting a head on the first flip.
B is the event of getting a tail on the second flip.
C is the event of a match between the two flips.
D is the elementary event of a head on both flips.

Find $Pr\{A\}$, $Pr\{B\}$, $Pr\{C\}$, $Pr\{D\}$, $Pr[A/B]$, and $Pr[C/D]$. Are A and B independent? Are C and D independent?

Solution

The events are defined by the following combination of outcomes,

$$\{A\} = \{HH, HT\}$$
$$\{B\} = \{HT, TT\}$$
$$\{C\} = \{HH, TT\}$$
$$\{D\} = \{HH\}$$

Therefore, as was done in Illustrative Example 3.1,

$$Pr\{A\} = Pr\{B\} = Pr\{C\} = \tfrac{1}{2}$$
$$Pr\{D\} = \tfrac{1}{4}$$

In order to find $Pr[A/B]$ and $Pr[C/D]$ we use Eq. (3.14),

$$Pr[A/B] = \frac{Pr\{A \text{ and } B\}}{Pr\{B\}}$$

and

$$Pr[C/D] = \frac{Pr\{C \text{ and } D\}}{Pr\{D\}}$$

The event $\{A \text{ and } B\}$ is $\{HT\}$.
The event $\{C \text{ and } D\}$ is $\{HH\}$.
Therefore,

$$Pr[A/B] = \frac{\frac{1}{4}}{\frac{1}{2}} = \frac{1}{2}$$

$$Pr[C/D] = \frac{\frac{1}{4}}{\frac{1}{4}} = 1$$

Since $Pr[A/B] = Pr\{A\}$, the event of a head on the first flip is independent of that of a tail on the second flip. Since $Pr[C/D] \neq Pr\{C\}$, the event of a match and that of two heads are not independent.

Illustrative Example 3.7

X and Y are each Gaussian random variables and they are independent of each other. What is their joint density?

Solution

$$p_X(x) = \frac{1}{\sqrt{2\pi}\sigma_1} \exp\left[-\frac{(x - m_1)^2}{2\sigma_1^2} \right] \tag{3.18}$$

$$p_Y(y) = \frac{1}{\sqrt{2\pi}\sigma_2} \exp\left[-\frac{(y - m_2)^2}{2\sigma_2^2} \right] \tag{3.19}$$

$$p(x, y) = \frac{1}{2\pi\sigma_1\sigma_2} \exp\left[-\frac{(x - m_1)^2}{2\sigma_1^2} \right] \exp\left[-\frac{(y - m_2)^2}{2\sigma_2^2} \right] \tag{3.20}$$

Functions of a Random Variable

"Everybody talks about the weather, but nobody does anything about it." We, as communication experts, would be open to the same type of accusation if all we ever did was make statements such as "there is a 42% probability of the noise being annoying." A significant part of communication engineering involves itself with changing noise from one form to another in the hopes that the new form will be less annoying than the old. We must therefore study the effects of processing upon random phenomena.

Consider a function of a random variable, $y = g(x)$, where X is a random variable with known density function. Since X is random, Y is also random. We therefore ask what the density function of Y will look like. The event $\{x_1 < X \le x_2\}$ corresponds to the event $\{y_1 < Y \le y_2\}$,[2] where

$$y_1 \overset{\Delta}{=} g(x_1) \quad \text{and} \quad y_2 \overset{\Delta}{=} g(x_2)$$

That is, the two events are identical since they include the same outcomes. (We are assuming for the time being that $g(x)$ is a single valued function.) Since the events are identical, their respective probabilities must also be equal,

$$Pr\{x_1 < X \le x_2\} = Pr\{y_1 < Y \le y_2\} \tag{3.21}$$

and in terms of the densities,

$$\int_{x_1}^{x_2} p_X(x)\, dx = \int_{y_1}^{y_2} p_Y(y)\, dy \tag{3.22}$$

If we now let x_2 get very close to x_1 Eq. (3.22) becomes

$$p_X(x_1)\, dx = p_Y(y_1)\, dy \tag{3.23}$$

and finally,

$$p_Y(y_1) = \frac{p_X(x_1)}{dy/dx} \tag{3.24}$$

If, on the other hand $y_1 > y_2$, (i.e., $dy/dx < 0$) we would find (do it)

$$p_Y(y_1) = -\frac{p_X(x_1)}{dy/dx} \tag{3.25}$$

We can account for both of these cases by writing

$$p_Y(y_1) = \frac{p_X(x_1)}{|dy/dx|} \tag{3.26}$$

Finally, writing $x_1 = g^{-1}(y_1)$, and realizing that y_1 can be set equal to any value, we have

$$p_Y(y) = \frac{p_X[g^{-1}(y)]}{|dy/dx|} \tag{3.27}$$

If the function $g(x)$ is not monotone, then the event $\{y_1 < Y \le y_2\}$ corresponds to several intervals of the variable X. For example, if $g(x) = x^2$, then the event $\{1 < Y \le 4\}$ is the same as the event $\{1 < X \le 2\}$ or $\{-1 > X \ge -2\}$. Therefore,

$$\int_1^2 p_X(x)\, dx + \int_{-2}^{-1} p_X(x)\, dx = \int_1^4 p_Y(y)\, dy$$

---

[2] We are assuming that $y_1 < y_2$ if $x_1 < x_2$ ($g(x)$ is monotonic increasing). That is, $dy/dx > 0$ in this range. If this is not the case, the inequalities must be reversed, or a single region in x can correspond to multiple regions in y.

In terms of the density functions this would mean that $g^{-1}(y)$ has several values. Denoting these values as x_a and x_b, then

$$p_Y(y) = \frac{p_X(x)}{|dy/dx|}\bigg|_{x=x_a} + \frac{p_X(x)}{|dy/dx|}\bigg|_{x=x_b} \tag{3.28}$$

Illustrative Example 3.8

A random voltage, v, is put through a full wave rectifier. v is uniformly distributed between -2 volts and $+2$ volts. Find the density of the output of the full wave rectifier.

Solution

Calling the output y, we have $y = g(v)$, where $g(v)$ and the density of V (letting the random variable be the same as the value of voltage) are sketched in Fig. 3.6.

FIGURE 3.6 $g(v)$ and $p_V(v)$ for Illustrative Example 3.7.

At every value of V, $|dg/dv| = 1$. For $y > 0$, $g^{-1}(y) = \pm y$. For $y < 0$, $g^{-1}(y)$ is undefined. That is, there are no values of v for which $g(v)$ is negative. Using Eq. (3.28), we find

$$p_Y(y) = p_X(y) + p_X(-y) \qquad (y > 0) \tag{3.29}$$

$$p_Y(y) = 0 \qquad (y < 0) \tag{3.30}$$

Finally, after plugging in values, we find $p_Y(y)$ as sketched in Fig. 3.7.

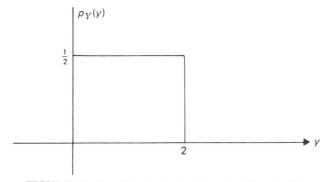

FIGURE 3.7   Resulting density for Illustrative Example 3.7.

Density Function of the Sum of Two Random Variables

Suppose that $z = x + y$, where x and y are random variables which are statistically independent of each other. We wish to find the density function of z in terms of the densities of x and y. The result will be useful in later work.

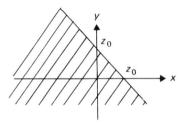

FIGURE 3.8   Region for which $x + y < z_0$.

We start by finding the distribution function of the sum variable, z. The probability that z is less than some quantity z_0 is the probability that the sum $x + y < z_0$. This is the probability that x and y fall within the shaded region of Fig. 3.8. This probability is given by the double integral of the joint density for x and y:

$$P_z(z_0) = \int_{-\infty}^{+\infty} \int_{-\infty}^{z_0-y} p(x, y)\, dx\, dy \qquad (3.31)$$

Since x and y are assumed to be independent, $p(x, y)$ is given by the product of the individual densities:

$$P_z(z_0) = \int_{-\infty}^{+\infty} \int_{-\infty}^{z_0-y} p_X(x) p_Y(y)\, dx\, dy \qquad (3.32)$$

$$= \int_{-\infty}^{+\infty} p_Y(y) \int_{-\infty}^{z_0-y} p_X(x)\, dx\, dy$$

We now differentiate to find the density function:

$$p_z(z_0) = \frac{dP_z(z_0)}{dz_0}$$

$$= \int_{-\infty}^{+\infty} p_Y(y) p_X(z_0 - y)\, dy \qquad (3.33)$$

Equation (3.33) has a very simple interpretation. That is, the density of the sum of two independent random variables is equal to the convolution of the two individual densities.

From Chapter 1, we know that if it were desired to avoid computation of the convolution, we could take Fourier Transforms (which, in this case, would be known as *characteristic functions*) and multiply these transforms together to get the transform of the resulting density function.

Expected Values

Picture yourself as a teacher who has just given an examination. How would you average the resulting grades?

You would add them all together and divide by the number of grades. Thus if an experiment is performed many times, the average of the random variable resulting would be found in the same way.

Let x_i, $i = 1, 2, \ldots, M$ represent the possible values of the random variable, and let n_i represent the number of times x_i occurs as the (random variable assigned to the) outcome. The average of the random variable after N performances of the experiment would therefore be

$$X_{avg} = \frac{1}{N} \sum_{i=1}^{M} n_i x_i \tag{3.34}$$

$$= \sum_{i=1}^{M} \frac{n_i}{N} x_i$$

Since x_i has ranged over all possible values of the random variable, we observe

$$\sum_{i=1}^{M} n_i = N$$

As N approaches infinity, n_i/N becomes the definition of $Pr\{x_i\}$. Therefore,

$$X_{avg} = \sum_{i=1}^{M} x_i Pr\{x_i\} \tag{3.35}$$

This average value is sometimes called the *mean* or *expected value* of X, and is often given the symbol m_x.

Now suppose that we wish to find the average value of a continuous random variable. We can use the previous result (Eq. (3.35)) if we first round off the continuous variable to the nearest multiple of Δx. Thus, if X is between $k\Delta x - \frac{1}{2}\Delta x$ and $k\Delta x + \frac{1}{2}\Delta x$, we shall round it off to $k\Delta x$. The probability of X being in this range is given by

$$\int_{k\Delta x - (1/2)\Delta x}^{k\Delta x + (1/2)\Delta x} p_X(x)\, dx \tag{3.36}$$

which, if Δx is small, is approximately equal to $p_X(k\Delta x)\Delta x$. Therefore, using Eq. (3.35), the expected value of X is

$$\sum_{k=-\infty}^{\infty} k\Delta x\, p_X(k\Delta x)\, \Delta x \approx X_{avg} \tag{3.37}$$

As Δx approaches zero, Eq. (3.37) becomes (by definition of the integral)

$$X_{avg} = m_x = \int_{-\infty}^{\infty} x p_X(x)\, dx \tag{3.38}$$

Now consider a function of the random variable, $y = g(x)$. Suppose we wish to find the expected value of Y. This is given by Eq. (3.38) as

$$Y_{avg} = \int_{-\infty}^{\infty} y p_Y(y)\, dy \tag{3.39}$$

which, by Eq. (3.27), becomes

$$Y_{avg} = \int_{-\infty}^{\infty} y p_X[g^{-1}(y)]\frac{dy}{|dy/dx|} \tag{3.40}$$

which, after substituting for y, becomes

$$[g(x)]_{avag} = \int_{-\infty}^{\infty} g(x) p_X(x)\, dx \tag{3.41}$$

This is an extremely significant equation. It tells us that in order to find the expected value of a function of X, we simply integrate the function weighted by the *density of* X.

We will switch between several different notational forms for the average value operator. These are

$$[g(x)]_{avg} = \overline{g(x)} = E\{g(x)\}$$

From Eq. (3.41), the expected value of x^2 is given by

$$E\{x^2\} = \int_{-\infty}^{\infty} x^2 p_X(x)\, dx \tag{3.42}$$

Illustrative Example 3.9

X is uniformly distributed as shown in Fig. 3.9. Find $E\{x\}$, $E\{x^2\}$, $E\{\cos x\}$, and $E\{(x - m_x)^2\}$.

Solution

$$E\{x\} = \int_{-\infty}^{\infty} x p_X(x)\, dx = \frac{1}{2\pi}\int_0^{2\pi} x\, dx = \pi \tag{3.43}$$

$$E\{x^2\} = \int_{-\infty}^{\infty} x^2 p_X(x)\, dx = \frac{1}{2\pi}\int_0^{2\pi} x^2\, dx = \frac{4}{3}\pi^2 \tag{3.44}$$

$$E\{\cos x\} = \int_{-\infty}^{\infty} \cos x\, p_X(x)\, dx = \frac{1}{2\pi}\int_0^{2\pi} \cos x\, dx = 0 \tag{3.45}$$

$$E\{(x - \pi)^2\} = \int_{-\infty}^{\infty} (x - \pi)^2 p_X(x)\, dx = \frac{1}{2\pi}\int_0^{2\pi} (x - \pi)^2\, dx = \frac{\pi^2}{3} \tag{3.46}$$

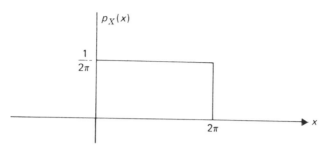

FIGURE 3.9   Density function of X for Illustrative Example 3.9.

The last expected value in Example 3.9 is called the variance of X, and it gives a measure of how far we can expect the random variable to deviate from its mean value. That is, if $E\{(x - m_x)^2\}$ is large, it means that on the average the square of the difference between X and its mean value will be large. This is the same as saying that the density function is spread out.

Thus if you were told that a variable X has mean 1 and variance 0.1, and that a variable Y has mean 1 and variance 10, you would intuitively say that on the average X will be closer to 1 than Y is. This is true even though the average of both variables is exactly equal to 1.

The concept of mean and variance of a random variable will prove to be a major building block in the work to follow.

Illustrative Example 3.10

Two independent random variables are uniformly distributed, each having a mean of 2 and a variance of $\frac{1}{3}$. Find the density of the sum of these two random variables.

Solution

It is first necessary to find the specific densities of the given functions. In order to yield a variance of $\frac{1}{3}$, the width of the uniform density must be 2. The sketch is shown in Fig. 3.10. When this density is convolved with itself, the triangular density shown in Fig. 3.11 results. The mean of this resulting density is 4, and the variance is found as follows:

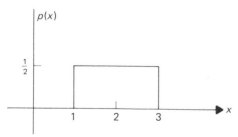

FIGURE 3.10   Density function for Illustrative Example 3.10.

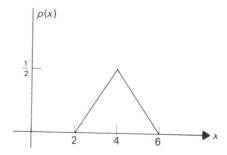

FIGURE 3.11   Result of convolving density of Fig. 3.10 with itself.

$$\sigma^2 = \int_{-\infty}^{\infty} (x - m)^2 p(x)\, dx = 2 \int_{4}^{6} (x - 4)^2 \left[\frac{1}{2} - \frac{x - 4}{4} \right] dx$$

Now, with a change of variables,

$$\sigma^2 = 2 \int_{0}^{2} (\tfrac{1}{2} - \tfrac{1}{4} x)\, x^2\, dx$$

$$= 2 \int_{0}^{2} \left(\frac{x^2}{2} - \frac{x^3}{4} \right) dx$$

$$= \frac{2}{3}$$

Note that the mean of the sum is the sum of the means and the variance is the sum of the variance. This is true whenever independent random variables are added together.

Central Limit Theorem

Extending the result of Illustrative Example 3.10 gives some hint of why the Gaussian density is of such great significance to us. When two independent uniform distributed variables are added together, the result is a variable with triangular density. If a third independent uniform variable were now added, an additional convolution would result in a parabolic (bell shaped) curve which resembles the Gaussian density to the naked eye. In fact, a powerful theorem formalizes this.

The *Central Limit Theorem* states that as the number of independent random variables being added together approaches infinity, the resulting density approaches Gaussian with mean equal to the sum of the individual means and variance equal to the sum of the variances. Only one broad restriction applies, that being that the individual variances are well behaved (The ratio of any particular variance to the variance of the sum must approach zero as the number of variables being summed approaches infinity). The individual densities need not be identical, and indeed, with proper alternative restrictions, we can even relax the independence requirement.

Thus, any real life phenomenon which can be thought of as resulting from the sum of many random parameters will be Gaussian distributed. For example, the static on a radio is caused by a sum of many factors including auto ignitions, spikes from lights being turned off and on, telephone dialing, lightning and other electrical disturbances, random signals generated by electron motion in components and devices (thermal noise), and leakage of other signals. Measurement of the density of this static reveals that it is approximately Gaussian distributed. In most communication applications, we will assume that the noise is Gaussian distributed.

Illustrative Example 3.11

X is a Gaussian distributed random variable with density function

$$p_x(x) = \frac{1}{\sqrt{2\pi}\,\sigma}\,e^{-(x-m)^2/2\sigma^2}$$

(a) Show that $p_x(x)$ integrates to unity.
(b) Find $E\{x\}$.
(c) Find $E\{(x - m_x)^2\}$.

   Solution

   (a) We first write the integral directly:

$$\frac{1}{\sqrt{2\pi}\,\sigma} \int_{-\infty}^{+\infty} e^{-(x-m)^2/2\sigma^2}\,dx$$

Making a change of variables, $y = (x - m)/\sigma$, we have

$$\frac{1}{\sqrt{2\pi}\,\sigma} \int_{-\infty}^{+\infty} e^{-(x-m)^2/2\sigma^2}\,dx = \frac{1}{\sqrt{2\pi}} \int_{-\infty}^{+\infty} e^{-y^2/2}\,dy$$

This integral cannot be evaluated in closed form for general limits of integration. In the case of infinite limits, the integral can be evaluated. Although the infinite limit integral can be found in most any table of integrals, it is instructive to show a "trick" for evaluating it. Define

$$I = \int_{-\infty}^{+\infty} e^{-y^2/2}\,dy \tag{3.47}$$

Then

$$I^2 = \int_{-\infty}^{+\infty} \int_{-\infty}^{+\infty} e^{-x^2/2}e^{-y^2/2}\,dx\,dy$$

$$= \int_{-\infty}^{+\infty} \int_{-\infty}^{+\infty} e^{-(x^2+y^2)/2}\,dx\,dy \tag{3.48}$$

Transforming to polar coordinates, we have

$$x^2 + y^2 = r^2$$

and

$$dx\,dy = r\,dr\,d\theta$$

$$I^2 = \int_0^{2\pi} \int_0^{\infty} e^{-r^2/2}\,r\,dr\,d\theta \tag{3.49}$$

Since the integrand is not a function of θ, Eq. (3.49) becomes

$$I^2 = 2\pi \int_0^{\infty} re^{-r^2/2}\,dr \tag{3.50}$$

The integrand of Eq. (3.50) is the derivative of $-e^{-r^2/2}$. Therefore,

$$I^2 = 2\pi(-e^{-r^2/2}\,|_0^{\infty}) = 2\pi$$

and

$$I = \sqrt{2\pi}$$

Finally,

$$\frac{1}{\sqrt{2\pi}} I = 1$$

which is the desired result.

(b)

$$E\{x\} = \frac{1}{\sqrt{2\pi}\,\sigma} \int_{-\infty}^{+\infty} x e^{-(x-m)^2/2\sigma^2} \, dx$$

$$= \frac{1}{\sqrt{2\pi}\,\sigma} \int_{-\infty}^{+\infty} (x-m) e^{-(x-m)^2/2\sigma^2} \, dx$$

$$+ \frac{m}{\sqrt{2\pi}\,\sigma} \int_{-\infty}^{+\infty} e^{-(x-m)^2/2\sigma^2} \, dx$$

$$= \frac{\sigma}{\sqrt{2\pi}} e^{-(x-m)^2/2\sigma^2} \Big|_{-\infty}^{+\infty} + m$$

Therefore the "m" in the formula for the Gaussian density is actually the mean value of the random variable. This should have been obvious from the symmetry displayed in Fig. 3.2.

(c)

$$E\{(x-m)^2\} = \frac{1}{\sqrt{2\pi}\,\sigma} \int_{-\infty}^{+\infty} (x-m)^2 e^{-(x-m)^2/2\sigma^2} \, dx$$

This can be integrated by parts. Let

$$u = x - m$$
$$dv = (x-m) e^{-(x-m)^2/2\sigma^2} \, dx$$

Then

$$du = dx$$
$$v = -e^{-(x-m)^2/2\sigma^2}$$

$$E\{(x-m)^2\} = \frac{1}{\sqrt{2\pi}\,\sigma} \int_{-\infty}^{+\infty} e^{-(x-m)^2/2\sigma^2} \, dx$$

$$- (x-m)\sigma^2 e^{-(x-m)^2/2\sigma^2} \Big|_{-\infty}^{+\infty}$$

$$= \sigma^2$$

Therefore, "σ^2" in the formula for the Gaussian density is the variance of the random variable.

A simple relationship may be derived involving the mean, variance, and second moment of any random variable. It is worth mentioning since one often has already calculated the mean and second moment and must find the variance. With the use of this relationship, no additional integration is required.

We start with the definition of variance:

$$\sigma_x^2 = E\{(x - E\{x\})^2\}$$

Expanding this, and using the fact that the expected value of a sum is the sum of the expected values, we have

$$\sigma_x^2 = E\{x^2 - 2E\{x\}E\{x\} + (E\{x\})^2\}$$
$$= E\{x^2\} - (E\{x\})^2 \tag{3.51}$$

Note that $E\{x\}$ is non-random and is treated as a constant in the above derivation. Equation (3.51) is the desired result. That is, the variance can be found by subtracting the square of the mean from the second moment.

3.3   STOCHASTIC PROCESSES

Until this point, we have considered primarily single random variables. All averages (and related parameters) were simply numbers. In this section we shall, in effect, add another dimension to the study; the dimension of time. Instead of talking of numbers only, we will now be able to characterize random functions. The advantages of such a capability should be obvious. Our approach to random functions' analysis will begin with the consideration of discrete-time functions (i.e., sampled), since these will prove to be a simple extension of random variables.

Imagine a single die being flipped 1,000 times. Let X_i be the random variable assigned to the outcome of the ith flip. Now list the 1,000 values of the random variables,

$$x_1 x_2 x_3, \ldots, x_{999} x_{1,000} \tag{3.52}$$

For example, if the random variable is assigned to be equal to the number of dots on the top face after flipping the die, a typical list might resemble that shown in Eq. (3.53):

$$463514253145, \ldots \tag{3.53}$$

Suppose that all possible sequences of random variables are now listed. We would have a collection of $6^{1,000}$ entries, each one resembling that shown in Eq. (3.53). This collection will be known as the *ensemble* of possible outcomes. This ensemble, together with associated statistical properties, forms a *stochastic process*. In this particular example, the process is discrete valued and discrete time.

If we were to view one digit, say the third entry, a random variable would result. In this example the random variable would represent that assigned to the outcome of the third flip of the die.

We can completely describe the above stochastic process by specifying the 1,000-dimensional probability density function,

$$p(x_1, x_2, x_3, \ldots, x_{999}, x_{1,000})$$

This 1,000-dimensional probability density function is used in the same way as a one-dimensional probability density function. Thus, the 1,000-dimensional integral of the probability density function is unity.

$$\int_{-\infty}^{\infty} \int_{-\infty}^{\infty} \cdots \int_{-\infty}^{\infty} p(x_1, x_2, \ldots x_{1,000})\, dx_1\, dx_2 \ldots dx_{1,000} = 1$$

The probability that the variables fall within any specified volume is the integral of the probability density function over that volume. The expected value of any single variable can be found by integrating the product of that variable with the probability density function. Thus, for example, the expected value of x_1 is

$$\int_{-\infty}^{\infty} \int_{-\infty}^{\infty} \cdots \int_{-\infty}^{\infty} x_1 p(x_1, x_2, \ldots x_{1,000})\, dx_1\, dx_2 \ldots dx_{1,000}$$

We will call this a *first order average*.

In a similar manner, the expected value of any multidimensional function of the variables is found by integrating the product of that function with the density function. Thus, for example, the expected value of the product $x_1 x_2$ is given by

$$E\{x_1 x_2\} = \int_{-\infty}^{\infty} \int_{-\infty}^{\infty} \cdots \int_{-\infty}^{\infty} x_1 x_2 p(x_1, x_2, \ldots x_{1,000})\, dx\, dx_2\, dx_{1,000}$$

This would be known as a *second order average*.

In many instances it will prove sufficient to specify only the first and second order averages (moments). That is, one would specify

$$E\{X_i\} \quad \text{and} \quad E\{X_i X_j\}$$

for all i and j.

We would now like to extend this example to the case of an infinite number of random variables in each list and an infinite number of lists in the ensemble. It may be helpful to refer back to the above simple example from time to time.

The most general form of the stochastic process would result if our simple experiment could yield an infinite range of values as the outcome, and if the sampling period approached zero (i.e., the die is flipped faster and faster).

Another way to arrive at a stochastic process is to perform a discrete experiment, but to each outcome assign a *time function* instead of a number. As a simple example, consider the experiment defined by picking a 6-volt DC generator from an infinite inventory in a warehouse. The generator voltage is then measured as a function of time on an oscilloscope. This waveform, $v(t)$, will be called a *sample function* of the process. The waveform will not be a perfect constant of 6 volts value, but will wiggle around due to imperfections in the generator construction and due to radio pickup when the wires act as an antenna. There are an infinite number of possible sample functions of the process. This infinite number of samples forms the ensemble. Each time we choose a generator and measure its voltage, a sample function from this infinite ensemble results.

If we were to sample the voltage at a specific time, say $t = t_0$, the sample, $v(t_0)$, can be considered as a random variable. Since $v(t)$ is assumed to be continuous with time, there are an infinity of random variables associated with the process.

Summing this up and changing notation slightly, let $x(t)$ represent a stochastic process. Then $x(t)$ can be thought of as being an infinite ensemble of all possible sample functions. For every specific value of time, $t = t_0$, $x(t_0)$ is a random variable.

As in the simple example which opened this section, we can first hope to characterize the process by a joint density function of the random variables. Unfortunately, since there are an infinite number of random variables this density function would be infinite dimensional. We therefore resort to first and second moment characterization of the process.

The first moment is given by the mean value.

$$m(t) \triangleq E\{x(t)\} \tag{3.54}$$

The second moments are given by

$$R_{xx}(t_1, t_2) \triangleq E\{x(t_1)x(t_2)\} \tag{3.55}$$

for all t_1 and t_2.

At the moment we must think of these averages as being taken over the ensemble of possible time function samples. That is, in order to find $m(t_0)$, we would average all members of the ensemble at time t_0. In practice, we could measure the voltage of a great number of the generators at time t_0 and average the resulting numbers. In the case of the DC generators, one would intuitively expect this average not to depend upon t_0. Indeed, most processes which we will consider have mean values which are independent of time.

A process with overall statistics that are independent of time is called a *stationary* process. If only the mean and second moment are independent of time, the process is *wide sense stationary*. Given a stationary process, $x(t)$, then the process described by $x(t - T)$ will have the same statistics independent of the value of T. Clearly, for a stationary process, $m(t_0)$ will not depend upon t_0.

If the process is stationary,

$$R_{xx}(t_1, t_2) = E\{x(t_1)x(t_2)\} = E\{x(t_1 - T)x(t_2 - T)\} \tag{3.56}$$

Since Eq. (3.56) applies for all values of T, let $T = t_1$. Then,

$$R_{xx}(t_1, t_2) = E\{x(0)x(t_2 - t_1)\}$$

This indicates that the second moment, $R_{xx}(t_1, t_2)$ does not depend upon the actual values t_1 and t_2, but only upon the difference, $t_2 - t_1$. That is, the left hand time point can be placed anywhere, and as long as the right hand point is separated from this by $t_2 - t_1$, the second moment will be $R_{xx}(t_1, t_2)$. We state this mathematically as

$$R_{xx}(t_1, t_2) = R_{xx}(t_2 - t_1) \tag{3.57}$$

for a stationary process.

The function $R_{xx}(t_2 - t_1)$ is called the *autocorrelation* of the process, $x(t)$. The word autocorrelation was used earlier in the context of power and energy. We shall soon see the reason for using the same word again.

In the work to follow, the argument, $t_2 - t_1$ will be replaced by "τ".

$R_{xx}(t_2 - t_1)$ tells us most of what we need to know about the process. In particular, it gives us some idea as to how fast a particular sample function can change with time. If t_2 and t_1 are sufficiently far apart such that $x(t_1)$ and $x(t_2)$ are independent random variables, the autocorrelation reduces to

$$R_{xx}(t_2 - t_1) = E\{x(t_1)x(t_2)\},$$
$$= E\{x(t_1)\}E\{x(t_2)\} = m^2 \tag{3.58}$$

For most processes encountered in communications, the mean value is zero ($m = 0$). In this case, the value of "t" at which $R_{xx}(t)$ goes to zero represents the time over which the process is "correlated." If two samples are separated by more than this length of time, one sample has no effect upon the other.

Illustrative Example 3.12

You are given a stochastic process, $x(t)$, with mean value m_x and autocorrelation, $R_{xx}(\tau)$. Obviously, the process is at least wide sense stationary. If not, the mean and autocorrelation could not have been given in this form. Find the mean and autocorrelation of a process $y(t)$, where

$$y(t) = x(t) - x(t - T)$$

Solution

To solve problems of this type, we need only recall the definition of the mean and autocorrelation and the fact that taking expected values is a linear operation. Proceeding, we have

$$m_y = E\{y(t)\} = E\{x(t) - x(t - T)\}$$
$$= E\{x(t)\} - E\{x(t - T)\} \tag{3.59}$$
$$= m_x - m_x = 0$$

$$R_{yy}(\tau) = E\{y(t)y(t+\tau)\}$$
$$= E\{[x(t) - x(t-T)][x(t+\tau) - x(t+\tau-T)]\}$$
$$= E\{x(t)x(t+\tau)\} - E\{x(t)x(t+\tau-T)\}$$
$$\quad - E\{x(t-T)x(t+\tau)\} + E\{x(t-T)x(t+\tau-T)\} \tag{3.60}$$
$$= R_{xx}(\tau) - R_{xx}(\tau-T) - R_{xx}(\tau+T) + R_{xx}(\tau)$$
$$= 2R_{xx}(\tau) - R_{xx}(\tau-T) - R_{xx}(\tau+T)$$

We note that the process $y(t)$ is wide sense stationary. If this were not the case, both m_y and R_{yy} would be functions of t.

Illustrative Example 3.13

Consider the experiment of starting a sinusoidal generator of deterministic frequency f_0 and amplitude A. The exact time of start is random. Thus,

$$x(t) = A \sin(2\pi f_0 t + \theta)$$

where the phase, θ, can be considered a random variable with a uniform density as shown in Fig. 3.12.

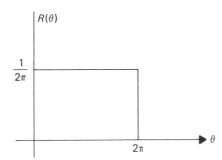

FIGURE 3.12 Density function of phase in Illustrative Example 3.13.

Find the autocorrelation of this random process.

Solution

The autocorrelation is found directly from the definition,

$$R(\tau) = E\{x(t)x(t+\tau)\}$$
$$= E\{A^2 \sin(2\pi f_0 t + \theta) \sin(2\pi f_0(t+\tau) + \theta)\}$$

By using trigonometric identities, this becomes

$$E\{\tfrac{1}{2} A^2 [\cos 2\pi f_0 \tau - \cos(2\pi f_0(2t+\tau) + \theta)]\}$$

The first term in this result is not random, so its expected value is the function itself. The expected value of the second term is found by integrating its product with the density of θ.

$$R(\tau) = \tfrac{1}{2} A^2 \cos 2\pi f_0 \tau - \frac{1}{2\pi} \int_0^{2\pi} \cos(2\pi f_0(2t + \tau) + \theta) \, d\theta$$

The second term represents the integral of a perfect cosine function over two entire periods. This integral is equal to zero. Thus,

$$R(\tau) = \tfrac{1}{2} A^2 \cos 2\pi f_0 \tau$$

Suppose you were asked to find the average value of the voltage in the DC generator example. You would have to measure the voltage (at any given time) of many generators and then compute the average. Once you were told that the process is stationary, you would probably be tempted to take one generator and average its voltage over a large time interval. You would expect to get 6 volts as the result of either technique. That is, you would reason that

$$m_v = E\{v(t)\} = \lim_{T \to \infty} \frac{1}{T} \int_{-T/2}^{T/2} v_i(t) \, dt \tag{3.61}$$

Here, $v_i(t)$ is one sample function of the ensemble. Actually this reasoning will not always be correct. Suppose, for example, that one of the generators was burned out. If you happened to choose this particular generator, you would find that $m_v = 0$, which is certainly not correct.

If any single sample of a stochastic process contains all of the information (statistics) about the process, we call the process *ergodic*. Most processes which we deal with will be ergodic. The generator example is ergodic as long as none of the generators is exceptional (e.g., burned out).

Clearly a process which is ergodic must also be stationary. This is true since, once we allow that the averages can be found from one time sample, these averages can no longer be a function of the time at which they are computed. Conversely, a stationary process need not be ergodic (e.g., the burned-out generator example . . . convince yourself this process would still be stationary).

If a process is ergodic, the autocorrelation can be found from any time sample,

$$R_{xx}(\tau) = E\{x(t)x(t + \tau)\} = \lim_{T \to \infty} \frac{1}{T} \int_{-T/2}^{T/2} x(t)x(t + \tau) \, dt \tag{3.62}$$

We can now see why the functional notation, $R(\tau)$, and the name "autocorrelation" were used. Equation (3.62) is exactly the same as the equation for $R(\tau)$ found in Section 2.9. The earlier equation held for deterministic finite power signals.

Since the equations are identical, we can borrow all of our previous results. Thus, the power spectral density of a stochastic process is given by

$$G(f) = \mathscr{F}[R(t)] = \int_{-\infty}^{\infty} R(t)e^{-j2\pi ft} \, dt \tag{3.63}$$

and the total average power by

$$P_{avg} = 2 \int_{0}^{\infty} G(f) \, df \tag{3.64}$$

Here, the word "average" takes on more meaning than it did for deterministic signals. Note that since

$$R(t) = \int_{-\infty}^{\infty} G(f)e^{j2\pi ft} \, df \tag{3.65}$$

then

$$P_{avg} = \int_{-\infty}^{\infty} G(f)df = R(0) \tag{3.66}$$

The observation that $R(0)$ is the average power makes a great deal of sense since

$$R(0) = E\{x^2(t)\} \tag{3.67}$$

If a stochastic process now forms the input to a linear system the output will also be a stochastic process. That is, each sample function of the input process yields a sample function of the output process. The autocorrelation of the output process is found as in Chapter 2. (*See* Fig. 3.13.)

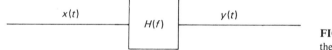

FIGURE 3.13   A stochastic process as the input to a linear system.

$$G_{yy}(f) = G_{xx}(f)|H(f)|^2 \tag{3.68}$$

Taking inverse transforms,

$$R_{yy}(t) = R_{xx}(t) * h(t) * h(-t) \tag{3.69}$$

We therefore have a way of finding the average power of the output of a system when the input is a stochastic process. Since most noise waveforms can be thought of as being samples of a stochastic process, the above analysis is critical to any noise reduction scheme analysis.

Illustrative Example 3.14

A received signal is made up of two components, signal and noise.

$$r(t) = s(t) + n(t)$$

The signal can be considered as a sample of a random process since random amplitude fluctuations are introduced by turbulence in the air. You are told that the autocorrelation of the signal process is

$$R_s(\tau) = 2e^{-|\tau|}$$

The noise is a sample function of a random process with autocorrelation,

$$R_n(\tau) = e^{-2|\tau|}$$

You are told that both processes have zero mean value, and that they are independent of each other.

Find the autocorrelation and total power of $r(t)$.

Solution

From the definition of autocorrelation,

$$
\begin{aligned}
R_r(\tau) &\overset{\Delta}{=} E\{r(t)r(t+\tau)\} \\
&= E\{[s(t)+n(t)][s(t+\tau)+n(t+\tau)]\} \\
&= E\{s(t)s(t+\tau)\} + E\{s(t)n(t+\tau)\} \\
&\quad + E\{s(t+\tau)n(t)\} + E\{s(t+\tau)n(t+\tau)\}
\end{aligned}
\tag{3.70}
$$

Since the signal and noise are independent,

$$E\{s(t+\tau)n(t)\} = E\{s(t+\tau)\}E\{n(t)\} = 0$$

and

$$E\{s(t)n(t+\tau)\} = E\{s(t)\}E\,n(t+\tau)\} = 0$$

Finally, Eq. (3.70) becomes

$$
\begin{aligned}
R_r(\tau) &= R_s(\tau) + R_n(\tau) \\
&= 2e^{-|\tau|} + e^{-2|\tau|}
\end{aligned}
\tag{3.71}
$$

The total power of $r(t)$ is simply $R_r(0) = 3$.

3.4   WHITE NOISE

Let $x(t)$ be a stochastic process with a constant power spectral density (*see* Fig. 3.14).

$G_{xx}(f) = K$

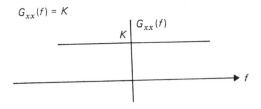

FIGURE 3.14   Power spectrum of white noise.

This process contains all frequencies to an equal degree. Since white light is composed of all frequencies (colors), the process described above is known as *white noise*.

The autocorrelation of white noise is the inverse Fourier Transform of K, which is simply $R_{xx}(t) = K\delta(t)$.

The average power of white noise, $R(0)$, is infinity. It therefore cannot exist in real life. However, many types of noise encountered can be assumed to be approximately white.

Illustrative Example 3.15

White noise forms the input to the RC circuit of Fig. 3.15 (approximation to a lowpass filter). Find the autocorrelation and power spectral density of the output of this filter.

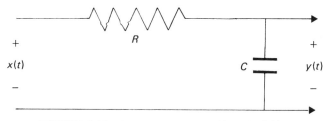

FIGURE 3.15   Circuit for Illustrative Example 3.15.

Solution

$$G_{yy}(f) = G_{xx}(f)\,|H(f)|^2 = \frac{K}{1 + (2\pi f)^2 C^2 R^2} \tag{3.72}$$

$$R_{yy}(\tau) = \frac{K}{2RC}\ \exp\left[\frac{-|\tau|}{RC}\right] \tag{3.73}$$

Illustrative Example 3.16

Repeat Illustrative Example 3.15 for an ideal lowpass filter with cutoff frequency, f_m, as shown in Fig. 3.16.

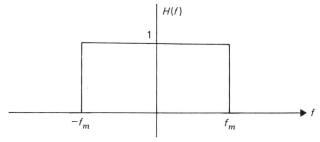

FIGURE 3.16   Lowpass filter for Illustrative Example 3.16.

Solution

$$G_{yy}(f) = G_{xx}(f)|H(f)|^2$$

$$= \begin{cases} K & |f| < f_m \\ 0 & \text{otherwise} \end{cases} \qquad (3.74)$$

This is shown in Fig. 3.17.

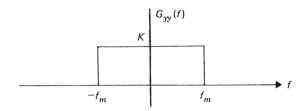

FIGURE 3.17   Output power spectrum for Illustrative Example 3.16.

The autocorrelation is the inverse transform of the power spectral density and is given by

$$R_{yy}(\tau) = K \frac{\sin 2\pi f_m \tau}{\pi \tau} \qquad (3.75)$$

A process with a power spectral density as shown in Fig. 3.17 is known as *bandlimited white noise*.

In the previous example, suppose that the input process had a power spectrum as shown in Fig. 3.18 with $f_1 \geq f_m$. The output process would be identical to that

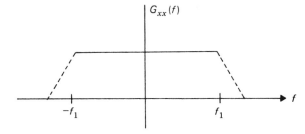

FIGURE 3.18   Spectrum of an input to the filter of Illustrative Example 3.15 which would yield the same output spectrum as white noise input did.

found in Illustrative Example 3.16. Therefore, if a processing system exhibits an upper cutoff frequency (all real systems approximately do) and the input noise has a flat spectrum up to this cutoff frequency, no error will be made by considering the input noise to be white. This will simplify the required mathematics considerably.

Since $R(t)$ is zero for $t \neq 0$, two samples of white noise are independent of each other even if they are taken very closely together.[3] That is, knowing the sample value

[3] Actually the zero correlation ($x(t)$ and $x(t + \tau)$ are "uncorrelated") is a necessary, but not sufficient, condition for independence of the two random variables. The only complete definition of independence relates to conditional probabilities or, equivalently, to the joint density function. However, one can show that two random variables which are jointly Gaussian are independent as long as they are uncorrelated. Therefore as long as we deal with Gaussian processes, we can use the words "uncorrelated" and "independent" interchangeably.

of white noise at one instant of time tells us absolutely nothing about its value an instant later. From a practical standpoint, this is a sad situation. It would seem to make the elimination of noise more difficult. Although this is true, the "independence of samples" assumption serves to greatly simplify the analysis of many complex processing systems.

Up to this point we have said nothing about the actual probability distributions of the process. We have only talked about the first and second moments. Each random variable, $x(t_0)$, has a certain probability distribution. For example, it can be either Gaussian or uniformly distributed. By only considering the means and second moments we are not telling the whole story. Two completely different random variables may possess the same mean and variance.

Perhaps this last idea will become clearer if we draw an analogy with simple dynamics. The equations for the first and second moment of a random variable are

$$E\{x\} = \int_{-\infty}^{\infty} xp(x)\,dx \qquad (3.76)$$

$$E\{x^2\} = \int_{-\infty}^{\infty} x^2 p(x)\,dx \qquad (3.77)$$

These are of the same form as the equations used to find the center of gravity (centroid) and the moment of inertia of a body. Certainly two bodies can be completely different in shape yet have the same center of gravity and moment of inertia.

We are rescued from this dilemma by the observation that most processes encountered in real life are Gaussian. This means that all relevant densities (joint and single) are Gaussian. Once we know that a random variable has a Gaussian distribution, the density function is completely specified by giving its mean and second moment.

Thermal and Shot Noise

The assumption of *white noise* is not too far off base for several common forms of noise. *Thermal noise* and *shot noise* are two physically encountered types of noise.

Thermal noise is produced by the random motion of electrons in a medium. The intensity of this motion increases with increasing temperature and is zero only at a temperature of absolute zero. If a resistor is hooked directly to the input of a sensitive oscilloscope, a random pattern will be displayed on the screen. The power spectral density of this random process can be shown to be of the form

$$G(f) = \frac{A\,|f|}{e^{B\,|f|} - 1} \qquad (3.78)$$

where A and B are constants that depend upon temperature and other physical constants. Figure 3.19 is a sketch of $G(f)$. For low frequencies, $G(f)$ is almost constant. We can therefore usually approximate thermal noise as white noise.

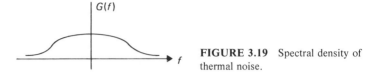

FIGURE 3.19 Spectral density of thermal noise.

Shot noise occurs since, although we think of current as being continuous, it is actually a discrete phenomenon. In fact, current occurs in discrete pulses each time an electron moves between two points. A plot of current as a function of time would show a waveform which wiggles around what we conventionally think of as the current plot. An example is shown in Fig. 3.20. This variation of current around the average value is known as shot noise. An analysis (see the references) would show that, for frequencies of interest to us, shot noise can usually be thought of as being white noise.

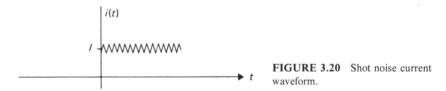

FIGURE 3.20 Shot noise current waveform.

3.5   NARROWBAND NOISE—QUADRATURE COMPONENTS

In studies of noise in communication systems, we encounter noise that is bandlimited to some range of frequencies. Such a noise waveform would result if any noise signal (white, for example), were passed through a bandpass filter. It will prove useful to us to have a trigonometric expansion for such so called narrowband noise signals. The actual form will be

$$n(t) = x(t) \cos 2\pi f_0 t - y(t) \sin 2\pi f_0 t \qquad (3.79)$$

where $n(t)$ is the noise waveform and f_0 is a frequency (often the center) within the band occupied by the noise. Since sine and cosine vary by $90°$, $x(t)$ and $y(t)$ are known as the "quadrature components" of the noise waveform.

Since the formula of Eq. (3.79) will be used many times in later studies, we shall spend some time examining the properties of the various time functions. The result of Eq. (3.79) can be derived by starting with the exponential notation

$$n(t) = Re\{r(t)e^{j2\pi f_0 t}\}$$

where $r(t)$ is a complex function with a low frequency bandlimited Fourier Transform, and the exponential has the effect of shifting the frequencies of $r(t)$ by f_0. Expanding the exponential using Euler's identity and letting $x(t)$ be the real part of $r(t)$ and $y(t)$ be the imaginary part yield

$$n(t) = Re\{(x(t) + jy(t))(\cos 2\pi f_0 t + j \sin 2\pi f_0 t)\}$$

$$= x(t) \cos 2\pi f_0 t - y(t) \sin 2\pi f_0 t$$

as in Eq. (3.79).

Explicit solution of Eq. (3.79) for $x(t)$ and $y(t)$ is not simple. One way to solve this is by using Hilbert Transforms. Since this concept will be useful to us again in the discussion of envelopes and in single sideband AM, we shall spend a few moments developing it.

Hilbert Transform

If all components of a time signal are shifted by -90 degrees in phase (and the amplitude is not changed), the result is a new time function known as the *Hilbert Transform*. Since the impulse response of a real system, $h(t)$, is real, the real part of $H(f)$ must be even and the imaginary part odd. Alternatively, the amplitude of $H(f)$ must be even and the phase odd. Thus, the system function of Figure 3.21 results.

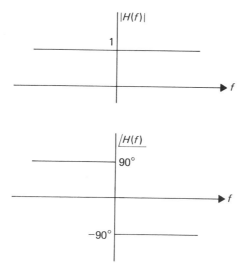

FIGURE 3.21   System function of Hilbert transformer.

$H(f)$ is then given by

$$H(f) = -j \, sgn(f) \tag{3.80}$$

where *sgn* is the "sign" function. The impulse response of this system is found by evaluating the inverse Fourier Transform of $H(f)$. A limiting process similar to that used in finding the Fourier Transform of the unit step function (see Section 1.8) yields

$$h(t) = \frac{1}{\pi t}$$

Thus the Hilbert Transform of a signal, $s(t)$, is given by the convolution of the signal with h(t).

$$\hat{s}(t) \triangleq \frac{1}{\pi} \int_{-\infty}^{\infty} \frac{s(\tau)}{t - \tau} d\tau$$

where the caret over the $s(t)$ is the notation used for the Hilbert Transform.

If we take the Hilbert Transform of a Hilbert Transform, the effect in the frequency domain is to multiply the signal transform by $H^2(f)$. We see from Eq. (3.80) that $H^2(f) = -1$. Thus, the inverse Hilbert Transform is given by

$$s(t) = -\frac{1}{\pi} \int_{-\infty}^{\infty} \frac{\hat{s}(\tau)}{t - \tau} d\tau$$

Illustrative Example 3.17

Find the Hilbert Transform of the following time signals.

(a) $s(t) = \cos(2\pi f_0 t + \theta)$

(b) $s_m(t) = \dfrac{\sin t}{t} \cos 2\pi x 100t$

(c) $s_m(t) = \dfrac{\sin t}{t} \sin 2\pi x 100t$

Solution

It is almost always easier to avoid time convolution by working with Fourier Transforms.

(a) The Fourier Transform of $s(t)$ is given by

$$S(f) = [\tfrac{1}{2}\delta(f - f_0) + \tfrac{1}{2}\delta(f + f_0)]e^{-j\theta/2\pi f_0}$$

Note that the phase shift of θ radians is equivalent to a time shift of $\theta/2\pi f_0$ seconds. We now multiply this by $-j\,\text{sgn}\,(f)$ to get

$$\hat{S}(f) = [-\tfrac{1}{2}j\,\delta(f - f_0) + \tfrac{1}{2}j\,\delta(f + f_0)]e^{-j\theta/2\pi f_0}$$

The quantity in square brackets is the Fourier Transform of a sine wave, so the result is

$$\hat{s}(t) = \sin(2\pi f_0 t + \theta)$$

This is not surprising, since the Hilbert Transform is a 90 degree phase shifting operation.

(b) We shall denote

$$s(t) = \frac{\sin t}{t}$$

The Fourier Transform of $s_m(t)$ is then given by

$$S_m(f) = \tfrac{1}{2}S(f - 100) + \tfrac{1}{2}S(f + 100)$$

Since $S(f)$ is bandlimited to $f = \pm 1$, the first term in $S_m(f)$ occupies frequencies between 99 and 101 Hz while the second term occupies frequencies between -101 and -99

Hz. When $S_m(f)$ is multiplied by $-j\,\text{sgn}(f)$ to get the Fourier Transform of the Hilbert Transform, we find

$$\hat{S}_m(f) = -\tfrac{1}{2}jS(f - 100) + \tfrac{1}{2}jS(f + 100)$$

The inverse transform of this yields

$$\hat{s}_m(t) = s(t)\sin 2\pi x 100 t = \frac{\sin t}{t}\sin 2\pi\, x 100t$$

(c) We can use the fact that the Hilbert Transform of a Hilbert Transform is the negative of the original function. Thus by inspection, the result is

$$-s(t)\cos 2\pi x 100t = -\frac{\sin t}{t}\cos 2\pi x 100t$$

We are now ready to return to the solution of Eq. (3.79). If $x(t)$ and $y(t)$ are assumed to be bandlimited to frequencies below f_0, we can take the Hilbert Transform of both sides of Eq. (3.79) to get

$$\hat{n}(t) = x(t)\sin 2\pi f_0 t + y(t)\cos 2\pi f_0 t \tag{3.81}$$

If Eq. (3.79) is multiplied by $\cos 2\pi f_0 t$ and Eq. (3.81) is multiplied by $\sin 2\pi f_0 t$, when the two expressions are added together, $y(t)$ is eliminated. This yields

$$n(t)\cos 2\pi f_0 t + \hat{n}(t)\sin 2\pi f_0 t = x(t)[\cos^2 2\pi f_0 t + \sin^2 2\pi f_0 t]$$

$$= x(t) \tag{3.82}$$

Similarly, if Eq. (3.79) is multiplied by $\sin 2\pi f_0 t$ and Eq. (3.81) by $\cos 2\pi f_0 t$, taking the difference eliminates $x(t)$. This yields

$$y(t) = \hat{n}(t)\cos 2\pi f_0 t - n(t)\sin 2\pi f_0 t \tag{3.83}$$

Illustrative Example 3.18

Show that the system of Figure 3.22 yields the quadrature components at the outputs.

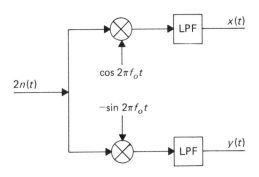

FIGURE 3.22   System to yield quadrature components.

Solution

The inputs to the lowpass filters are given by

$$n_1(t) = 2x(t) \cos^2 2\pi f_0 t - 2y(t) \sin 2\pi f_0 t \cos 2\pi f_0 t$$
$$= x(t) + x(t) \cos 4\pi f_0 t - y(t) \sin 4\pi f_0 t$$

$$n_2(t) = -2x(t) \cos 2\pi f_0 t \sin 2\pi f_0 t + 2y(t) \sin^2 2\pi f_0 t$$
$$= -x(t) \sin 4\pi f_0 t + y(t) - y(t) \cos 4\pi f_0 t$$

Using the modulation theorem, we see that the Fourier Transform of $x(t) \cos 4\pi f_0 t$ occupies a range around a frequency of $2f_0$, as is true of the other terms with $4\pi f_0$ in the above expression. The lowpass filter is designed to pass the frequencies of $x(t)$ and $y(t)$, so it will reject this high frequency term. The outputs are therefore as shown on the diagram.

The autocorrelation of $x(t)$ and $y(t)$ can now be derived from Eqs. (3.82) and (3.83).

$$R_x(\tau) = R_y(\tau) = R_n(\tau) \cos \omega_0 \tau + \left[R_n(\tau) * \frac{1}{\pi t} \right] \sin 2\pi f_0 \tau \qquad (3.84)$$

Finally, from Eq. (3.81) and using the modulation theorem we get the power spectral density relationship,

$$G_x(f) = G_y(f) = G_n(f - f_0) + G_n(f + f_0) \qquad (3.85)$$
$$f_0 - f_m < |f| < f_0 + f_m$$

Equation (3.85) is the key result which will enable us to calculate the effects of noise upon AM and FM communication systems.

Illustrative Example 3.19

Express the following three narrowband noise processes in quadrature form using f_0 as the center frequency.

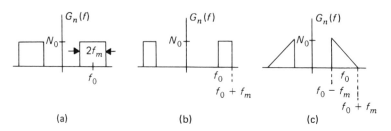

(a)   (b)   (c)

Solution

Using Eq. (3.85), we can immediately sketch the power spectral densities of $x(t)$ and $y(t)$, where the noise is expressed as

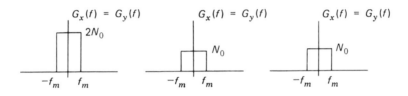

$$n(t) = x(t) \cos 2\pi f_0 t - y(t) \sin 2\pi f_0 t$$

3.6   SIGNAL TO NOISE RATIO

In most communication studies, the probabilistic parameter of interest to us is the *average power*. By itself, this quantity would not tell very much, since we could always modify the average power of a signal by putting it through an amplifier or an attenuator. The problem arises since the received waveform usually consists of a desired signal portion plus additive noise. If we amplify, or attenuate, the total signal we will do the same thing to the noise as we do to the signal. The parameter of interest then is not the signal power, but the ratio of this power to the power of the unwanted noise. This ratio is, quite logically, called the *signal to noise ratio* and is abbreviated as *S/N* or *SNR*.

In a good number of communication situations, the purpose of processing a received signal is to increase the signal to noise ratio. That is, we receive a signal including unwanted noise and we wish to somehow process this function so as to increase the S/N.

Figure 3.23 shows the block diagram of a system with the input S/N and output S/N indicated. The ratio of these two S/Ns gives some measure of the effectiveness of

$(S/N)_{in}$ ──── $H(f)$ ──── $(S/N)_{out}$

FIGURE 3.23   System for definition of signal to noise improvement.

the system. This ratio, designated ΔSNR, is defined as the *signal to noise improvement* of the system:

$$\Delta SNR = \frac{(S/N)_{in}}{(S/N)_{out}} \tag{3.86}$$

The ratio is often expressed in *decibels*, or dB, defined by

$$\Delta SNR_{dB} = 10 \log_{10} (\Delta SNR)$$

Illustrative Example 3.20

A signal is given by

$$r(t) = s(t) + n(t)$$

where

$$s(t) = 5 \cos 2\pi x 1{,}000t + 10 \cos 2\pi x 1{,}100t$$

The noise, $n(t)$, is white noise with a power spectral density of 0.05 W/Hz.

The total received signal is put through a bandpass filter with pass band between 990 and 1,110 Hz. Find the signal to noise ratio at the filter output.

Solution

We can assume that, in the steady state, the entire signal portion, $s(t)$, appears at the output of the filter. Thus, the output power is $25/2 + 100/2 = 62.5$ watts. The noise power at the output of the filter is found from

$$P_n = 2 \int_{990}^{1110} G_n(f)\, df = 12 \text{ watts}$$

The signal to noise ratio is then given by

$$S/N = 62.5/12 = 5.21 = 7.17 \text{ dB}$$

Illustrative Example 3.21

The input to the bandpass filter shown in Fig. 3.24 is the sum of a signal and noise. The power spectral densities of the signal and of the noise are as shown. Find the signal to noise improvement of the filter.

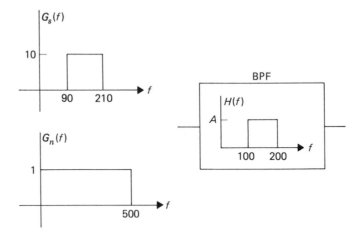

FIGURE 3.24 System and input spectra for Illustrative Example 3.21.

Solution

The input powers are found by integrating the corresponding densities.

$$\text{Signal Power In} = 2400 \text{ watts}$$

$$\text{Noise Power In} = 1000 \text{ watts}$$

$$S/N \text{ In} = 2.4 = 3.8 \text{ dB}$$

The output power spectral densities, which are illustrated in Fig. 3.25, result from multiplying the input densities by the magnitude squared of the filter transfer function. Integration of these densities results in the following output powers.

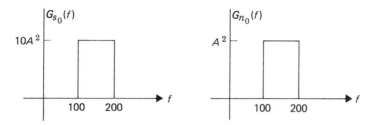

FIGURE 3.25   Output spectral densities for Illustrative Example 3.21.

$$\text{Signal Power Out} = 2000A^2 \text{ watts}$$

$$\text{Noise Power Out} = 200A^2 \text{ watts}$$

$$\text{S/N Out} = 10 = 10 \text{ dB}$$

The signal to noise improvement is given by

$$\text{SNR} = 10/2.4 = 4.17 = 6.2 \text{ dB}$$

Note that we could have simply subtracted the input S/N from the output S/N when both are expressed in dB. Further note that the answer is independent of A. This is true because both the noise and the signal are multiplied by A, so the ratio is unaffected.

PROBLEMS

3.1.   Three coins are tossed at the same time. List all possible outcomes of this experiment. List five representative events. Find the probabilities of the following events:

{all tails}   {one head only}   {three matches}

3.2.   If a perfectly balanced die is rolled, find the probability that the number of spots on the face turned up is greater than or equal to 2.

3.3.   An urn contains 4 white balls and 7 black balls. An experiment is performed in which three balls are drawn out in succession without replacing any. List all possible outcomes and assign probabilities to each one.

3.4.   In Problem 3.3 you are told that the first two draws are white balls. Find the probability that the third is also a white ball.

3.5.   An urn contains 4 red balls, 7 green balls and 5 white balls. Another urn contains 5 red balls, 9 green balls and 2 white balls. One ball is drawn from each urn. What is the probability that both balls will be of the same color?

3.6.   Three people, A, B, and C live in the same neighborhood and use the same bus line to go to work. Each of the three has a probability of $\frac{1}{4}$ of making the 6:10 bus, a probability of $\frac{1}{2}$ of making the 6:15 and a probability of $\frac{1}{4}$ of making the 6:20 bus. Assuming independence, what is the probability that they all take the same bus?

3.7. The probability density function of a certain voltage is given by

$$p_V(v) = ve^{-v}U(v)$$

where $U(v)$ is the unit step function.

(a) Sketch this probability density function.
(b) Sketch the distribution function of v.
(c) What is the probability that v is between 1 and 2 volts?

3.8. A Gaussian random variable has a mean value of 2. The probability that the variable lies between 2 and 5 is 0.3. Find the probability that the variable is between 1 and 3.

3.9. A Gaussian random variable has a variance of 9. The probability that the variable is greater than 5 is 0.1. Find the mean of the random variable.

3.10. A zero-mean Gaussian random variable has a variance of 4. Find x_0 such that

$$Pr(|x| > x_0) < 0.001$$

3.11. A random variable is Rayleigh distributed as follows.

$$p(x) = \frac{x}{4} \exp\left(\frac{-x^2}{8}\right)U(x)$$

Find the probability that the random variable is between 1 and 3.

3.12. Find the density of $y = |x|$ given that $p(x)$ is as shown below. Also find the mean and variance of both y and x.

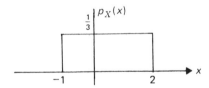

3.13. x is a Gaussian random variable with zero mean and unit variance. A new variable is defined by

$$y = |x|$$

(a) Find the density of y.
(b) Find the mean value of y.
(c) Find the variance of y.

3.14. A function is defined by

$$y = e^{-2x}$$

Find the density of $y, p(y)$, if
(a) x is uniformly distributed between 0 and 3.
(b) $p(x) = e^{-x}U(x)$.

3.15. Find the mean and variance of x where the density of x is shown below.

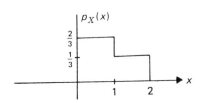

3.16. The density function of x is shown below. A random variable, y, is related to x as shown. Determine $p_Y(y)$.

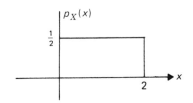

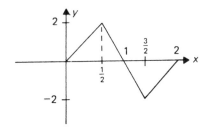

3.17. Find the expected value of $y = \sin x$ if x is uniformly distributed between 0 and 2π.

3.18. You are told that the mean rainfall per year is $10''$ in California and that the variance in the amount of rain per year is 1. Can you tell, from this information, what the density of rainfall is? If so, roughly sketch this density function.

3.19. The random variable x is Rayleigh distributed as in Eq. (3.11). Find the mean, second moment, and variance.

3.20. The random variable x is uniformly distributed between $x = -1$ and $x = +1$. The random variable y is uniformly distributed between $y = 0$ and $y = +5$. x and y are independent of each other. A new variable, z, is formed as the sum of x and y.
(a) Find the probability density function of z.
(b) Find the mean and variance of x, y, and z.
(c) Find a general relationship among the means of x, y, and z.
(d) Repeat part (c) for the variance.

3.21. (a) Find and sketch the density of the sum of two independent random variables as follows. One of the variables is uniformly distributed between -1 and $+1$. The second variable is triangularly distributed between -2 and $+2$.
(b) Compare the result to a Gaussian density function, where the variance should be chosen to make the Gaussian variable as close as possible to your answer to part (a). (You will have to start by defining "closeness.")

3.22. A random variable x has the following density function:

$$p_X(x) = Ae^{-2|x|}$$

(a) What is the value of A?
(b) Find the probability that x is between -3 and $+3$.
(c) Find the variance of x.

3.23. A voltage is known to be Gaussian distributed with mean value of 4. When this voltage is impressed across a 4-ohm resistor, the average power dissipated is 8 watts. Find the probability that the voltage exceeds 2 volts at any instant of time.

3.24. k is a random variable. It is uniformly distributed in the interval between -1 and $+1$. Sketch several possible samples of the process

$$x(t) = k \sin 2t$$

3.25. $x(t)$ is a stationary process with mean value 1 and autocorrelation

$$R(\tau) = e^{-|\tau|}$$

Find the mean and autocorrelation of the process $y(t) = x(t-1)$.

3.26. $x(t)$ is a stationary process with zero mean value. Is the process

$$y(t) = tx(t)$$

stationary? Find the mean and autocorrelation of $y(t)$.

3.27.   Given a stationary process, $x(t)$, find the autocorrelation of $y(t)$ in terms of $R_x(\tau)$, where

$$y(t) = x(t-1) + \sin 2t$$

3.28. $x(t)$ is stationary with mean value "m." Find the mean value of the output, $y(t)$, of a linear system with $h(t) = e^{-t}U(t)$ and $x(t)$ as input. That is,

$$y(t) = \int_0^t h(\tau)x(t-\tau)\, d\tau$$

3.29.   A random process is defined by

$$x(t) = K_1 t + K_2$$

where K_2 is a deterministic constant and K_1 is uniformly distributed between 0 and 1.

(a) Sketch several samples of this process.
(b) Find the mean of the process.
(c) Write an expression for the process autocorrelation.
(d) Is the process stationary?
(e) If K_1 is now uniformly distributed between -1 and $+1$, what changes occur in your answers to (b), (c) and (d)?

3.30.   Which of the following could not be the autocorrelation function of a process?

$$\text{(a) } R(\tau) = \begin{cases} 1 - |\tau| & |\tau| < 1 \\ 0 & |\tau| > 1 \end{cases}$$

$$\text{(b) } R(\tau) = 5 \sin 3\tau$$

$$\text{(c) } R(\tau) = \begin{cases} 1 & |\tau| < 2 \\ 0 & |\tau| > 2 \end{cases}$$

$$\text{(d) } R(\tau) = \frac{\sin \tau}{\tau}$$

3.31. $x(t)$ is white noise with autocorrelation $R_x(\tau) = \delta(\tau)$. It forms the input to an ideal lowpass filter with cutoff frequency f_m. Find the average power of the output, $E\{y^2(t)\}$.

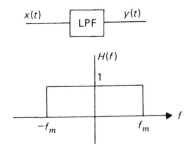

3.32. Use the result of Illustrative Example 3.12 in order to find the autocorrelation of the process $y(t) = dx/dt$ in terms of the autocorrelation of $x(t)$. (Hint: You will need the definition of the derivative,

$$\frac{dx}{dt} \overset{\Delta}{=} \lim_{T \to \infty} \frac{x(t) - x(t - T)}{T}$$

and l'Hospital's rule.)

3.33. Use an approach similar to that of Problem 3.32 in order to find

$$E\{x'(t)x(t + \tau)\} \quad \text{and} \quad E\{x'(t + \tau)x(t)\}$$

where $x'(t) = dx/dt$.

3.34. Find $R_e(\tau)$ in terms of $R_i(\tau)$ where $i(t)$ and $e(t)$ are related by the following circuit: That is, $e(t) = Ri(t) + L \, di/dt$.

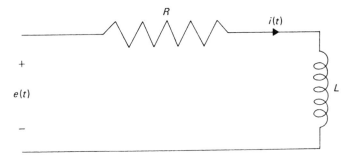

3.35. Given a constant, "a," and a random variable, "f," with density $p_F(f)$, form

$$x(t) = ae^{j2\pi ft}$$

Find $R_x(\tau)$ and show that $G_x(f) = |a|^2 p_f(f)$. (Note: For complex $x(t)$, use $R_x(\tau) = E\{x(t + \tau)x^*(t)\}$.)

3.36. Find the autocorrelation and power spectral density of the wave

$$v(t) = A \cos(2\pi f_c t + \theta)$$

where A and f_c are not random, and θ is uniformly distributed between 0 and 2π. (Hint: Use the definition of autocorrelation and the definition of expected value. You may also need the identity $\cos x \cos y = \frac{1}{2}\cos(x + y) + \frac{1}{2}\cos(x - y)$.)

3.37.   Prove that the inverse Fourier Transform of a Hilbert Transform is as shown below.

$$\mathscr{F}^{-1}(-j\,sgn(f)) = \frac{1}{\pi t}$$

3.38.   Find and sketch the Hilbert Transform of the function shown below.

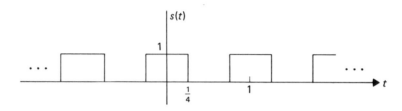

3.39.   Express the following narrowband noise processes in quadrature form. For each process, choose two different center frequencies and sketch the power spectral density of $x(t)$ and $y(t)$.

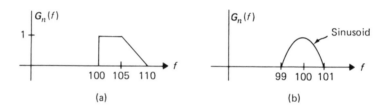

3.40.   In a given communication system, the signal is $f(t) = 20 \cos 2\pi t$. Noise of power spectral density $G_n(f) = e^{-3|f|}$ is added to the signal, and the resulting sum forms the input to a filter, $H(f)$.
(a) Find the S/N at the input to the filter.
(b) If the filter is lowpass with $H(f) = 1$ for $|f| < 2$, find the signal to noise improvement and express this in decibels.
(c) If the filter is bandpass with $H(f) = 1$ for $0.9 < |f| < 1.1$, find the signal to noise improvement and express this in decibels.

3.41.   You are given the following system composed of two cascaded filters:

$s(t)$ and $n(t)$ are as given in Problem 3.40. $H_1(f)$ is a lowpass filter with $H_1(f) = 1$ for $|f| < 2$. $H_2(f)$ is a bandpass filter with $H_2(f) = 1$ for $0.9 < |f| < 1.1$. Find the signal to noise improvement of $H_1(f)$ and the signal to noise improvement of $H_2(f)$, and express these in decibels. Now find the overall signal to noise improvement of the entire system and compare this to the individual signal to noise improvement of the two filters.

3.42.   Repeat Illustrative Example 3.21 if the bandpass filter passes from 90 to 210 Hz instead of 100 to 200 Hz. Compare the results.

Chapter 4
Amplitude Modulation

In the majority of current communication situations, the information to be transmitted is in the form of signals suitable for human ears (no reference to taste or censorship is intended). This is beginning to change as data transmission becomes prevalent.

Figure 4.1 shows a possible sample of a typical speech waveform. One can observe that, in general, it is a mess. Since the exact waveform which we wish to transmit is not known, how can anything be said about the system required to transmit it? Asked in a different way, does any similarity exist among various information signals which will help us to design a universal transmission system?

FIGURE 4.1 A representative speech waveform.

In the case of the speech (or any audio) waveform, the answer to this question comes in the form of recalling something learned in high school physics. The human ear can only respond to (hear) signals with frequencies below about 15,000 Hertz. Thus, if our eventual goal is the reception of audio type signals, it is fair to assume that the Fourier Transform of the time signal is zero for $|f| > 15,000$ Hz. That is,

$$S(f) = 0 \qquad (|f| > f_m)$$

where

$$f_m = 15,000 \text{ Hz}$$

One might argue that the vocal chords, or other sound generators, are capable of generating frequency components above 15,000 Hz, even though the ear cannot hear them. This is indeed correct. However, if one of these signals were passed through a lowpass filter with cutoff frequency of 15,000 Hz, the output of the filter, if fed into a loudspeaker, would sound exactly the same as the input. We are therefore justified in assuming that our "information" signals are bandlimited to an upper frequency of 15,000 Hz. While this may seem to be an insignificant observation, this one assumption will prove sufficient for the design of reasonably sophisticated communications systems.

"What about data transmission?" you ask. In that case, the upper frequency limitation is supplied by amplifiers and other electronic devices. Although this upper frequency limitation is usually considerably higher than that imposed by the human ear, it is nevertheless present and definable. The signal frequency cutoff restriction therefore applies to all information signals of interest, though the actual upper cutoff frequency depends upon the source and the eventual application of the signal.

For all of the future discussion, we will therefore assume that the signal to be transmitted (the "information" signal) has a Fourier Transform which is zero above a certain frequency. We will call this upper frequency limit, f_m, for "f-maximum." Even if f_m is high (the word "high" has no meaning by itself, so let us interpret it as "high with respect to 15,000 Hz"), we will call signals of this type "baseband signals."

In the case of short range transmission, as in the path between the pre-amp and amplifier or between the amplifier and speakers of a hi-fi, these low frequency signals are usually sent through wires. For longer distances, this is often not desirable since wires require "rights of way" meaning that towers must be built or trenches dug. Also, in the case of transmission through wires, one must specify the location of every terminal. For example, in the case of television, the wire would have to terminate in the home of every prospective viewer.

For these reasons, transmission through the air is often used instead of wire transmission. If a low frequency signal were simply applied to the input of an antenna, it would not radiate very far. In fact, in order to efficiently transfer the signal to the air (i.e., impedance match), the antenna would have to be many kilometers long for audio signals!

Even if one could efficiently transmit this signal through the air, a serious problem would arise. Suppose it were desirable to transmit more than one signal at a time, as it certainly is. That is, suppose that more than one of the dozens of local radio stations wished to transmit broadcasts simultaneously. They would each have antennas several miles long on top of their studios, and they would pollute the air with many audio signals. The poor listener would erect a receiving antenna several kilometers high, and get some sort of weighted sum (depending upon relative distances from the different transmitting antennas to the receiving antenna) of all of the signals. Since the only information the listener (receiver) has about the signals is that they are all low frequency bandlimited to the same upper cutoff frequency, there would be absolutely no way of separating one station from all of the others.

For the above two reasons, that is, efficient radiation and station separation, it is desirable to modify the low frequency signal before sending it from one point to another. An added bonus arises if the modified signal is less susceptible to noise than the original signal, but we're getting ahead of the game.

The most common method of accomplishing this modification is to use the low frequency signal to modulate (modify the parameters of) another signal. Most commonly, this other signal is a pure sine wave.

The following sections will analyze several different possibilities. During each analysis, the accomplishment of the primary objectives of efficient radiation and station separation will become obvious.

4.1   MODULATION

We start with a pure sinusoid, $s_c(t)$, called the *carrier* wave. It is given this name since it is used to carry the information signal from the transmitter to the receiver.

$$s_c(t) = A \cos(2\pi f_c t + \theta) \tag{4.1}$$

If f_c is properly chosen, this carrier wave can be efficiently transmitted. For example, suppose you were told that frequencies in the range between 1 MHz and 3 MHz propagate in a mode which allows them to be transmitted reliably in a range up to about 250 kilometers. If you chose the carrier frequency, f_c, to be in this range, then the pure sinusoidal carrier would transmit efficiently. The wavelength of transmissions in this range of frequencies is of the order of 100 meters, and reasonable length antennas can be used.

We now ask the question whether this pure sinusoidal carrier waveform can somehow be altered in a way that (a) the altered waveform still transmits efficiently and (b) the information we wish to send is somehow superimposed on the new waveform in a way that it can be recovered at the receiver. We are asking whether there is some way that the sinusoid can "carry" the information along.

Successful ways have been discovered to superimpose the information on the carrier. The techniques involve varying, or modulating one of the parameters of the carrier with the information signal.

Examination of Eq. (4.1) illustrates that there are three parameters which may be varied; the amplitude, A; the frequency, f_c; and the phase, θ. Using the information signal to vary A, f_c, or θ leads to amplitude modulation, frequency modulation, or phase modulation respectively.

For each of these three cases we will show that the two objectives of modulating are achieved. In addition we will have to illustrate a third property. That is, the information signal which we shall denote $s(t)$ must be uniquely recoverable from the modulated carrier wave. It wouldn't be of much use to modify a carrier waveform with $s(t)$ for efficient transmission if we could not reproduce $s(t)$ accurately at the receiver.

4.2   AM USING TRIGONOMETRY

Prior to a thorough analysis of AM using the Fourier Transform, it may be instructive to give the subject a once-over using only trigonometry.

Suppose first that a talented performer is whistling a single tone at a frequency of 1 kHz. It is desired that this (uninteresting) signal be transmitted through the air. From Section 4.1, we know that, without modification, the signal will transmit very poorly. We decide that shifting of the frequency to something near 1 MHz would improve the transmission capabilities. Since it is unreasonable to ask the performer to whistle at a frequency of 1 MHz, we investigate electronic techniques for performing this frequency shift.

You may recall a trigonometric identity:

$$\cos (A \pm B) = \cos A \ \cos B \mp \sin A \ \sin B \qquad (4.2)$$

From this, we see that (perform the derivation)

$$\cos 2\pi f_1 t \ \cos 2\pi f_2 t = \tfrac{1}{2} \cos 2\pi (f_1 + f_2)t + \tfrac{1}{2} \cos 2\pi (f_1 - f_2)t \qquad (4.3)$$

Therefore, if we wish to shift our 1-kHz sinusoid to a frequency near 1 MHz, we need simply multiply the sinusoid by another sine wave of about 1 MHz and then filter out the undesired term (*see* Fig. 4.2). This process is known as *heterodyning*.

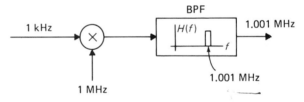

FIGURE 4.2   Heterodyning.

Let's simplify things and try eliminating the filter in Fig. 4.2. That is, we consider transmitting the product of the 1-kHz sinusoid with the 1-MHz sinusoid. This new signal has frequencies of 1.001 and 0.999 MHz. Both of these new frequencies will transmit efficiently.

One important job remains if we are to be convinced that this is a useful way of transmitting our 1-kHz whistle. We must be able to recover the original whistle at the receiver.

Since the AM wave is formed by shifting the frequency of the sound signal, the original signal is recovered by performing yet another shift. We must shift the received frequency *down* by 1 MHz, and we do this by multiplying the received signal by a sinusoid of frequency 1 MHz. (*See* Fig. 4.3.) The result of this multiplication is the signal

$$g(t) = \cos (2\pi \times 10^3 t) \cos^2 (2\pi \times 10^6 t) \qquad (4.4)$$

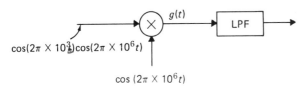

$\cos(2\pi \times 10^3 t)\cos(2\pi \times 10^6 t)$

$\cos(2\pi \times 10^6 t)$

FIGURE 4.3   Recovering original
signal from AM.

But from Eq. (4.3) the square of a cosine is equal to $\frac{1}{2}$ plus a cosine of twice the frequency. We therefore have

$$g(t) = \tfrac{1}{2}\cos(2\pi \times 10^3 t) + \tfrac{1}{2}\cos(2\pi \times 10^3 t)\cos(2\pi \times 2 \times 10^6 t) \quad (4.5)$$

The second term in Eq. (4.5) is composed of two frequencies—2.001 and 1.999 MHz. If a lowpass filter is added as shown in Fig. 4.3, these two relatively high frequencies are easily rejected, leaving only the original 1-kHz whistle.

We have therefore shown that a sinusoid of low frequency can be transmitted through the air by first multiplying it by a sinusoid of high frequency. The product is an AM wave, and it can be transmitted. At the receiver, the reverse operation is performed to recover the original signal.

A more complex audio signal than a whistle can be thought of as being composed of a sum of sinusoids. In this manner, the above trigonometric analysis can be extended to the general case.

The next section considers the general case but does so using the Fourier Transform rather than trigonometric identities. The results are more general, and, indeed, with a little bit of comfort with Fourier Transforms, one can extend the analysis more readily.

4.3   DOUBLE SIDEBAND SUPPRESSED CARRIER AMPLITUDE MODULATION

The case of amplitude modulation is characterized by a modulated signal waveform of the form $s_m(t)$.

$$s_m(t) = A(t)\cos(2\pi f_c t + \theta) \quad (4.6)$$

f_c and θ are constants and $A(t)$, the amplitude, varies somehow in accordance with $s(t)$.

In the following analysis, we shall assume that $\theta = 0$ for simplicity. This will not affect any of the basic results since the angle actually corresponds to a time shift of $\theta/2\pi f_c$. That is, we can rewrite $s_m(t)$ as

$$s_m(t) = A(t)\cos 2\pi f_c\left(t + \frac{\theta}{2\pi f_c}\right) \quad (4.7)$$

A time shift is not considered as distortion in a standard communications system since the exact time of arrival of a transmitted signal is usually of no importance.

If somebody asked you how to vary $A(t)$ in accordance with $s(t)$, the simplest answer you could suggest would be to make $A(t)$ equal to $s(t)$. Thus, the modulated signal would be of the form

$$s_m(t) = s(t) \cos 2\pi f_c t \qquad (4.8)$$

This type of modulated signal is given the name *double sideband suppressed carrier amplitude modulation* for reasons that will become clear in Section 4.7.

This simple guess for $A(t)$ does indeed satisfy the criteria demanded of a communication system. The easiest way to illustrate this fact is to express $s_m(t)$ in the frequency domain, that is, to find its Fourier transform.

Suppose that we call the Fourier Transform of $s(t)$, $S(f)$. Note that we are requiring nothing more of $S(f)$ other than its being equal to zero for frequencies above some cutoff frequency, f_m. We sketch $S(f)$ as in Fig. 4.4 since this represents an easy

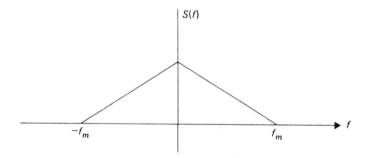

FIGURE 4.4   General form of $S(f)$.

sketch to reproduce. This does not mean that $S(f)$ must be of the shape shown. The sketch is meant only to indicate the transform of a general low frequency bandlimited signal.

From Section 1.7 (the modulation theorem), it should be clear that the transform of $s_m(t)$, $S_m(f)$, is given by

$$S_m(f) = \mathcal{F}[s(t) \cos 2\pi f_c t] = \mathcal{F}\left[\frac{s(t)e^{-j2\pi f_c t}}{2} + \frac{s(t)e^{+j2\pi f_c t}}{2} \right]$$

$$S_m(f) = \tfrac{1}{2}[S(f+f_c) + S(f-f_c)] \qquad (4.9)$$

This transform is sketched as Fig. 4.5.

Observation of Fig. 4.5 and the acceptance of a result carried over from electromagnetic field theory indicates that the first objective of a communication system has been met. The observation is that the frequencies present in $s_m(t)$ fall in the range between $f_c - f_m$ and $f_c + f_m$. The result from electromagnetics is that an efficient radiator (antenna) must have a length that is of the order of the wavelength of the signal to be transmitted. This is what we used to get the figure of many kilometers

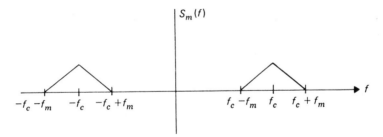

FIGURE 4.5 $S_m(f)$, the transform of $s_m(t)$.

for an audio signal antenna. Since wavelength is inversely proportional to frequency, we can make the wavelength as short as desired by raising the frequency of a wave. Since the lowest frequency present in $s_m(t)$ is $f_c - f_m$, and this can be made as large as desired by choosing f_c large, the signal can be tailored to any length antenna. In addition, as f_c increases, the range of frequencies occupied by $s_m(t)$ (i.e., the bandwidth) becomes smaller relative to f_c. This also leads to simpler antenna construction.

As an example, in the audio case even though $s(t)$ has frequencies between 0 and 15,000 Hz, $s_m(t)$ can be made to have only those frequencies between 985,000 and 1,015,000 Hz by choosing a carrier frequency of 1 MHz. This would lend itself to efficient transmission.

The second objective is that of channel separability. We see that if one information signal modulates a sinusoid of frequency f_{c1} and another information signal modulates a sinusoid of frequency f_{c2}, the Fourier Transforms of the two modulated carriers will be separated in frequency provided that f_{c1} and f_{c2} are not too close together. Figure 4.6 shows a sketch of a typical two channel case. Here, two

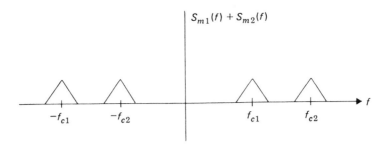

FIGURE 4.6   An example of two-channel AM.

modulated signals have been added together. Since taking the Fourier Transform of a function represents a linear operation, the transform of the sum of the two waves is the sum of the two individual transforms. If the frequencies of the two modulated waveforms are not too widely separated, both signals can even be transmitted via the same antenna. That is, although the optimum antenna length is not the same for both channels, the total bandwidth can be made relatively small compared to the carrier frequencies. In practice, the antenna is useable over a range of frequencies rather than just being effective at a single frequency. If the latter were the case, radio broadcasting could not exist.

As an example, you don't have to readjust your car antenna when you tune across the AM dial. The effectiveness of the antenna does not vary greatly from one frequency limit to the other. However, if you switch to the FM band which encompasses a much higher range of carrier frequencies than AM, the radio manufacturer recommends shortening the car antenna to about 75 centimeters. (Don't try doing this if you have the type of antenna which is sandwiched within the windshield).

We have repeatedly emphasized that signals can be separated if they are non-overlapping in either time or frequency. If they are non-overlapping in time, gates or switches can be used to effect the separation. If they are non-overlapping in frequency, the signals can be separated from each other by means of bandpass filters. Thus, a system such as that shown in Fig. 4.7 could be used to separate the two modulated carriers from each other in the case presented in Fig. 4.6.

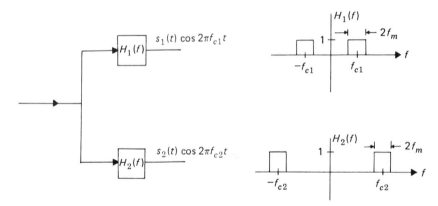

FIGURE 4.7   Separating the two channels of Fig. 4.6

The extension of this system to more than two channels should be obvious. Therefore, even if many modulated signals were transmitted over the same stretch of air (i.e., added together at the transmitter) they could be separated at the receiver by using bandpass filters which accept only those frequencies present in the desired modulated signal. This is true provided that the separate carrier frequencies are wide enough apart to prevent overlapping of the transforms. Observation of Fig. 4.6 indicates that if each information signal is low frequency bandlimited to f_m, the adjacent carrier frequencies must be separated by at least $2f_m$ to avoid overlapping.

The process of stacking many stations in separate "frequency slots" is called *frequency division multiplexing* (*FDM*), and is the method used in all standard broadcast transmission.

Before going on, we should again note that all of our sketches of Fourier Transforms seem to indicate that they are real functions of f. This would require that the corresponding time functions all be even functions, a very severe restriction indeed. However, examination of the formulae and derivations will indicate that this assumption was never made. It is simply used in the sketches to avoid clouding the picture with phase diagrams. Indeed, the sketches are included only to give an intuitive feel for what is happening. If the reader feels that absolute precision is needed throughout, it can be assumed that all of our drawings are actually plots of the magnitudes of the Fourier Transforms and that the phase plots were inadvertently omitted.

Illustrative Example 4.1

An information signal, $s(t) = (\sin t)/t$ is used to amplitude modulate a carrier of frequency $10/\pi$ Hz. Sketch the AM wave and its Fourier Transform.

Solution

The AM wave is given by the time function

$$s_m(t) = s(t) \cos 20t$$

$$= \frac{\sin t}{t} \cos 20t \tag{4.10}$$

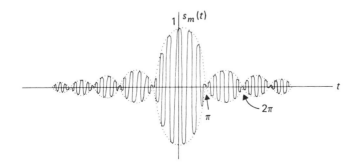

FIGURE 4.8 $S_m(t)$ for Illustrative Example 4.1.

This is sketched in Fig. 4.8.

We note that when the carrier, $\cos 20t$, is equal to 1, $s_m(t) = s(t)$, and when $\cos 20t = -1$, $s_m(t) = r(t)$. Therefore, the information signal, $s(t)$, and its mirror image, $-s(t)$, can be used as a type of outline to guide in the sketching of the waveform.

The transform of $s(t) = (\sin t)/t$ is as sketched in Fig. 4.9. This is found from the table in the Appendix.

From Eq. (4.9), the transform of the modulated wave is

$$S_m(f) = \tfrac{1}{2} [S(f - 10/\pi) + S(f + 10/\pi)] \tag{4.11}$$

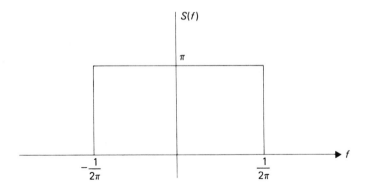

FIGURE 4.9 $S(f)$, the transform of $(\sin t)/t$.

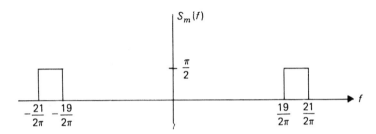

FIGURE 4.10 $S_m(f)$ for $S(f)$ of Fig. 4.9.

This is sketched as Fig. 4.10.

Illustrative Example 4.2

Define the dot product of $s_1(t)$ with $s_2(t)$ as the convolution of the two functions.

$$s_1(t) \cdot s_2(t) \overset{\Delta}{=} s_1(t) * s_2(t) \qquad (4.12)$$

Using this definition of the dot product, show that the two modulated carrier signals illustrated in Fig. 4.6 are orthogonal to each other.

Solution

We shall define

$$s_{m1}(t) = s_1(t) \cos 2\pi f_{c1} t$$

$$s_{m2}(t) = s_2(t) \cos 2\pi f_{c2} t$$

$$s_{m1}(t) * s_{m2}(t) = \int_{-\infty}^{\infty} s_{m1}(\tau)\, s_{m2}(t - \tau)\, d\tau$$

Since we do not know the exact time functions, we cannot evaluate this convolution integral. However, for this special case, the evaluation is trivial via the time convolution theorem.

$$s_{m1}(t) * s_{m2}(t) \leftrightarrow S_{m1}(f)S_{m2}(f)$$

In Fig. 4.6 we see that the product of the two transforms is zero since, at values of f for which $S_{m1}(f)$ is not equal to zero, $S_{m2}(f)$ is zero and vice versa. We are assuming that the carrier frequencies are wide enough apart so that the individual transforms do not overlap.

Therefore,

$$s_{m1}(t) * s_{m2}(t) = \mathscr{F}^{-1}[0] = 0 \qquad (4.13)$$

and $s_{m1}(t)$ is orthogonal to $s_{m2}(t)$. Actually, this analogy with the vector case can be carried much further. The set of time functions with transforms shown in Fig. 4.11 can be thought of as

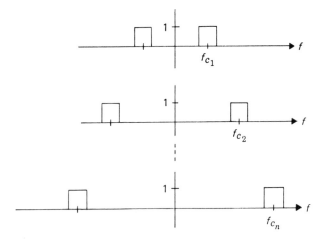

FIGURE 4.11   Unit vectors for Illustrative Example 4.2.

analogous to unit vectors. The actual $s_{m1}(t)$ and $s_{m2}(t)$ can be regarded as the coefficients. The results of Section 1.1 can therefore indicate how to send many signals at once, and how to recover one signal from the total sum (i.e., evaluate one of the coefficients).

The details will not be covered here as we will take a different approach. This example was meant to stimulate the interested student to thinking about the meaning and various applications of the principle of orthogonality. This principle will exhibit itself many times throughout the study of communications, but particularly in the area of noise reduction systems.

We have indicated that an AM wave of the type discussed could be transmitted efficiently, and that more than one information signal could be sent at the same time and still allow for separation at the receiver. An important property that must still be verified is that the information signal, $s(t)$, can be recovered from the modulated waveform, $s_m(t)$. We proceed to do that now.

Since $S_m(f)$ was derived from $S(f)$ by shifting all of the frequency components of $s(t)$ by f_c, we can recover $s(t)$ from $s_m(t)$ by shifting the frequencies again by the same amount, but this time in the opposite direction.

   The frequency shifting theorem tells us that multiplication of $s(t)$ by a sinusoid shifts the Fourier Transform both up and down in frequency. Thus multiplying $s_m(t)$ by $\cos 2\pi f_c t$ will shift the Fourier Transform back down to its low frequency position (sometimes called its DC position since direct current represents a zero frequency sinusoid). This multiplication will also shift the transform of $s(t)$ up to a position centered about a frequency of $2f_c$, but this part can easily be rejected by using a lowpass filter. This process is illustrated in Fig. 4.12.

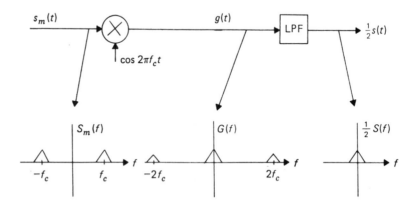

FIGURE 4.12   Recovery of $S(t)$ from $S_m(t)$.

   Mathematically, we have just described the following:

$$[s_m(t)] \cos 2\pi f_c t = [s(t) \cos 2\pi f_c t] \cos 2\pi f_c t$$
$$= s(t) \cos^2 2\pi f_c t \qquad\qquad (4.14)$$
$$= \tfrac{1}{2}[s(t) + s(t) \cos 4\pi f_c t]$$

   The output of the lowpass filter is therefore $\frac{1}{2} s(t)$ which represents an undistorted version of $s(t)$. Quite logically, this process of recovering $s(t)$ from the modulated waveform is known as *demodulation*. We have taken this moment now to discuss demodulation rather than waiting until Section 4.5 where we will have more to say about it. Indeed, if $s(t)$ could not be recovered from $s_m(t)$, there would be no reason to go on to Section 4.4.

4.4   MODULATORS

   Although this text is primarily intended to present system theory rather than practical realizations, it is instructive to consider a few actual techniques of building a modulator or demodulator. We say this is instructive since the modulator represents the first time we have come across a system that is *not* linear time invariant.

   Why is modulation not linear time invariant? We can use the basic properties of systems to illustrate that this must be the case. Recall that any linear time invariant system has an output whose Fourier Transform is the product of the Fourier Transform of the input with $H(f)$. Thus, if the Fourier Transform of the input, $S(f)$, is zero at some value of f, the Fourier Transform of the output must also be zero at this frequency. Indeed, a general property of linear time invariant systems is that they cannot generate any output frequencies that do not appear in the input. Going one step further, linear systems can always be described by linear differential equations. No matter how many times we integrate or differentiate a sine wave, it never changes frequency. Thus a linear time invariant is incapable of creating any new frequencies that do not exist as part of its input.

   We now ask whether any linear time invariant system can have $s(t)$ as an input and $s_m(t)$ as an output. In other words, is there any $H(f)$ for which $S_m(f) = S(f)H(f)$ as in Fig. 4.13?

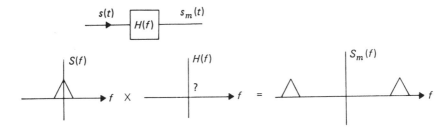

FIGURE 4.13   Can a linear time invariant system modulate?

   Since $S_m(f)$ is non-zero at frequencies for which $S(f) = 0$, there exists no $H(f)$ that can be used as a modulator.

   The problem arises because the form of AM we are discussing is linear, but not time invariant. To show linearity, we note

$$s_1(t) \rightarrow s_1(t) \cos 2\pi f_c t$$

$$s_2(t) \rightarrow s_2(t) \cos 2\pi f_c t$$

$$as_1(t) + bs_2(t) \rightarrow [as_1(t) + bs_2(t)] \cos 2\pi f_c t$$

$$= as_1(t) \cos 2\pi f_c t + bs_2(t) \cos 2\pi f_c t$$

We shall see that this linearity makes the analysis much simpler. (This will become clear when we study FM, a type of modulation which is not even linear.)

   Synthesis of a system which is not linear time invariant is, in general, extremely complicated. We shall attempt to simplify the synthesis, at least in the case of modulation, to a point at which the physical system realization seems intuitively realizable.

   We state the problem of amplitude modulation as follows: Given $s(t)$ and $\cos 2\pi f_c t$ we wish to form the product, $s(t) \cos 2\pi f_c t$.

   The methods of accomplishing this can be divided into three general areas. We shall assign the names gated, square law, and waveshape modulators to these three classes.

The *gated modulator* takes advantage of an observation that can be made from the sampling theorem. That is, if $s(t)$ is sampled (multiplied) with a periodic wave of fundamental frequency f_c, the Fourier Transform of the sampled wave will consist of the original Fourier Transform shifted and repeated at multiples of f_c in frequency. In the case of the sampling theorem, we were interested in the Fourier Transform "hump" centered about $f = 0$, and we rejected the others with a lowpass filter in order to recover $s(t)$. In amplitude modulation, we are interested in the component of the Fourier Transform centered about $f = f_c$, and we shall reject all of the other components using a bandpass filter.

As a special case, consider the sampling function, $p(t)$, as shown in Fig. 4.14. If

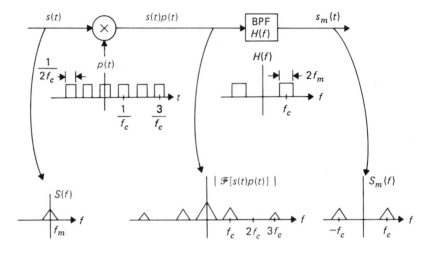

FIGURE 4.14   The gated modulator.

$s(t)$ is multiplied by $p(t)$, we have

$$s(t)p(t) = s(t) \left[\frac{1}{2} + \frac{2}{\pi} \cos 2\pi f_c t - \frac{2}{3\pi} \cos 6\pi f_c t + \cdots \right] \qquad (4.15)$$

where we have replaced $p(t)$ by its Fourier Series representation. When the above expression is multiplied out, each term will be of the form

$$a_n s(t) \cos n2\pi f_c t \qquad (4.16)$$

which is actually an amplitude modulated wave whose carrier frequency is nf_c. A bandpass filter of bandwidth $2f_m$ and centered about f_c will sift out the desired term from all of the undesired terms (harmonics). For this particular $p(t)$, the term will be $(2/\pi)s(t) \cos 2\pi f_c t$ which is, indeed, the desired AM wave. The only requirement is that of the sampling theorem to insure that the separate sections of the Fourier Transform do not overlap. That is, $f_c > 2f_m$. We will see later that for all cases of interest, f_c is intentionally chosen to be much much greater than f_m, so this restriction is easily met. For example, in standard AM broadcast, f_m is 5000 Hz and f_c is something in the vicinity of 1,000,000 Hz, the exact value depending upon the desired station.

The entire gated modulating process is illustrated in Fig. 4.14.

A reasonable question would now be "So what? What have we accomplished?" Observation of Fig. 4.14 indicates that we must somehow generate $p(t)$ from $\cos 2\pi f_c t$ (what we were given), perform a multiplication, and also build a bandpass filter. Why did we not just take $s(t)$ and $\cos 2\pi f_c t$ and put them into a multiplier as shown in Fig. 4.15?

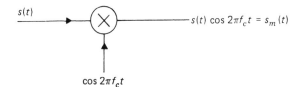

FIGURE 4.15   An AM modulator.

The answer is that there are very few devices which can take two continuous analog signals and multiply them together accurately.

How then does the gated modulator avoid the multiplication problem? Since one of the functions, $p(t)$, is always either equal to zero or one, the multiplication process can be thought of as a gating process. That is, whenever $p(t) = 1$, the output is equal to the input. When $p(t) = 0$, the output is zero. Multiplication by an $p(t)$ of this type corresponds to periodically opening and closing a gate, or switch in a circuit. Consider the circuit shown in Fig. 4.16.

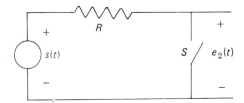

FIGURE 4.16   A gating circuit

If the operator sits at the circuit and periodically opens and closes switch S, one can see that $e_2(t)$ will be of the desired form, $s(t)p(t)$.

Unfortunately, the switch has to be opened and closed at a rate which is often greater than a million times every second (f_c times per second). Mechanical switches are therefore usually ruled out. The obvious alternative is to use an electrical switch.

In practice, one often uses a solid state device operating between cutoff and saturation as an effective electronic switch. For our purpose here, a simple diode is sufficient. The circuit of Fig. 4.17 is known as a *diode bridge modulator*. It accomplishes the gating in the same way as the circuit of Fig. 4.16.

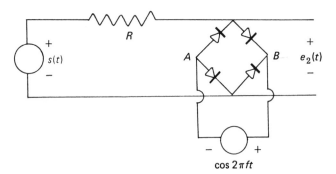

FIGURE 4.17   The diode bridge modulator.

When $\cos 2\pi ft > 0$, the point "B" is at a higher potential than point "A." In this condition, all four diodes are open circuited, and the circuit is equivalent to that of Fig. 4.16 with switch S open. When $\cos 2\pi ft < 0$, point "A" is at a higher potential than point "B," and all four diodes are short circuits. In that case, the circuit is equivalent to that of Fig. 4.16 with switch S closed. Therefore the $\cos 2\pi ft$ source essentially opens and closes the switch at a rate of f times per second. The only limit to this rate is imposed by the nonidealness of practical diodes (i.e., capacitance).

There is another type of amplitude modulator which is conceptually simpler than the gated modulator. This is known as the *square law modulator*. The square law modulator takes advantage of the fact that the square of the sum of two functions contains their product as the cross term in the expansion.

$$[s_1(t) + s_2(t)]^2 = s_1^2(t) + s_2^2(t) + 2s_1(t)s_2(t) \qquad (4.17)$$

Therefore, a summer and squarer could be used to multiply two functions together provided that the $s_1^2(t)$ and the $s_2^2(t)$ terms could be rejected from the result.

Considering the special case where $s_1(t) = s(t)$, the information signal, and $s_2(t) = s_c(t)$, the carrier sinusoid, we have

$$[s(t) + s_c(t)]^2 = s^2(t) + s_c^2(t) + 2s(t)s_c(t) \qquad (4.18)$$

The third term in Eq. (4.18) is the desired modulated signal, and if the other two terms can be rejected, the modulation is accomplished.

We have already seen that functions whose Fourier Transforms occupy different intervals along the frequency axis can always be separated from each other. The mechanism of this separation is the bandpass filter. We would therefore be overjoyed if it could somehow be shown that the three terms on the right side of Eq. (4.18) have non-overlapping Fourier Transforms.

We note that $s_c^2(t)$ can be explicitly written,

$$s_c^2(t) = \cos^2 2\pi f_c t = \tfrac{1}{2}[1 + \cos 4\pi f_c t] \qquad (4.19)$$

Figure 4.18 shows the sum of the transforms of $s_c^2(t)$ and $2r(t)s_c(t)$, both of which have been found in previous work.

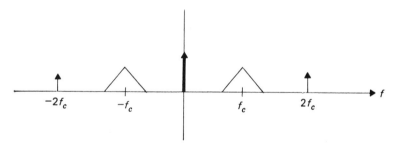

FIGURE 4.18   Sketch of transform of $s_c^2(t) + 2r(t)s_c(t)$.

The transform of $s(t)$ is not known exactly, since we do not wish to restrict this to any particular information signal. How can we then hope to know anything about the transform of $s^2(t)$? The answer lies in the convolution theorem. Since $s^2(t)$ is the product of $s(t)$ with itself, the transform of this will be the convolution of $S(f)$ with $S(f)$.

$$s^2(t) \leftrightarrow [S(f) * S(f)] \tag{4.20}$$

Illustrative Example 1.6 showed that, if $S(f)$ is low frequency bandlimited to f_m, then $S(f)$ convolved with itself is low frequency band limited to $2f_m$.

$$S(f) * S(f) = 0 \qquad (|f| > 2f_m) \tag{4.21}$$

We can now sketch the transform of all terms of Eq. (4.18). This is done in Fig. 4.19. There has been no attempt to show the exact shape of $S(f) * S(f)$, just as the sketch of $S(f)$ is not meant to represent its exact shape. The range of frequency occupied is the only important parameter for this study.

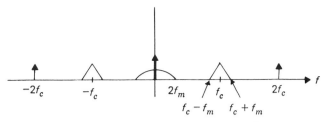

FIGURE 4.19   Transform of $[s(t) + s_c(t)]^2$.

As long as $f_c > 3f_m$, the components of Fig. 4.19 do not overlap, and the desired AM wave can be recovered using a bandpass filter centered at f_c with bandwidth equal to $2f_m$. This yields a square law modulator block diagram as shown in Fig. 4.20.

The obvious question now arises as to whether or not one can construct this device.

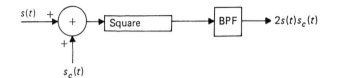

FIGURE 4.20   The square law modulator.

Summing devices are trivial to construct since any linear circuit obeys superposition.

Square law devices are not quite as simple. Any practical non-linear device has an output over input relationship which can be expanded in a Taylor series. This assumes that no energy storage is taking place. That is, the output at any time depends only upon the value of the input at that same time, and not upon any earlier input values. This is illustrated in Fig. 4.21.

FIGURE 4.21   A nonenergy storing non-linear device.

Thus,

$$y(t) = F[x(t)]$$
$$= a_0 + a_1 x(t) + a_2 x^2(t) + a_3 x^3(t) + \cdots$$

The term we are interested in is the $a_2 x^2(t)$ term in the above expansion. If we could somehow find a way to separate this term from all of the others, it would then be possible to use any non-linear device in place of the squarer in Fig. 4.20. We now optimistically investigate this possibility of separation.

It would be beautiful if the transforms of each of the terms in the Taylor series were non-overlapping in frequency. It would even be sufficient if the transform of each term other than that of the square term fell outside of the interval $f_c - f_m$ to $f_c + f_m$ on the frequency axis.

Beautiful as this might be, it is not the case. The transform of the linear term, assuming $x(t) = s(t) + s_c(t)$, has an impulse sitting right in the middle of the interval of interest. The cube term contains $s^2(t) \cos 2\pi f_c t$, which consists of $S(f) * S(f)$ shifted up around f_c in the frequency domain. Indeed, every higher order term will have a component sitting right on top of the desired transform, and therefore inseparable from it by means of filters.

All we have indicated is that not just any non-linear device can be used in our modulator. The device must be an essentially pure squarer. Fortunately, such devices do exist approximately in real life. A simple semiconductor diode can be used as a squarer when confined to operate in certain ranges. Indeed, the modulator shown in Fig. 4.22 is often used (with minor modifications).

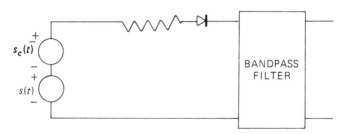

FIGURE 4.22 A diode square law modulator.

The previous discussion has shown that the modulator of Fig. 4.20 requires an essentially pure square law device. This was true since the means of separating desired and undesired terms was the bandpass filter. The terms therefore had to be non-overlapping in frequency. More exotic techniques do exist for separating two signals. One such technique allows the non-linear device of Fig. 4.21 to contain odd powers in addition to the square term.

Suppose, for example, we build the square law modulator and a cube term is present. We already found that an undesired term of the form $s^2(t) \cos 2\pi f_c t$ appears in the output. Suppose that we now build another modulator, but use $-s(t)$ as the input instead of $+s(t)$. The term $s^2(t) \cos 2\pi f_c t$ will remain unchanged, while the desired term, $s(t) \cos 2\pi f_c t$ will change sign. If we subtract the two outputs, the $s^2(t) \cos 2\pi f_c t$ will be eliminated to leave the desired modulated waveform. Some thought will convince one that this technique will eliminate all product terms resulting from odd powers in the non-linear device. Such a modulator is called a *balanced modulator*, and one form is sketched in Fig. 4.23.

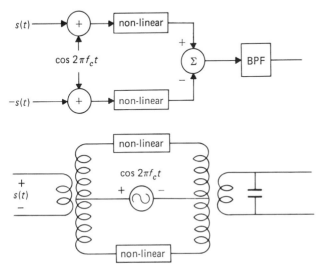

FIGURE 4.23 A balanced AM modulator.

The third method of constructing a modulator is more direct than the previous two techniques. We shall call this the *waveshape modulator*. The modulation is performed as an actual waveshape multiplication of two signals in an electronic circuit.

One possible realization is drawn in Fig. 4.24. In this case, the carrier sinusoid is introduced into the emitter of a transistor. The transistor is in the common base

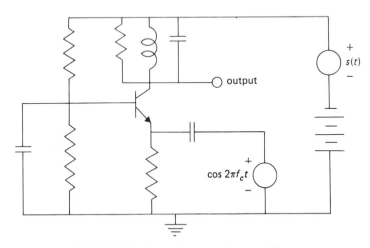

FIGURE 4.24   A transistor waveshape modulator.

configuration. The information signal, $s(t)$, is introduced as a variation in the B^+ supply voltage, E_{cc}. The relatively slow changes in E_{cc} have the effect of moving the load line and changing the quiescent point of the circuit.

Without any further analysis, it should not be too surprising to find that if the transistor is not permitted to saturate, the output will contain the desired AM waveform.

4.5   DEMODULATORS

We have previously stated that $s(t)$ is recovered from $s_m(t)$ by "remodulating" $s_m(t)$ and then passing the result through a lowpass filter. This yields the demodulator system block diagram of Fig. 4.25.

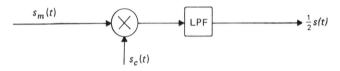

FIGURE 4.25   An AM demodulator.

Since the multiplier in the above figure looks no different from the multiplier used in the modulator, one would expect that the gated and square law modulators can again be used.

Illustrative Example 4.3

Show that the gated modulator can be used as a demodulator.

Solution

The system is illustrated in Fig. 4.26.

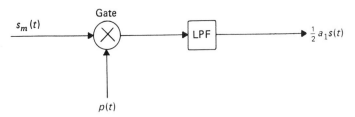

FIGURE 4.26   The gated demodulator.

We wish to consider $s_m(t)p(t)$, where $p(t)$ is periodic and given by a Fourier Series expansion,

$$p(t) = \sum_{n=0}^{\infty} a_n \cos n2\pi f_c t \qquad (4.22)$$

Expanding the product, we have

$$s_m(t)p(t) = s(t) \cos 2\pi f_c t \sum_{n=0}^{\infty} a_n \cos n2\pi f_c t$$

$$= s(t) \sum_{n=0}^{\infty} a_n \cos 2\pi f_c t \cos n2\pi f_c t \qquad (4.23)$$

$$= s(t) \sum_{n=0}^{\infty} a_n \tfrac{1}{2}[\cos(n-1)2\pi f_c t + \cos(n+1)2\pi f_c t]$$

Writing out the first few terms, and sketching their transforms, we have

$n = 0 \rightarrow a_0 s(t) \cos 2\pi f_c t$

$n = 1 \rightarrow \tfrac{1}{2} a_1 s(t)[1 + \cos 4\pi f_c t]$

$n = 2 \rightarrow \tfrac{1}{2} a_2 s(t)[\cos 2\pi f_c t + \cos 6\pi f_c t]$

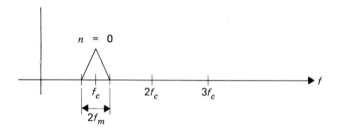

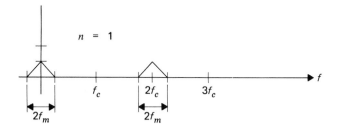

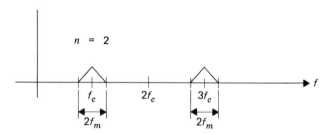

FIGURE 4.26A   Uncaptioned part of Illustrative Example 4.3

Viewing the transform of each of these parts, we see that each one represents $S(f)$ shifted to some multiple of f_c, except for the term $\frac{1}{2} a_1 s(t)$. This has as its transform

$$\tfrac{1}{2} a_1 S(f)$$

and it can be separated by use of a lowpass filter. We have therefore accomplished demodulation.

Illustrative Example 4.4

Show that the square law modulator *cannot* be used to demodulate AM.

Solution

We wish to show that $[s_m(t) + s_c(t)]^2$ does not contain a separable term of the form, $s(t)$. Expanding the squared sum, we have

$$[s_m(t) + s_c(t)]^2 = [s(t) \cos 2\pi f_c t + \cos 2\pi f_c t]^2$$
$$= \cos^2 2\pi f_c t[1 + s(t)]^2$$

$$= \tfrac{1}{2}[1 + \cos 4\pi f_c t][1 + 2s(t) + s^2(t)]$$
$$= \tfrac{1}{2}[1 + 2s(t) + s^2(t) + \cos 4\pi f_c t$$
$$+ 2s(t)\cos 4\pi f_c t + s^2(t)\cos 4\pi f_c t] \tag{4.24}$$

The desired term in Eq. (4.24) is $r(t)$. It should be clear from Fig. 4.18 that the term $\tfrac{1}{2}s^2(t)$ has a transform which lies right on top of $S(f)$, and can therefore not be filtered out. A square law modulator can therefore not be used for demodulation. (In Section 4.7, we will see a modification that can be made which will permit square law demodulation).

Illustrative Example 4.5

This example should give some practice in applying the principles of modulation and demodulation to a real system. We have already demonstrated that modulators and demodulators can be built, so we will again revert to block diagrams.

Consider the system sketched in Fig. 4.27, with $R_1(f)$ and $R_2(f)$ as shown.

(a) Sketch the transform of $r_3(t)$.
(b) Sketch the transform of $g(t)$.
(c) If $r_1(t)$ and $r_2(t)$ were to represent the left and right channels of a stereo broadcast, sketch one possible realization of a complete receiver.

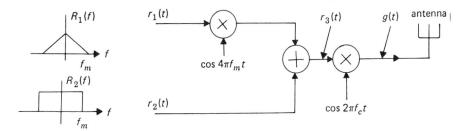

FIGURE 4.27   System for Illustrative Example 4.5.

Solution

$r_3(t)$ is given by

$$r_3(t) = r_1(t)\cos 4\pi f_m t + r_2(t) \tag{4.25}$$

The transform of this is sketched in Fig. 4.28. $g(t)$ is given by

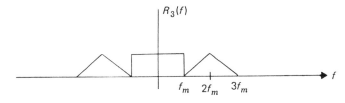

FIGURE 4.28 $R_3(f)$ for Illustrative Example 4.5.

$$g(t) = r_3(t) \cos 2\pi f_c t \tag{4.26}$$

and its transform is sketched in Fig. 4.29. We note that, while $r_3(t)$ could not be efficiently transmitted through the air, $g(t)$ can be, with proper choice of f_c.

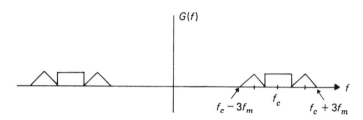

FIGURE 4.29 $G(f)$ for Illustrative Example 4.5.

The demodulation system is sketched in Fig. 4.30. It essentially reverses the operations of the original system, with the addition of filtering to separate the two "channels." The student should be able to sketch the Fourier Transform of the signal at each point in the system of Figure 4.30.

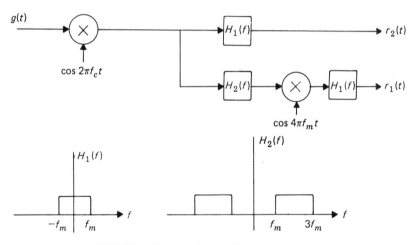

FIGURE 4.30 Receiver for Illustrative Example 4.5.

In many ways, this system is similar to the system used for FM stereo multiplex broadcasting. We will say more about this in Chapter 5.

Even though the waveshape and gated demodulation schemes appear to be acceptable, significant problems are encountered in practice. In general, we have essentially said "just multiply $s_m(t)$ by $\cos 2\pi f_c t$ and lowpass filter in order to recover $r(t)$ from the modulated waveform." We now state that this type of demodulator is given the name, *synchronous demodulator* (or detector). The word "synchronous" is used since the receiver must possess an oscillator which is at exactly the same frequency and phase as the carrier oscillator which is located in the transmitter.

   In actual operation, the receiver and transmitter are usually separated by many miles, and this matching becomes difficult if not impossible.

   Let us first see what happens if either the phase or the frequency of the oscillator in the receiver differs by a small amount from that of the transmitter.

4.6   PROBLEMS IN SYNCHRONOUS DEMODULATION

In synchronous demodulation, $s_m(t)$, the received modulated waveform is multiplied by a sinusoid of the same frequency and phase as the carrier sinusoid. We now consider the case of multiplying $s_m(t)$ by a cosine waveform which deviates in frequency by Δf, and in phase by $\Delta\theta$ from the desired values of frequency and phase. Performing the necessary trigonometric operations, we have

$$s_m(t) \cos [2\pi(f_c + \Delta f)t + \Delta\theta]$$

$$= s(t) \cos 2\pi f_c t \cos [2\pi(f_c + \Delta f)t + \Delta\theta]$$

$$= s(t) \left[\frac{\cos (2\pi\Delta f t + \Delta\theta)}{2} + \frac{\cos [2\pi(2f_c + \Delta f)t + \Delta\theta]}{2} \right] \tag{4.27}$$

Since the expression in Eq. (4.27) forms the input to the lowpass filter of the synchronous demodulator, the output of this filter will be that given in Eq. (4.28). This is true since the second term of Eq. (4.27) has a Fourier Transform which is centered about a frequency of $2f_c$, and is therefore rejected by the lowpass filter. The term in Eq. (4.28) has frequency components extending to $f_m + \Delta f$, so we are assuming that the filter doesn't cut off exactly at f_m.

$$\text{Filter output} = s(t) \frac{\cos (2\pi\Delta f t + \Delta\theta)}{2} \tag{4.28}$$

If Δf and $\Delta\theta$ are both equal to zero, Eq. (4.28) simply becomes $\frac{1}{2}s(t)$, and the demodulator gives the desired output. Assuming that these deviations cannot be made equal to zero, we are stuck with the output given by Eq. (4.28).

   To see the physical manifestations of this, let us first assume that $\Delta\theta = 0$. The output then becomes

$$s(t) \frac{\cos (2\pi\Delta f t)}{2} \tag{4.29}$$

Δf is usually (one hopes) small, and the result will be a slowly varying amplitude (beating) of $s(t)$. If, for example, $s(t)$ were an audio signal, this effect would be highly annoying since the volume would periodically vary from zero to a maximum and back to zero again. One could simulate this in a standard radio by continually turning the volume control clockwise and then counterclockwise once every $1/\Delta f$ seconds.

As a more specific example, if $f_c = 10^6$ Hz, one would certainly not consider an error in adjusting the frequency of the demodulator oscillator as being unreasonably large if it were only 1 Hz. If anything, the opposite is true. That is, 1 Hz is an optimistically small deviation. This would occur if the receiver oscillator were set to either 999,999 or 1,000,001 Hz instead of the desired 1,000,000 Hz of the transmitter oscillator. The result would be heard as a continuous variation in volume from maximum to zero *twice every second*. This is clearly an intolerable situation. Even if one could adjust the demodulation oscillator until no beating is heard, it is not unusual for an oscillator to drift in frequency over an interval of time. The example cited above would represent a drift of only 0.0001% in frequency!

The above analysis is enough to doom synchronous demodulation to highly limited use. However, for completeness, we shall examine the effects of $\Delta\theta$. This can best be done separately from the effects of Δf. We therefore assume that the frequency is miraculously adjusted perfectly, and $\Delta f = 0$. The output of the demodulator is then given by

$$s(t) \, \frac{\cos(\Delta\theta)}{2} \qquad (4.30)$$

This is not as annoying as the previous problem since $\cos(\Delta\theta)$ is a constant. We can usually compensate for this factor by varying the volume control (amplification) of the receiver. However, a deviation in phase can end up being just as deadly as a deviation in frequency. If the phase factor gets too small (i.e., $\Delta\theta$ approaches 90°), we may not be able to add enough amplification, and might therefore lose the signal. A more serious problem will turn out to be that of the signal getting so small that it gets lost in the background noise, but this is getting ahead of the game.

Is there any way to get around these problems? Wouldn't it be nice if the so-called local carrier (i.e., the cosine wave required at the receiver to demodulate) could somehow be derived from, or generated by, the incoming signal? If this were possible, we would be assured of having the frequency and phase properly adjusted.

Let's be intuitive for a minute. The Fourier Transform of the arriving modulated waveform is redrawn in Fig. 4.31.

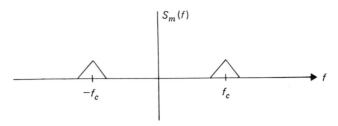

FIGURE 4.31   Transform of AM wave.

Stare at this transform for a while, and ask yourself whether it possesses enough information to uniquely specify f_c.

Looking at the portion lying around $f = f_c$, we see that it is symmetrical about $f = f_c$. This is true since the original $S(f)$ was an even function (i.e., symmetrical about $f = 0$). Therefore, in terms of $S_m(f)$, one can find f_c as the point at which $S_m(f)$ can be folded upon itself. That is, the symmetry mentioned above serves to point out f_c as its midpoint. We therefore claim that, intuitively, the incoming waveform possesses sufficient information to allow unique determination of the exact frequency, f_c. The construction of a system to "detect" this symmetry point is not at all simple.

Pay close attention now as we hand-wave our way through one possible system (Fig. 4.32). The "hand-waving" description of this system is not critical to the continuity of this chapter, and the timid student may therefore skip the next few pages.

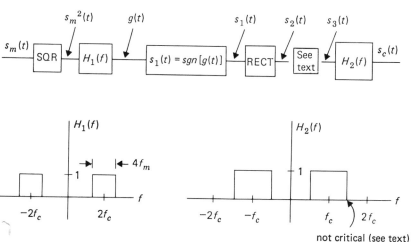

FIGURE 4.32   A system for generating $s_c(t)$ at the receiver, given $s_m(t)$.

Given $s_m(t) = s(t) \cos 2\pi f_c t$, we wish to somehow generate the term $\cos 2\pi f_c t$. Let us first square $s_m(t)$:

$$s_m^2(t) = s^2(t) \cos^2 2\pi f_c t \tag{4.31}$$

$$= \frac{s^2(t)}{2} + \frac{s^2(t) \cos 4\pi f_c t}{2} \tag{4.32}$$

If this waveform is passed through a bandpass filter with center frequency $2f_c$ and bandwidth $4f_m$, the $\frac{1}{2} s^2(t)$ term will be rejected to yield $g(t)$.

$$g(t) = \frac{s^2(t) \cos 4\pi f_c t}{2} \tag{4.33}$$

Figure 4.33 sketches this and other significant waveforms to follow for a representative example.

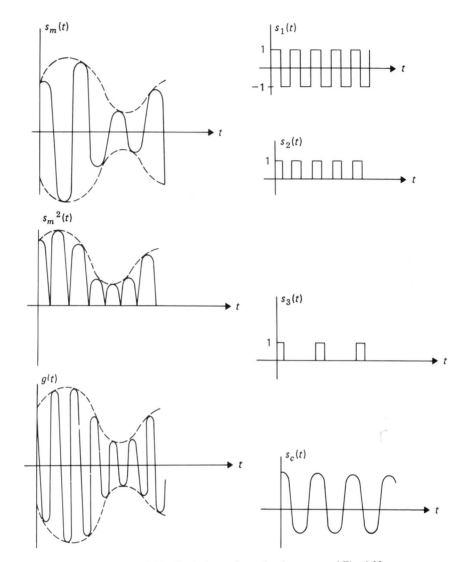

FIGURE 4.33 Typical waveforms for the system of Fig. 4.32.

We note that, since $s^2(t)$ is always positive, $g(t)$ will be positive when $\cos 4\pi f_c t$ is positive, and negative when $\cos 4\pi f_c t$ is negative. We now define the "sign function" as follows,

$$sgn(t) = \begin{cases} 1 & t > 0 \\ -1 & t < 0 \end{cases} \qquad (4.34)$$

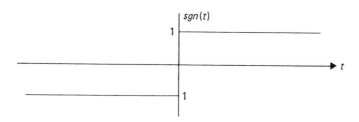

From Fig. 4.33, it should now be clear that $sgn[g(t)]$ will be a square wave of frequency $2f_c$. The student can accept the fact that the operation of taking $sgn[g(t)]$ is relatively easy to perform. It is sometimes called "infinite clipping" since it is like clipping the waveform at a very low level and then amplifying the result.

We must now halve the fundamental frequency of this square wave. This can be done by using a device known as a bistable multivibrator (flip-flop). In case the reader is unfamiliar with this, we will describe a slightly modified approach. $s_1(t)$ is half-wave-rectified to yield the pulse train shown as $s_2(t)$. We then put $s_2(t)$ through a system which ignores every second pulse. The result is the pulse train $s_3(t)$. The Fourier Transform of this pulse train is a train of impulse functions at frequencies which are multiples of f_c. A bandpass filter which passes f_c but rejects $2f_c$ will yield a sinusoid of exactly f_c frequency. This resulting signal can now be used in the synchronous detector with no worries about Δf being anything other than identically zero.

4.7   AM TRANSMITTED CARRIER

The previous system, while quite simple in theory, is reasonably complex to build. If standard broadcast radio had to resort to it, the garden variety pocket or table radio would probably be much more expensive than it is today. The question arises as to whether or not there exists a simpler method of demodulating an amplitude modulated signal.

Since the major difficulty appears to be caused by the absence of an accurate reproduction of the carrier at the receiver, why not send a pure carrier along with the AM signal? That is, suppose we add a pure carrier term to the AM signal as shown in Fig. 4.34.

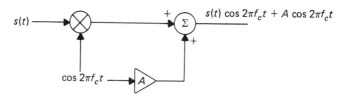

FIGURE 4.34 Addition of a carrier term.

We have continually preached that signals can be separated if they are non-overlapping in time or non-overlapping in frequency. Clearly the carrier and AM waveform are overlapping in time, so our only hope is to examine frequencies. Figure 4.35 shows the Fourier Transform of the AM wave with the added carrier term. It

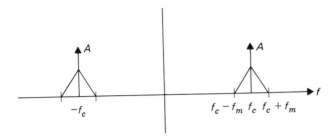

FIGURE 4.35   Fourier Transform of AM wave with added carrier.

would appear that we are in trouble, since the two portions do overlap in frequency.

We are saved by two observations. First, the portion of the signal overlapping the carrier results from the low frequency terms of the original information signal, $s(t)$. That is, the energy lying around the carrier frequency, f_c, results from shifting that portion of $S(f)$ lying around zero. Most information signals we will be dealing with have very little energy at frequencies near zero (e.g., the human voice has little energy below about 25 Hz). Therefore, there will be a small gap in the transform of the AM wave around f_c. But even if this were not true, we are saved by another observation.

All of the energy of the added carrier term is concentrated as a single frequency point, while the energy of the modulated wave is distributed across a range of frequencies. Thus, if we place a very narrow bandpass filter at the carrier frequency, all of the carrier term will pass through it in the steady state. On the other hand, the narrower the filter, the less of the AM waveform will appear at the output. Analyzing this in the Fourier Transform domain, we find the output transform of the filter shown in Fig. 4.36 is given by Eq. (4.35).

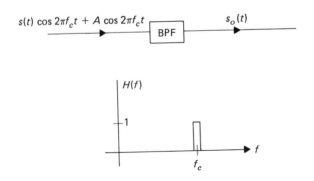

FIGURE 4.36   Narrowband filter used to recover carrier.

$$S_0(f) = \frac{1}{2}[S(f-f_c) + S(f+f_c) + A\delta(f-f_c) + A\delta(f+f_c)]$$

$$f_c - \frac{1}{2}BW < |f| < f_c + \frac{1}{2}BW$$

$$s_0(t) = A\cos 2\pi f_c t + \frac{1}{2}\int_{f_c-\frac{BW}{2}}^{f_c+\frac{BW}{2}} S(f-f_c)\cos 2\pi ft\, df \qquad (4.35)$$

The integral of Eq. (4.35) can be bounded by

$$\frac{1}{2}\int_{f_c-\frac{BW}{2}}^{f_c+\frac{BW}{2}} S(f-f_c)\cos 2\pi ft\, df \le \frac{1}{2\pi t} S_{\max}(f)\, BW \qquad (4.36)$$

Thus, the output of the filter can be made as close to the pure carrier term as is desired by making the bandwidth approach zero. Of course, the narrower the filter, the longer the output takes to reach steady state.

The pure carrier at the output of the filter can be used in the synchronous demodulator, as shown in Fig. 4.37.

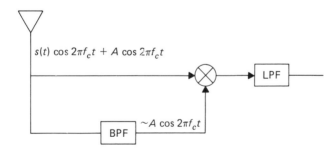

FIGURE 4.37   Recovering carrier term for use in synchronous demodulator.

The system illustrated in Fig. 4.37 will work provided the term at the output of the bandpass filter is a faithful reproduction of the pure carrier signal. This will require proper matching of the filter bandwidth to the carrier power transmitted. Thus, if the output is not close enough to the pure sinusoid, either the carrier power, A, can be raised at the transmitter, and/or the filter bandwidth can be narrowed. If the added carrier is made sufficiently large in amplitude, several simpler demodulation techniques become available. But before examining these techniques, we must establish some concepts relating to transmitted carrier waveforms.

Figure 4.38 shows what might be a typical amplitude modulated waveform. Since the carrier frequency, f_c, is usually much greater than f_m, the maximum

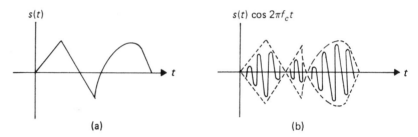

FIGURE 4.38 An AM waveform.

frequency component of the information, one can conclude that the "outline" of the waveform in Fig. 4.38(b) varies much more slowly than the cosine carrier wave does (in practice, the carrier would be of a much higher frequency than that indicated in Fig. 4.38(b)). We would therefore be justified in sketching the modulated waveform as an outline with the interior regions just shaded in. That is, the cosine carrier varies up and down so much faster than does its amplitude, the individual cycles of the carrier are indistinguishable in an accurate sketch.

If, instead of representing a voltage waveform as a function of time, this curve represented the actual shape of a wire or stiff piece of cord, one can envisage a cam (any object) moving along the top surface. If the cam is attached by means of a shock absorber, or viscous damper device, it will approximately follow the upper outline of the curve (*see* Fig. 4.39). This is true since the shock absorber will not allow the cam to respond to the rapid carrier oscillations. This is much the same as the behavior of an automobile suspension system while the car is travelling on a bumpy road.

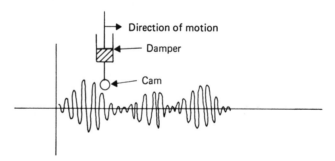

FIGURE 4.39 A mechanical outline follower.

The higher the carrier frequency, the more smoothly the cam will describe the upper outline provided that it can respond fast enough to follow the shape of the outline itself. This upper outline, or boundary of the curve, is called the *envelope* of the waveform. As long as f_c is much greater than f_m, our intuition allows us to define this new parameter. Actually, the definition of the envelope of any waveform can be made quite exact and rigorous (*see* Section 4.9) using very straightforward Fourier Transform analysis techniques which we have already developed. Since for modulation applications f_c is always much greater than f_m, we need not resort to this rigorous approach, but can depend upon our intuition.

At this point, we accept the fact that the electrical analog to the mechanical cam system does exist and is very simple to construct. The critical question is whether the envelope of the modulated signal represents the information signal that we wish to recover.

It should be quite clear from Fig. 4.38 that the envelope of $s(t) \cos 2\pi f_c t$ is given by $|s(t)|$, the absolute value of $s(t)$. Taking the absolute value of a waveform represents a very severe form of distortion. If you don't believe this, consider the simple example of a pure sine wave. Examination of Fig. 4.40 shows that the absolute

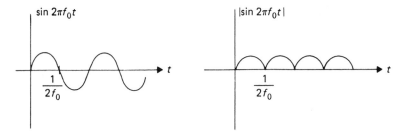

FIGURE 4.40   The absolute value of a sine wave.

value of a pure sine wave is a periodic function whose period is one half of the period of the original sine wave. Fourier Series analysis tells us that it also contains components at multiples of $2f_c$ in frequency. If the original sinusoid represented a pure musical tone, the absolute value would sound like a "raspy" tone one octave higher than the original tone.

Suppose now that $s(t)$ happened to always be non-negative? If this were the case, the absolute value of $s(t)$ would be the same as $s(t)$, and the envelope of $s_m(t)$ would equal $s(t)$. Unfortunately, most information signals of interest have an average value of zero. They must therefore possess both positive and negative excursions.

All is not lost. One can always add a constant to $s(t)$, where the constant is chosen large enough so that the sum is always non-negative. Therefore, instead of using $s(t)$ as the information signal, we use

$$\hat{s}(t) = s(t) + A$$

where A is chosen such that

$$[s(t) + A] \geq 0 \qquad \text{for all } t \tag{4.37}$$

The modulated waveform is now of the form

$$s_m(t) = [A + s(t)] \cos 2\pi f_c t \tag{4.38}$$

This is sketched in Fig. 4.41 for the same example as shown in Fig. 4.38.

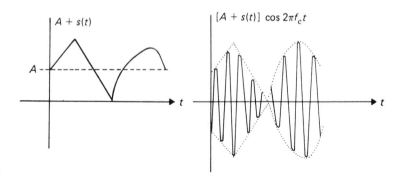

FIGURE 4.41   Example of Fig. 4.38 with a constant added to $s(t)$.

If we use a device which follows the envelope in order to demodulate $s_m(t)$, the result will be a signal of the form $A + s(t)$. Since the actual desired information signal was simply $s(t)$, we may wish to reject the "A" term. It is always quite simple to get rid of a constant term since it represents zero frequency. For example, a capacitor looks like an open circuit to DC (zero frequency), so a series capacitor will reject any DC term. Equation (4.38) can be rewritten as

$$s_m(t) = A \cos 2\pi f_c t + s(t) \cos 2\pi f_c t$$

The addition of the constant to $s(t)$ can therefore be viewed as the addition of an unmodulated carrier term to $s_m(t)$. That is, since the generation of a sinusoid exactly like the carrier was difficult to perform at the receiver, we have essentially resorted to sending it along with the modulated signal. The only difference between this and the carrier injection system of the previous section is that we are now setting a minimum amplitude for the added carrier as specified by Eq. (4.37). Since part of the transmitted signal is the carrier itself, this form of amplitude modulation is called *double sideband AM transmitted carrier*. Until we encounter an example of modulation which is not "double sideband," we will omit the words. Thus we will call this type of signal *AM transmitted carrier*, or *AMTC*. We can now digress and mention that the form of modulation that we were talking about until now is called *AM suppressed carrier*, or *AMSC*, since no carrier term was explicitly transmitted.

Since synchronous detection is so difficult to perform and we have prematurely accepted the fact that envelope demodulation is easy to perform, why not always transmit the carrier, that is, use AMTC instead of AMSC?

The answer lies in efficiency. The power of the signal, $s_m(t)$ increases with increasing A. It is a minimum when $A = 0$ (i.e., for suppressed carrier transmission). The usable information power remains the same regardless of the value of A. One would therefore like to use the minimum value of A consistent with the envelope restriction. That is, choose A such that

$$\min_{t} [A + s(t)] = 0 \tag{4.39}$$

Without knowing the actual waveshape of $s(t)$, it is impossible to say anything about the efficiency when A is chosen to be this minimum value, since we don't know the minimum value of $s(t)$.

Illustrative Example 4.6

Let the information signal in an AMTC scheme be a pure sine wave. Calculate the efficiency of transmission as a function of A, and find the maximum possible value of efficiency.

Solution

Let the information signal, $s(t)$, be given by

$$s(t) = K \cos 2\pi f_m t$$

and the carrier frequency be f_c Hz. The AMTC waveform is given by Eq. (4.38),

$$\begin{aligned} s_m(t) &= [A + K \cos 2\pi f_m t] \cos 2\pi f_c t \\ &= A \cos 2\pi f_c t + \tfrac{1}{2} K[\cos 2\pi (f_c + f_m)t + \cos 2\pi (f_c - f_m)t] \end{aligned} \tag{4.40}$$

The Fourier Transform of $s_m(t)$ is sketched in Fig. 4.42.

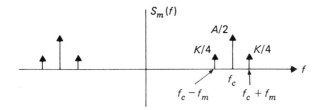

FIGURE 4.42 $S_m(f)$ for Illustrative Example 4.6.

Using the techniques of Chapter 2, the power of this waveform is given by

$$\begin{aligned} P_{\text{avg}} &= \frac{A^2}{2} + \frac{1}{2}\left[\frac{K}{2}\right]^2 + \frac{1}{2}\left[\frac{K}{2}\right]^2 \\ &= \frac{A^2}{2} + \frac{K^2}{4} \end{aligned} \tag{4.41}$$

The usable signal power is contained in the term

$$r(t) \cos 2\pi f_c t$$

and in this case, is equal to $K^2/4$. The system efficiency is therefore given by the ratio of usable power to total power. It is often denoted by the symbol the Greek letter "eta" (η).

$$\eta = \frac{\tfrac{1}{4} K^2}{\tfrac{1}{2} A^2 + \tfrac{1}{4} K^2} \tag{4.42}$$

η increases as A decreases. We must now ask what the minimum value of A can be. Since the maximum negative excursion of $s(t)$ is $s(t) = -K$, the minimum value of A that will guarantee that $s(t) + A$ is always non-negative is K. For this choice of A, the efficiency is

$$\eta = \frac{\frac{1}{4}K^2}{\frac{1}{2}K^2 + \frac{1}{4}K^2} = \frac{1}{3} = 33\% \tag{4.43}$$

Even though the maximum possible efficiency (for a pure sine wave information signal) is only 33% as compared with 100% for the AMSC case, AMTC is used in almost all standard broadcasting. Envelope demodulation (detection) must certainly be an excellent salesperson!

Rectifier Detector

At this point, you should be convinced that a synchronous demodulator can recover the information signal from the AM waveform. If the transmitted waveform contains no explicit carrier term, the local carrier must be reconstructed either through trial and error, or through a complex extraction system. If the received waveform contains a carrier term, that term can be separated out using a narrow bandpass filter. If the amplitude of the added carrier term is such that the envelope of the AM wave is a faithful reproduction of the information signal, simpler techniques become available. The simpler techniques are collectively known as *incoherent* demodulation, since they do not require reconstruction of the precise carrier at the receiver.

We start by observing that the synchronous demodulator of Fig. 4.43 would recover the information signal. Here, we have replaced the locally generated sinusoid by a square wave of the same fundamental frequency.

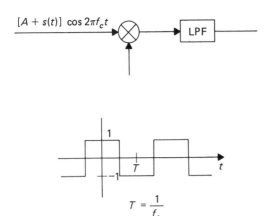

FIGURE 4.43   Synchronous demodulator using square wave instead of sine wave.

The proof that this modified demodulator works is left as an exercise at the end of this chapter. The general case was dealt with in Illustrative Example 4.3. It would

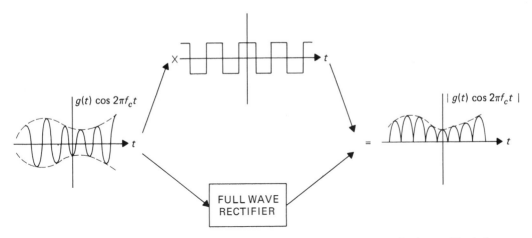

FIGURE 4.44 Multiplication by square wave and full wave rectification are identical.

seem that things are getting more complicated instead of simpler. Close observation of Fig. 4.44 will indicate that all is not lost.

In Fig. 4.44, we show an AMTC waveform being multiplied by a square wave as in the system of Fig. 4.43. Now note that if $A + s(t)$ never goes negative, multiplication by the square wave is the same as full wave rectification. The multiplier of Fig. 4.43 can therefore be replaced by a rectifier. The beauty of using rectification is that no effort must be expended to reproduce the exact carrier frequency—it comes out automatically.

Now that you can see how the rectifier can be used to build a detector which is emulating the synchronous demodulator, we will proceed to the analysis of the rectifier detector.

Figure 4.45 shows a block diagram of the rectifier detection system. The

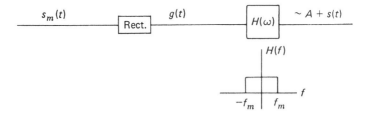

FIGURE 4.45 Detector for AMTC.

rectifier in Fig. 4.45 can either be full wave or half wave. We shall analyze both cases simultaneously. For the full wave rectifier,

$$g(t) = \begin{cases} s_m(t) & s_m(t) > 0 \\ -s_m(t) & s_m(t) < 0 \end{cases} \qquad (4.44)$$

This input/output relationship is sketched below.

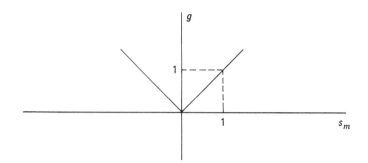

For the half wave rectifier,

$$g(t) = \begin{cases} s_m(t) & s_m(t) > 0 \\ 0 & s_m(t) < 0 \end{cases} \qquad (4.45)$$

This relationship is sketched below.

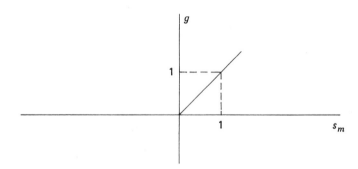

The output of the full wave rectifier is given by

$$|s_m(t)| = |[A + s(t)] \cos 2\pi f_c t|$$
$$= |A + s(t)| |\cos 2\pi f_c t| \qquad (4.46)$$
$$= [A + s(t)] |\cos 2\pi f_c t|$$

We have eliminated the absolute value sign around $A + s(t)$, since we assume that A is chosen so that this term is never negative. Thus its absolute value is equal to the function itself.

The only difference if a half wave rectifier were used would be that the second term in Eq. (4.46) would be the half wave rectified version of $\cos 2\pi f_c t$ instead of the full wave version. Both the half wave and full wave rectified forms of $\cos 2\pi f_c t$ are

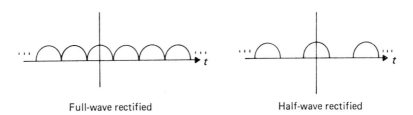

Full-wave rectified   Half-wave rectified

FIGURE 4.46   Rectified forms of cos $2\pi f_c t$.

sketched in Fig. 4.46. Both functions shown in Fig. 4.46 are periodic and can therefore be expressed as Fourier Series expansions. The half wave version has a fundamental frequency of f_c, and the full wave, $2f_c$, since the period has been cut in half. The output of the rectifier is therefore given by $g_1(t)$ for the full wave case, and by $g_2(t)$ for the half wave case.

$$g_1(t) = [1A + s(t)] \sum_{n=0}^{\infty} a_n \cos 2\pi n f_c t$$

(4.47)

$$g_2(t) = [1A + s(t)] \sum_{n=0}^{\infty} a'_n \cos 2\pi n f_c t$$

The prime notation is used for a_n in $g_2(t)$ to indicate that the Fourier Series coefficients are not the same as those of $g_1(t)$. If we write out the first few terms of each of these series, it should be clear that a lowpass filter can be used to reject all but the desired term.

$$g_1(t) = a_0[A + s(t)] + a_1[A + s(t)] \cos 4\pi f_c t$$
$$+ a_2[A + s(t)] \cos 8\pi f_c t + \cdots$$

(4.48a)

$$g_2(t) = a'_0[A + s(t)] + a'_1[A + s(t)] \cos 2\pi f_c t$$
$$+ a'_2[A + s(t)] \cos 4\pi f_c t + \cdots$$

(4.48b)

If $g_1(t)$ is put through a lowpass filter with cutoff frequency f_m, the output is $a_0[A + s(t)]$. If $g_2(t)$ were the input to the same filter, the output would be $a'_0[A + s(t)]$. In both cases, the output of the lowpass filter is in the form of a constant multiplying the desired signal, $A + s(t)$. We have therefore built an effective demodulator using only a rectifier and lowpass filter.

We make an interesting observation at this point. While the above system has simulated what appears to be the envelope of $s_m(t)$, this is not exactly the case. It has actually yielded a_0 times the exact envelope (a'_0 in the half wave rectifier case). While this subtle difference does not represent distortion of the signal, it does indicate that the exact mechanism of operation is basically different from that of the mechanical cam.

Envelope Detector

If you were asked to construct the rectifier detector of Fig. 4.45, you might make a lucky mistake, as the first person to attempt this problem might have done. You may, in effect, build an actual analog to the cam instead of the system of Fig. 4.45.

In Fig. 4.47 we illustrate a simple half wave rectifier and a simple (non-ideal) lowpass filter. One might just blindly put these two together and think that the system

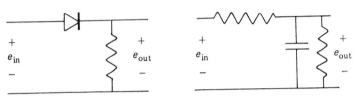

FIGURE 4.47   A half wave rectifier and lowpass filter.

which we just analyzed was being constructed. Unfortunately, the idea of just putting two systems together to develop the composite properties is totally wrong. It doesn't even apply for simple linear circuits (can you prove that the rectifier is non-linear from the definition of linearity?). As a trivial example, consider the two circuits of Figs. 4.48(a) and 4.48(b). The composite is shown as Fig. 4.48(c). The first circuit multiplies the input by $\frac{1}{2}$, as does the second circuit. The composite, instead of multiplying the input by $\frac{1}{4}$, possesses the relationship $e_{out}/e_{in} = \frac{1}{5}$. The concept of cascading systems without any regard to isolation is equally inapplicable to more complicated circuits.

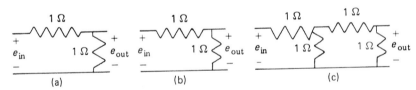

FIGURE 4.48   Incorrect cascading of systems.

In any case, a simple circuit constructed using the fallacious reasoning turned out to be the cheapest, simplest, and easiest to build envelope detector that could be dreamed possible. It can be built for a few pennies and is used in all standard household receivers. Unfortunately, it is virtually impossible to analyze in detail. Fortunately, a detailed analysis is totally unnecessary.

The device of which we speak is shown in Fig. 4.49.

We claim that the circuit in Fig. 4.49 is an exact electrical analog of the cam system of Fig. 4.39. The capacitor represents the mass of the cam. The resistor is analogous to the viscous damping, or shock absorber. The diode takes into account

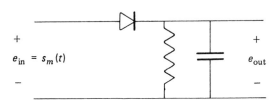

FIGURE 4.49   The envelope detector.

the fact that the wire in Fig. 4.39 (bumpy road?) can only push up on the cam. It cannot pull the cam down.

Now that our intuition should be satisfied, we shall attempt a partial analysis of the circuit. As a first step in the approximate analysis of the envelope detector, consider the well known *peak detector* circuit shown in Fig. 4.50. If the input to this

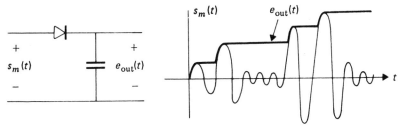

FIGURE 4.50   The peak detector.

peak detector is $s_m(t)$, the output will be as shown. That is, $s_m(t)$ can never become greater than $e_{out}(t)$ since this would imply a forward voltage across the diode. Similarly, the capacitor voltage, $e_{out}(t)$, can never decrease since there is no path through which the capacitor can discharge. The output, $e_{out}(t)$, is therefore always equal to the maximum past value of the input, $s_m(t)$. Note that, in order to exaggerate the effect, the carrier frequency has been shown much lower than it would actually be in practice.

If a "discharging" resistor is now added (*see* Fig. 4.51) the output voltage will exponentially decay toward zero, until such time as the input tries to exceed the

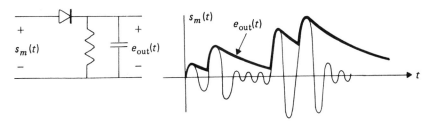

FIGURE 4.51   Addition of a discharging resistor.

output. Again, the carrier has been sketched with a very low frequency. We can see that, as the carrier frequency gets high, $e_{out}(t)$ will approach the envelope of $s_m(t)$.

Ideally, we would like the output, $e_{out}(t)$, to be able to decay as fast as the envelope of $s_m(t)$ changes. Any faster decay will cause $e_{out}(t)$ to start resembling the fast carrier oscillations.

Since the shortest time in which the envelope can go from its maximum to its minimum value is $\frac{1}{2} f_m$, the time constant of the RC circuit should be made to be the same order of magnitude as this value (usually about $\frac{1}{10} f_m$). This analysis is admittedly sloppy, but the large difference between f_c and f_m permits considerable leeway. In practice, it is better to choose the RC time constant to be a little too short rather than too long. A time constant which is too large would result in $e_{out}(t)$ completely missing some of the peaks of $s_m(t)$, as it does in Fig. 4.50. On the other hand, if the time constant is too small, the result is that $e_{out}(t)$ has a ripple around the actual envelope waveshape. This ripple represents a high frequency wave since it occurs at the peaks of the carrier. It is therefore composed of frequencies around f_c, the carrier frequency. An exact analysis would show that it is quite simple to remove this ripple with a lowpass filter. In practice one doesn't even bother since the radio speaker or, indeed, the human ear could never respond to frequencies of the order of f_c.

Illustrative Example 4.7

Design an envelope detector to demodulate the following AM signal:

$$5(\cos 100t + 1) \cos 2\pi x 10^6 t$$

Solution

It is first necessary to choose the time constant of the RC circuit. Several approaches will be presented, not all of which are appropriate to the specific given time function.

The capacitor discharges starting at the peak of the carrier, and it continues to discharge until the waveform intersects that of the modulated carrier. If the capacitor voltage is greater than the modulated carrier at the next peak, the envelope detector output will miss that peak (see Fig. 4.51). If this were to occasionally happen, it would probably not be noticed in a communication system. Nonetheless, it is more serious to design the envelope detector with too large a time constant than with one that is too small. This is true since the only consequence of too small a time constant is a ripple term at the carrier frequency. This term is easy to filter out. For this reason, our first approach will be to design the detector for the worst case of just following the maximum negative slope of the envelope. This is illustrated in Fig. 4.52.

Let's be intuitive for a second. Since the carrier amplitude in the given waveform is just large enough to allow envelope demodulation, the envelope does periodically touch the zero axis. Since a decaying exponential doesn't reach zero until time infinity, it will be impossible to design this detector to hit every peak of the modulated carrier. Nonetheless, we will develop the appropriate formulae for use in those cases where the carrier amplitude is larger than needed.

If the notation of Fig. 4.52 is used, the capacitor voltage at the second peak is given by

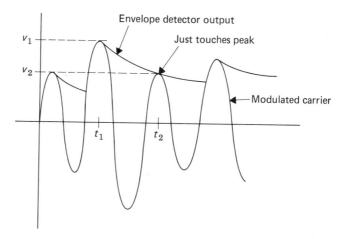

FIGURE 4.52   Envelope detector designed to respond to peaks of carrier.

$$v_2 = v_1 e^{-T/RC}$$

where $T = 1$ microsecond for the given carrier frequency. It is now necessary to express v_1 and v_2 using the given function for the envelope. In order to arrive at a solution without numerical techniques, we shall make some approximations. Since the carrier period, 1 microsecond, should be short compared to the resulting time constant, the exponential can be approximated by a straight line. Thus

$$\frac{v_2}{v_1} = 1 - \frac{T}{RC}$$

The slope of this line is $-v_1/RC$. This slope must be more negative than the maximum negative slope of the envelope. Taking the derivative of the envelope, and plugging in T_1 from the given function, we find

$$\frac{5 + 5 \cos 100t}{RC} > 5000 \sin 100t \tag{4.49}$$

$$RC < \frac{1 + \cos 100t}{1000 \sin 100t} \tag{4.50}$$

Unfortunately, as t approaches $\pi/100$, the right side of Eq. (4.50) approaches zero. This confirms the intuitive observation made earlier. That is, we have placed an impossible constraint upon the design, since the envelope does go to zero periodically.

   If instead the error in missing peaks is minimized, it would seem reasonable to design for the maximum slope of the envelope. This occurs when the envelope goes through an amplitude of 5, at which point the slope is $-5{,}000$. Thus, using Eq. (4.49), we have

$$5000 < \frac{5}{RC}$$

$$RC < 0.001$$

A third approach to the design looks only at the global peaks of the envelope, and not at the individual peaks of the carrier. Thus, it takes $\pi/100$ seconds for the envelope to go from a maximum to a minimum. Assuming that the RC circuit reaches its final value after about 4 time constants, a time constant of about 8 milliseconds would seem acceptable. Such a choice would allow the detector to follow the general movement of the envelope, but it would not assure an accurate reproduction of the shape.

We now proceed with the design using the earlier value of time constant equal to 1 millisecond.

The choice of the diode depends upon two observations. The first is voltage level. Practical diodes are not identical to ideal diodes. When forward biased, the practical diode sustains a voltage on the order of 0.3 to 0.8 volts. With total voltage swings of ± 10 volts for the given signal, this should not prove to be a serious problem, and any type of diode could be used. A second consideration is that of frequency of operation. As frequencies increase, diodes look more and more like capacitors. Again, with frequencies in the range of 1 MHz (relatively low), any "junk-box" diode could be chosen for this application.

Since the time constant is the product of R and C, it would appear that one of these two parameter values is arbitrary. This would be true except that R is normally chosen with a view toward input and output impedance matching. For purposes of the input to the envelope detector, we would like R to be as large as possible to avoid loading of the previous circuitry. For purposes of this example, let us choose $R = 1$ kΩ. Then, to achieve a time constant of 1 msec, C would be 1 microfarad.

Illustrative Example 4.8

Design an envelope detector for use in demodulating an AM Transmitted Carrier waveform. The carrier frequency is 1 MHz and the information signal is a voice waveform.

Solution

Once the diode is selected, all that is required in the design of the envelope detector is to choose the value of R and C in the circuit of Fig. 4.49.

The highest frequency of the envelope of the AM waveform is the highest frequency of the information signal. Since this was given to be a voice signal, we will assume a maximum frequency of 5 kHz, as used in broadcast AM. The envelope detector must be capable of responding to the fastest possible changes in this signal. At the maximum frequency of 5 kHZ, the envelope can go from a maximum to a minimum in about 0.1 milliseconds. This is one half of the period of a 5 kHz sinusoid. The envelope detector must be capable of following this. In dealing with exponential responses, one normally considers the final value to have been reached after about 4 time constants. This is true since $\exp(-4) = 0.018$, so the exponential is within 2% of its final value after four time constants. Using the numbers given above, we see a time constant of less than 25 microseconds would be sufficient to respond to even the fastest changes in the envelope. On the other hand, the 1 MHz carrier goes from its peak to zero in 0.25 microseconds, so with a time constant of 25 microseconds the RC circuit would have only 1% of one time constant to start responding to the carrier. Since $\exp(-0.01) = 0.99$, the exponential only would have time to traverse about 1% of its excursion to match the carrier.

The choice of a time constant equal to 25 microseconds would therefore appear to be an acceptable one, except that such a choice would not guarantee that the envelope detector output would hit all of the peaks of the carrier. Since a certain amount of ripple at 1 MHz will not hurt the system, we could afford to shorten the time constant. For example, a time constant of 10 microseconds would allow the signal to come within 0.005% of the final value in tracking the fastest envelope frequency, and the carrier response would only decrease to 0.975 of the peak. Even though the envelope is very rarely at the maximum frequency for any sustained period of time (i.e., how many speakers squeak at 5 kHz except for an instant?), we can certainly afford to play it safe and design for a rather short time constant. The fact that a system is designed for audio does not mean that all transmitted signals will be audio—only that they will occupy audio frequencies. (We will later see that a MODEM transmits voice-frequency signals which sustain a fixed frequency.)

Once the time constant is chosen, it is necessary to specify the diode type, the resistance and capacitance. The considerations were discussed in the previous example, and will not be repeated here. If a resistor value of 1 kΩ is chosen, the capacitor would have to equal 0.01 microfarads in order to achieve a time constant of 10 microseconds.

Square Law Detector

In Illustrative Example 4.4, we showed that a square law device could not be used to demodulate suppressed carrier AM. We now reexamine this situation for transmitted carrier AM.

The output of a square law device with AMTC at the input is

$$[A + s(t)]^2 \cos^2 2\pi f_c t = [A + s(t)]^2 [\tfrac{1}{2} + \cos 4\pi f_c t]$$

If we now put this signal into a lowpass filter which passes all frequencies of $[A + s(t)]^2$, the output will be

$$\frac{1}{2}[A + s(t)]^2$$

Note that the lowpass filter would have to pass frequencies up to $2f_m$, twice the highest frequency of $s(t)$. Now since $A + s(t) \geq 0$, we can unambiguously take the positive square root to get

$$\frac{1}{\sqrt{2}}[A + s(t)]$$

Thus, the system of Fig. 4.53 can be used for demodulation of transmitted carrier AM, provided the carrier is large enough to make the envelope non-negative.

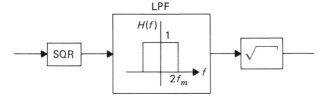

FIGURE 4.53 The square law detector.

The problems at the back of this chapter illustrate that other types of non-linearities can be used for demodulation of transmitted carrier AM.

If the carrier amplitude, A, is much larger than the information signal, $s(t)$, the square root operation in Fig. 4.53 can be eliminated. That is, the output of the lowpass filter is

$$\frac{1}{2}A^2 + As(t) + \frac{1}{2}s^2(t)$$

but if A is much larger than $s(t)$, this can be approximated by

$$\frac{1}{2}A^2 + As(t)$$

For this case, the lowpass filter need only pass frequencies up to f_m instead of $2f_m$.

4.8   THE SUPERHETERODYNE AM RECEIVER

We are now in a position to fulfill the dream of most students when they first start studying communication theory. That is, we can understand the operation of the standard radio receiver. We probably could not repair it (a technician with a familiarity of electronics is needed for this) but we certainly can understand a block diagram of it. This section will analyze the superheterodyne AM receiver from a system's approach standpoint (block diagram). The entire section can be skipped on a first reading without loss of continuity.

Several basic operations can be identified in any broadcast receiver. The first is station separation. We must pick out the one desired signal and reject all of the other signals (stations). The second operation is that of amplification. The signal picked up by a radio antenna is far too weak to drive the cone of a loudspeaker without first being amplified many times. The third, and final operation is that of demodulation. The incoming signal is amplitude modulated, and contains frequencies centered around the carrier frequency. The carrier is usually near one million Hertz in frequency (standard broadcast), and the signal received must therefore be demodulated before it can be fed into a speaker.

In an actual receiver, these three operations are sometimes combined, and the actual order in which they are performed is somewhat flexible.

In standard broadcast AM the highest frequency component of the information signal (f_m) is set at 5,000 Hz. The adjacent carrier signals are separated by 10 kHz (the minimum separation allowed) so that the transform of the signal received by the antenna would look something like that shown in Fig. 4.54.

We see that to receive one signal and reject all others requires a very accurate filter with a sharp frequency cutoff characteristic.

Assuming that the listener wanted the capability of choosing any station (a reasonable assumption), this filter would have to be tuneable. That is, the band of frequencies which it passes must be capable of variation.

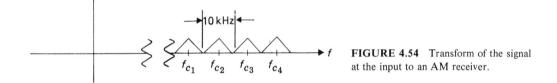

FIGURE 4.54   Transform of the signal at the input to an AM receiver.

We have seen how one can build filters which approximate ideal bandpass filters to any desired degree of accuracy. However, when we require that the pass band of the filters be capable of moving all over the frequency axis while maintaining sharp bandpass characteristics, we have made the practical construction of the filter virtually impossible. Early AM radio receivers contained bandpass filters that could be tuned. The tuning required the careful changing of a number of capacitors in just the right way. "Bringing in" a new station was an art of no minor difficulty.

We are rescued from this dilemma by recalling that multiplication of the incoming signal by a sinusoid shifts all frequencies up and down by the frequency of the sinusoid. Because of this, station selection can be accomplished by building a fixed bandpass filter and shifting the input frequencies so that the station of interest falls in the pass band of the filter. That is, we sort of construct a viewing "window" on the frequency axis, and shift the desired station so that it sits in the window. This shifting process (multiplication by a sinusoid) is called *heterodyning*. The receiver which we shall describe is called a "superheterodyne" receiver. Although we will only discuss the AM version of this receiver, the principle of superheterodyning is equally applicable to other forms of modulation (e.g., FM). We shall now analyze the total block diagram of such a receiver. (*See* Fig. 4.55.)

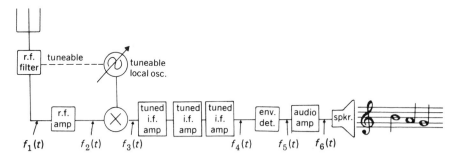

FIGURE 4.55   The superheterodyne receiver.

Starting at the antenna, a signal consisting of the sum of all broadcasted signals (stations) is received. We then amplify all stations in a radio frequency (r.f.) amplifier. The tuneable r.f. filter shown will be described later. The signal, $f_2(t)$ is then shifted up and down in frequency by multiplying it by a sinusoid generated by a "local" oscillator. The output of this heterodyner is applied to the almost ideal bandpass filter (usually part of the i.f. amplifiers). In standard AM, this fixed filter has a bandwidth of 10 kHz, and is centered at 455 kHz, known as the intermediate frequency (i.f.) of the receiver. In most receivers, the filter is made up of three tuned circuits which are often aligned so as to generate the Butterworth filter poles. These three stages of filtering are usually combined with three stages of i.f. amplification. Thus, at $f_4(t)$, we have a modulated signal whose carrier frequency has been shifted to 455 kHz, and which has already been amplified and sifted out from all of the other signals.

The receiver then simply envelope detects $f_4(t)$ and then amplifies once more before applying the signal to a loudspeaker. More expensive receivers may add a bass (lowpass filter) or treble (highpass filter) control. Inexpensive radios may eliminate one or more stages of amplification from the above block diagram.

It remains to justify the presence of the tuneable r.f. filter. One problem was not mentioned in the above analysis. The heterodyning process shifts the incoming frequencies both up and down in frequency. Therefore, while the "up-shift" might place the desired signal in the viewing window (pass band) of the i.f. filter, the downshift might place an undesired signal right on top of that deposited during the upshift. This undesired signal is called an "image" signal. Its elimination is not very difficult to perform as we shall presently see.

In order to continue, we will need to know how one chooses the frequency of the local oscillator. In standard broadcast AM, we may wish to listen to any station with carrier frequency between 550 kHz and 1,600 kHz. This is the band of frequencies which the FCC (Federal Communications Commission) has allocated to AM transmission.

For example, suppose that we wished to listen to a station with carrier frequency of 550 kHz. Since the i.f. filter frequency in standard AM is 455 kHz, the local oscillator must be tuned to either 95 kHz or 1,005 kHz. (Convince yourself that multiplication of a sinusoid of frequency 550 kHz by one of either 95 kHz or 1,005 kHz results in a sinusoidal term at 455 kHz.) We choose the higher of the two oscillator frequencies for the following practical reason.

Reception of the standard AM band requires that the local oscillator tune over a range of 1,050 kHz, the difference between the lowest and highest carrier frequency. It is far easier to construct an oscillator whose frequency must vary from 1,005 kHz to 2,055 kHz than it is to construct one whose frequency varies from 95 kHz to 1,145 kHz. This is due to the fact that the bandwidth of the latter oscillator is very large compared to the frequencies involved.

Returning now to the image problem, if the incoming signal is multiplied by a sinusoid of 1,005 kHz, the signal with carrier frequency 550 kHz is shifted into the pass band of the i.f. filter (455 kHz). The station with carrier frequency 1,460 kHz is also shifted into the pass band of the filter! This is the *image station*.

It turns out to be relatively simple to reject this unwanted signal. Had we first placed the incoming r.f. signal through a very sloppy bandpass filter that passed the signal around 550 kHz but rejected that around 1,460 kHz, we would have eliminated the image signal.

The difference between the local oscillator frequency, f_{lo}, and the desired station carrier frequency, f_d, is the i.f. frequency, $f_{\text{i.f.}}$. Likewise, the difference between the image carrier frequency, f_{image}, and the local oscillator frequency is the i.f. frequency. Thus,

$$f_d - f_{\text{image}} = 2f_{\text{i.f.}}$$

The image station is therefore separated from the desired station by 910 kHz. A bandpass filter with a bandwidth of 1,820 kHz would accomplish image rejection (see Fig. 4.56). Since the actual form of this filter is not critical (i.e., we don't really care

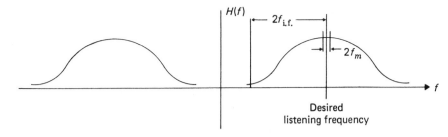

FIGURE 4.56   The rejection filter need not be ideal.

what it does to signals lying between the desired and image station in frequency), it is not difficult to construct even though it must be tuneable.

In practice, when the tuning dial on a receiver is turned, this sloppy r.f. rejection filter is being tuned and the frequency of the local oscillator is simultaneously being changed.

4.9   ᵉENVELOPES AND PRE-ENVELOPES OF WAVEFORMS

The previous section has led us to an intuitive definition of the envelope of a restricted class of waveforms. Given $s(t)\cos 2\pi f_c t$, we said that if the maximum frequency component of $s(t)$ is much less than f_c, then the envelope of $s(t)\cos 2\pi f_c t$ is given by the absolute value of $s(t)$. We made this intuitive definition on the basis of a sketch of the waveform and an intuitive feel for what we wanted to mean by envelope.

This definition is not very satisfying to anyone who desires a firm mathematical development of the analysis. We therefore rigorously define what is meant by the envelope of any general waveform. Since this is being labelled as a definition, it requires no further justification. However, we shall show that the mathematical and intuitive definition are identical for that class of function for which the intuitive approach applies.

This technique of defining something in a rigorous way such that it agrees with intuition for special cases is very common. We will see it again in our approach to frequency modulation, where a rigorous definition of frequency will be required.

The *pre-envelope* (sometimes called "complex envelope" or "analytic function") of a waveform is defined as the complex time function whose Fourier Transform is given by

$$2U(f)R(f) \tag{4.51}$$

where $U(f)$ is the unit step function and $R(f)$ is the transform of the original waveform. That is, the transform of the pre-envelope is zero for negative f and is equal to twice the Fourier Transform of the original function for positive f. The symbol, $z(t)$, is commonly used to represent the pre-envelope function. It should be clear from the properties of the transform that $z(t)$ could not possibly be a nonzero real time function.

Illustrative Example 4.9

Find the pre-envelope of $r(t) = \cos 2\pi f_0 t$.

Solution

The Fourier Transform of $r(t) = \cos 2\pi f_0 t$ is given by

$$R(f) = \frac{1}{2}[\delta(f - f_0) + \delta(f + f_0)]$$

The transform of the pre-envelope is therefore given by

$$Z(f) = \delta(f - f_0)$$

and the pre-envelope, denoted as $z(t)$, is given by $e^{j2\pi f_0 t}$, the inverse transform of $Z(f)$.

In elementary sinusoidal steady state analysis, a system input of $\cos 2\pi f_0 t$ is rarely carried through equations. Instead, circuits are solved for an input of $e^{j2\pi f_0 t}$, and the real part of the resulting output is taken. The reasons for this approach are not relevant to our present discussion. We simply note that this technique amounts to the substitution of the pre-envelope of the input for the actual input.

We now wish to express $z(t)$ in the time domain in terms of a general $r(t)$. Starting with $Z(f) = 2R(f)U(f)$, we use the time convolution theorem to get

$$z(t) = 2r(t) * \mathscr{F}^{-1}[U(f)]$$

The inverse Fourier Transform of $U(f)$ is found in much the same way as the transform of $U(t)$ was found (Eq. 1.93). It is perfectly acceptable to look this up in the table of transform pairs in Appendix III of this text.

$$U(f) \leftrightarrow \frac{1}{2}\delta(t) + \frac{-1}{2\pi jt}$$

Finally,

$$z(t) = 2r(t) * \frac{1}{2}\delta(t) + 2r(t) * \frac{-1}{2\pi jt}$$

$$(4.52)$$

$$z(t) = r(t) + \frac{j}{\pi} \int_{-\infty}^{\infty} \frac{r(t)}{t - \tau} d\tau$$

We note that the real part of $z(t)$ is the original time function, $r(t)$, just as the real part of $e^{j2\pi f_0 t}$ is $\cos 2\pi f_0 t$.

The imaginary part of $z(t)$ is given by the convolution of $r(t)$ with $1/\pi t$. This convolution is recognized as the Hilbert Transform of $r(t)$.

We now *define* the envelope of $r(t)$ as the magnitude of the pre-envelope, $z(t)$.

Illustrative Example 4.10

Find the envelope of $r(t) = \cos 2\pi f_0 t$.

Solution

The pre-envelope of $r(t)$ was found in Illustrative Example 4.8 to be

$$z(t) = e^{j2\pi f_0 t}$$

The magnitude of this is equal to 1. Therefore, the envelope of $r(t) = \cos 2\pi f_0 t$ is a constant, 1. We already knew this from our intuitive definition of envelope since this time function represents an unmodulated carrier wave.

Illustrative Example 4.11

Find the envelope of $s_m(t) = s(t) \cos 2\pi f_c t$

Solution

We first must find the pre-envelope of $s(t) \cos 2\pi f_c t$

$$z_m(t) = s_m(t) + j \left[s_m(t) * \frac{1}{\pi t} \right]$$

$$(4.53)$$

We shall perform the indicated convolution in the frequency domain by multiplying the two transforms together. The transform of $1/\pi t$ is $-j\,\text{sgn}(f)$. Therefore, the transform of the second term in Eq. (4.53) (i.e., the imaginary part of $z_m(t)$) is given by the product of $S_m(f)$ with $-j\,\text{sgn}(f)$. Recall that

$$S_m(f) = \frac{S(f - f_c) + S(f + f_c)}{2}$$

The product of this transform with $-j\,\text{sgn}(f)$ is given by

$$\frac{S(f - f_c) - S(f + f_c)}{2j} \qquad (f_m \leq f_c) \tag{4.54}$$

as shown in Fig. 4.57. This is recognized as the transform of $js(t) \sin 2\pi f_c t$ (verify this statement!). The pre-envelope is therefore given by

$$z_m(t) = s(t) \cos 2\pi f_c t + js(t) \sin 2\pi f_c t \tag{4.55}$$

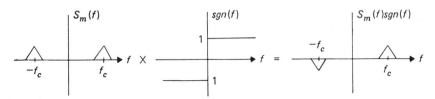

FIGURE 4.57 Construction of transform of the imaginary part of $z_m(t)$ for Illustrative Example 4.11.

The envelope is defined as the magnitude of $z_m(t)$ and is therefore given by

$$\begin{aligned}
|z_m(t)| &= \sqrt{s^2(t) \cos^2 2\pi f_c t + s^2(t) \sin^2 2\pi f_c t} \\
&= \sqrt{s^2(t) (\cos^2 2\pi f_c t + \sin^2 2\pi f_c t)} \\
&= \sqrt{s^2(t)} = |s(t)|
\end{aligned} \tag{4.56}$$

This agrees with the result accepted using our intuitive definition. Note that this rigorous definition only requires that $f_m \leq f_c$. Previously, we required that the maximum frequency component of $s(t)$ be much less than f_c. The mathematical definition of envelope is less restrictive than the intuitive one.

We therefore have a rigorous definition of the envelope of any function of time as the magnitude of the pre-envelope of the time function.

It would be a worthwhile piece of work if one could analyze the envelope detector circuit in the light of the definition of envelope presented here. Since the envelope detector has not been accurately analyzed, this task would seem to be extremely difficult.

The intuitive definition of envelope which applies when the envelope is slowly varying will be sufficient for all work that we will encounter in amplitude modulation theory. The concept of pre-envelope will prove useful in detection theory (e.g., radar).

4.10   SINGLE SIDEBAND

In the standard AM systems which we have been describing, the range of frequencies required to transmit a signal, $r(t)$, is that band between $f_c - f_m$ and $f_c + f_m$. This represents a bandwidth of $2f_m$ Hz. It was shown that, if we frequency multiplex two signals, their carrier frequencies must be separated by at least $2f_m$ Hz in order for the signals not to have overlapping frequency components. This was necessary to allow separation of the various signals at the receiver.

In addition to smog and atomic fallout, our air is becoming polluted with electromagnetic signals. Considerations relating to antenna design tend to limit the range of usable frequencies for transmission. Therefore, the number of channels which can be frequency multiplexed is limited due to practical considerations.

Certain ranges (bands) of frequencies prove better than others for transmitting over long distances, as in the case of transoceanic transmission. This is due to ionospheric skip conditions. These frequencies are obviously at a premium.

For the above reasons, the Federal Communications Commission (FCC) regulates the frequencies allocated to each particular class of user. Wouldn't it be nice if each channel required a smaller portion of the frequency band than $2f_m$? We could then stack adjacent channels closer together.

Single sideband transmission is a technique which allows adjacent carriers to be closer together than the separation required in double sideband transmission (the type of AM which we have been discussing until now).

As in Fig. 4.58 we define that portion of $S_m(f)$ which lies in the band of frequencies above the carrier as the *upper sideband*. Similarly, that portion which lies in the band of frequencies below the carrier is called the *lower sideband*. A double

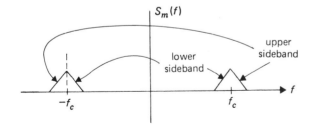

FIGURE 4.58   Definition of upper and lower sideband.

sideband AM suppressed carrier wave is therefore made up of a lower and an upper sideband. A double sideband AM transmitted carrier wave is composed of upper and lower sidebands plus a carrier term.

If the information signal, $s(t)$, is a real function of time, the magnitude of $S(f)$ must be even in f. It should therefore be clear that $S(f-f_c)$ is symmetrical about f_c. That is, the upper sideband is a mirror image of the lower sideband. Since the upper sideband can be exactly derived from the lower sideband, and vice versa, all of the information about the modulated waveform is contained in either the lower or the upper sideband. Why should we transmit both sidebands?

In the past, we transmitted both sidebands since the corresponding modulators and demodulators were easy to construct. We automatically got the double sideband version with each modulation scheme we examined. If, however, frequency bands are

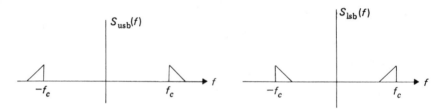

FIGURE 4.59   Upper and lower sideband transforms.

really at a premium in a particular application, the extra work we will find necessary
to send only one sideband may prove profitable.

Figure 4.59 shows the Fourier Transforms of the upper, and lower sideband
versions of $s_m(t)$, denoted by $s_{usb}(t)$ and $s_{lsb}(t)$, respectively.

Since the double sideband suppressed carrier waveform, $s_m(t)$, is composed of
upper and lower sidebands, it should be clear that

$$s_m(t) = s_{usb}(t) + s_{lsb}(t) \qquad (4.57)$$

This can be seen from Fig. 4.58 since the sum of the two transforms shown is the
suppressed carrier version of $S_m(f)$.

SSB Modulators and Demodulators

Since the upper and lower sidebands are separated in frequency, our old standby of
using filters to reject signals can be used as a method of generating single sideband
signals. We simply form the double sideband suppressed carrier waveform as
discussed previously and pass it through a bandpass filter. The filter should only
transmit the sideband of interest. Figures 4.60 and 4.61 show such systems to

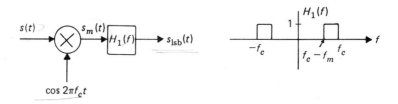

FIGURE 4.60   Single sideband generator for lower sideband.

generate the lower and the upper sideband signals, respectively. In Fig. 4.61, we
either use a bandpass filter which passes the upper sideband, or we can take
advantage of Eq. (4.57) to write

$$s_{usb}(t) = s_m(t) - s_{lsb}(t) \qquad (4.58)$$

Both of these methods are illustrated in Fig. 4.6.

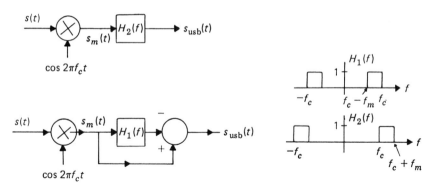

FIGURE 4.61   Single sideband generators for upper sideband.

The systems shown in Figs. 4.60 and 4.61 are not easy to construct. They require filters whose bandpass characteristics are perfect. For example, in the lower sideband transmitter, just below a frequency of f_c the filter must transmit unattenuated while just above f_c it must completely reject or attenuate the signals. This is fine for an ideal filter, but practical filters could never accomplish this task. We present a system which shifts the problem from that of designing a very good bandpass filter to that of designing a very good phase shifter. For certain applications, this is a better approach, although the basic problem is not any simpler. In the process, we will develop an expression for the SSB waveform in the time domain. This will prove useful in later work.

We note that if the information signal, $s(t)$, is a pure sinusoid, $\cos 2\pi f_m t$, the single sideband waveform is easy to derive explicitly. The double sideband wave is given by

$$s_m(t) = s(t) \cos 2\pi f_c t = \cos 2\pi f_m t \cos 2\pi f_c t$$

which, using trigonometric identities, becomes

$$s_m(t) = \tfrac{1}{2}\left[\cos 2\pi(f_c + f_m)t + \cos 2\pi(f_c - f_m)t\right] \qquad (4.59)$$

The transform of this waveform is shown in Fig. 4.62. The lower sideband wave is given by

$$s_{\text{lsb}}(t) = \tfrac{1}{2}\cos 2\pi(f_c - f_m)t \qquad (4.60)$$

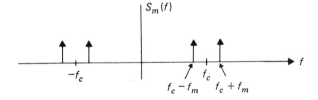

FIGURE 4.62   Double sideband transform for $r(t) = \cos 2\pi f_m t$.

which, by trigonometric identity, can be written as

$$s_{\text{lsb}}(t) = \tfrac{1}{2}\cos 2\pi f_c t \cos 2\pi f_m t + \tfrac{1}{2}\sin 2\pi f_c t \sin 2\pi f_m t \qquad (4.61)$$

Since a sine wave can be thought of as the corresponding cosine wave shifted by $-90°$ in phase, the $s_{\text{lsb}}(t)$ given in Eq. (4.61) can be generated by the system shown in Fig. 4.63.

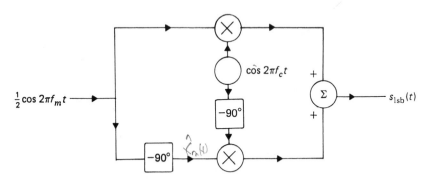

FIGURE 4.63   LSB generator for $r(t) = \cos 2\pi f_m t$.

Before examining this system in more detail, we would like to derive an equivalent system which will generate a single sideband signal from an arbitrary $s(t)$. That is, we ask the question whether any modifications must be made to the system shown in Fig. 4.63 so that it generates a single sideband wave when the input is not a pure sinusoid? The derivation follows. We could wave our hands and say that any $s(t)$ can be expressed as a sum of sinusoids, and then use a linearity argument. We will not do so as this text intends to promote some familiarity and comfort with Fourier Transform analysis techniques.

The Fourier Transform of $s_{\text{lsb}}(t)$ is given by $S_m(f)H(f)$, where

$$H(f) = U(f + f_c) - U(f - f_c) \qquad (4.62)$$

This is shown in Fig. 4.64.

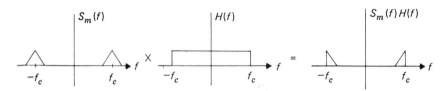

FIGURE 4.64 $S_{\text{isb}}(f)$ derived from $S_m(f)$.

Writing $S_m(f)$ as

$$S_m(f) = \tfrac{1}{2}[S(f - f_c) + S(f + f_c)]$$

we recognize that (*see* Fig. 4.65)

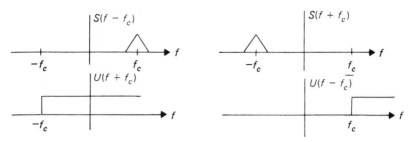

FIGURE 4.65   Products appearing in Eq. (4.63).

$$U(f+f_c)S(f-f_c) = S(f-f_c) \qquad (4.63\text{a})$$

and

$$U(f-f_c)S(f+f_c) = 0 \qquad (4.63\text{b})$$

Using these, we have

$$
\begin{aligned}
S_{\text{lsb}}(f) &= S_m(f)H(f), \\
&= \tfrac{1}{2}\, S(f+f_c)U(f+f_c) - \tfrac{1}{2}\, S(f-f_c)U(f-f_c) \\
&\quad + \tfrac{1}{2}\, S(f-f_c)
\end{aligned}
\qquad (4.64)
$$

Note that the unit step function can be rewritten in the following form,

$$U(f+f_c) = \frac{1 + sgn(f+f_c)}{2} \qquad (4.65\text{a})$$

and

$$1 - U(f-f_c) = \frac{1 - sgn(f-f_c)}{2} \qquad (4.65\text{b})$$

Now defining $\hat{s}(t)$ as the Hilbert Transform of $s(t)$,

$$\hat{S}(f) = \frac{S(f)\, sgn(f)}{j} \qquad (4.66)$$

where $\hat{s}(t) \leftrightarrow \hat{S}(f)$.

Substituting Eqs. (4.65) and (4.66) into Eq. (4.64) we get the desired result,

$$S_{\text{lsb}}(f) = \frac{1}{2}\left[\frac{S(f+f_c) + S(f-f_c)}{2} + \frac{\hat{S}(f-f_c) - \hat{S}(f+f_c)}{2j} \right] \qquad (4.67)$$

Both of the ratios of Eq. (4.67) look familiar. The first is the form of the Fourier Transform of an AM wave. The positive and negative frequency shift represents a multiplication of $s(t)$ by a cosine in the time domain. We can find the inverse transform of Eq. (4.67).

$$s_{lsb}(t) = \tfrac{1}{2}s(t) \cos 2\pi f_c t + \tfrac{1}{2}\hat{s}(t) \sin 2\pi f_c t \qquad (4.68)$$

The second identity used in Eq. (4.68) is easy to prove. We used it once before in the pre-envelope discussion. That is, if

$$s(t) \leftrightarrow S(f)$$

then

$$s(t) \sin 2\pi f_c t \leftrightarrow \frac{1}{2j}[S(f-f_c) - S(f+f_c)] \qquad (4.69)$$

The proof of this is left as an exercise for the student.

Equation (4.68) can be implemented with the system of Fig. 4.66. Note that the Hilbert Transform is derived from $s(t)$ using a 90° phase shifter.

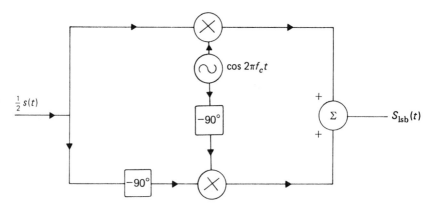

FIGURE 4.66   A lower sideband SSB generator.

If we wished to build a transmitter to generate the upper sideband instead of the lower sideband, we would simply subtract the lower sideband signal from $s_m(t)$. That is,

$$s_{usb}(t) = s_m(t) - s_{lsb}(t)$$

$$s_{usb}(t) = s(t) \cos 2\pi f_c t - [\tfrac{1}{2} s(t) \cos 2\pi f_c t + \tfrac{1}{2} \hat{s}(t) \sin 2\pi f_c t] \qquad (4.70)$$

$$= \tfrac{1}{2} s(t) \cos 2\pi f_c t - \tfrac{1}{2} \hat{s}(t) \sin 2\pi f_c t$$

The block diagram of a modulator which forms the upper sideband version is shown in Fig. 4.67. We note that it differs from the lower sideband generator by a simple change of sign in the output summer.

We comment that if one were to sketch the time waveform of a single sideband signal, the sketch would reveal no obvious relationship to $s(t)$. That is, while for double sideband transmission, the envelope was a replica of $s(t)$, the envelope of a SSB signal does not resemble $s(t)$.

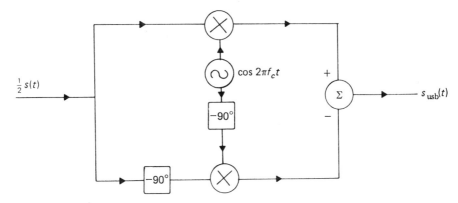

FIGURE 4.67 An upper sideband SSB generator.

Demodulation of SSB

Let us first investigate what the output of a synchronous detector would be if the input were a single sideband instead of a double sideband waveform. The synchronous demodulator was shown earlier, and is repeated as Fig. 4.68.

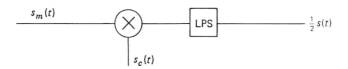

FIGURE 4.68 An AM demodulator.

Multiplication of the modulated signal by $\cos 2\pi f_c t$ has the effect of shifting the Fourier Transform up and down in frequency by f_c. This is illustrated for the upper sideband signal in Fig. 4.69. When this heterodyned signal is put through the

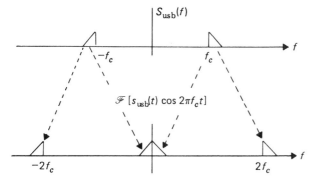

FIGURE 4.69 Heterodyning of SSB waveform.

synchronous detector's lowpass filter, we see that the desired information signal, $s(t)$, is recovered.

The same synchronous demodulator could be used for lower sideband waveforms, as is shown in Fig. 4.70.

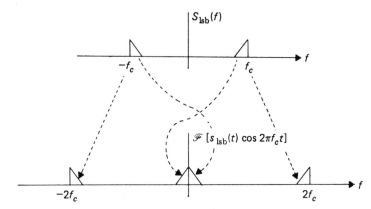

FIGURE 4.70   Heterodyning of SSB LSB waveform.

The inquiring scientist should not be happy until the previous graphical results can be proven mathematically. Consider the upper sideband waveform treated earlier. We found that its transform could be expressed as the standard double sideband transform multiplied by the system function of an ideal bandpass filter which passed frequencies between f_c and $f_c + f_m$. (We repeat Fig. 4.60 as 4.71.) That is,

$$S_{usb}(f) = S_m(f)H(f) = \tfrac{1}{2} H(f)[S(f-f_c) + S(f+f_c)] \qquad (4.71)$$

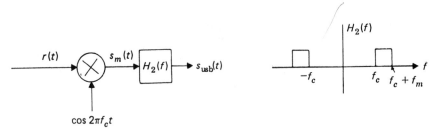

FIGURE 4.71   One type of SSB generator.

where

$$H(f) = \begin{cases} 1 & f_c < |f| < f_c + f_m \\ 0 & \text{otherwise} \end{cases}$$

If we now introduce $s_{usb}(t)$ into a synchronous demodulator, we first multiply it by $\cos 2\pi f_c t$.

$$s_{usb}(t) \cos 2\pi f_c t \leftrightarrow \tfrac{1}{4} \{ [H(f + f_c)][S(f - f_c + f_c) + S(f + 2f_c)]$$
$$+ [H(f - f_c)][S(f - 2f_c) + S(f + f_c - f_c)] \}$$
$$= \tfrac{1}{4} S(f)[H(f + f_c) + H(f - f_c)] \tag{4.72}$$
$$+ \tfrac{1}{4} S(f + 2f_c)H(f + f_c) + \tfrac{1}{4} S(f - 2f_c)H(f - f_c)$$

If we now pass this product, $s_{usb}(t) \cos 2\pi f_c t$ through a lowpass filter, the terms centered in frequency about $2f_c$ and about $-2f_c$ will be rejected, and the output will have the following transform.

$$\tfrac{1}{4} S(f)[H(f + f_c) + H(f - f_c)] \tag{4.73}$$

We note from Fig. 4.72 that $[H(f + f_c) + H(f - f_c)] = 1$, for f between $-f_m$ and

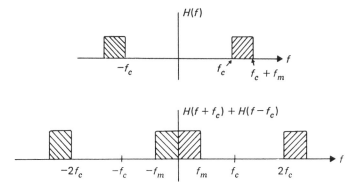

FIGURE 4.72 $H(f + f_c) + H(f - f_c)$ for Eq. (4.75).

$+f_m$. That is, it resembles the transfer function of an ideal lowpass filter. Therefore,

$$\tfrac{1}{4} S(f)[H(f + f_c) + H(f - f_c)] = \tfrac{1}{4} S(f) \leftrightarrow \tfrac{1}{4} s(t) \tag{4.74}$$

Figure 4.72 illustrates this "lowpass filter" characteristic of the sum, $H(f + f_c) + H(f - f_c)$. Note that $S(f) = 0$ for $|f| > f_m$. Thus the sections shown in Fig. 4.72 at $-2f_c$ do not affect the final result since they are multiplied by $S(f) = 0$.

The mathematical proof that a synchronous detector can also be used for the lower sideband case is similar to that given above for the upper sideband. The details are left to the exercises at the end of this chapter.

We therefore see that a synchronous detector will work as well with SSB as it does with double sideband signals. Unfortunately, this is not very well as you may recall.

In the double sideband case, we talked our way through a system which would use the incoming signal to generate a local carrier term. This was desirable since any slight error in the frequency of the sinusoid generated in the receiver proved disastrous. This is also true in the SSB case, although the physical manifestations of this error are not the same as in the double sideband case.

While the distortion in the double sideband case exhibited itself as a beating of the amplitude of the demodulated wave, in the single sideband case it results in total garbling of the message. Anybody who is an amateur radio operator can tell you how difficult it is to "tune in" a SSB station, and of the funny-sounding doubletalk that comes out whenever the local oscillator drifts in frequency. Slight frequency differences can be tolerated if all that is desired is to understand human speech.

If we view the transform of a SSB wave, it seems intuitively reasonable that sufficient information does *not* exist to derive the carrier from the SSB signal. The symmetry (redundancy) that was present in the AM double sideband wave no longer exists. One might be tempted to say (for the upper sideband case) that f_c can be found just by looking for the lowest frequency component present in the incoming modulated waveform. There are two fallacies in this reasoning. First, other stations may be occupying the frequency slot just below the signal of interest. There is no way of distinguishing the modulated portion due to the high information frequencies of one channel from that due to the low frequencies of an adjoining channel. (*See* Fig. 4.73(a).) Second, we cannot be sure that the information signal has frequencies that go right down to zero. It is more likely that the signal cuts off at some non-zero low

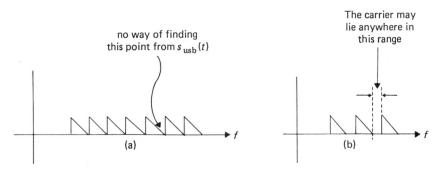

FIGURE 4.73 Carrier frequency cannot be derived from the incoming SSB waveform.

frequency. We would not want to falsely label this lowest frequency as the carrier (*see* Fig. 4.73(b)). It would therefore appear that we must simply "fish" for the proper carrier frequency and readjust it whenever the oscillator drifts slightly in frequency.

Suppose now, just as in the double sideband case, one decided to transmit a carrier term in the hopes that envelope detection might be used. In the lower sideband case, this would result in the following.

$$s_{\text{lsb}}(t) + A \cos 2\pi f_c t = [A + \tfrac{1}{2} s(t)] \cos 2\pi f_c t + \tfrac{1}{2} \hat{s}(t) \sin 2\pi f_c t \quad (4.75)$$

Performing a trigonometric identity, we find

$$s_{\text{lsb}}(t) + A \cos 2\pi f_c t = B(t) \cos(2\pi f_c t + \sigma(t))$$ (4.76)

where

$$B(t) = \sqrt{[A + \tfrac{1}{2} s(t)]^2 + [\tfrac{1}{2} \hat{s}(t)]^2}$$

and

$$\sigma(t) = \tan^{-1} \left[\frac{\tfrac{1}{2} \hat{s}(t)}{A + \tfrac{1}{2} s(t)} \right]$$

The output of an envelope detector with this as an input is the term "$B(t)$." In general, $\hat{s}(t)$ looks completely different from $s(t)$ since every frequency component is shifted by the same phase shift. Recall that distortionless transmission corresponded to a phase shift which was linear with frequency. It would therefore appear that the envelope detector output would not resemble the desired output, $s(t)$. This is indeed the case unless "A" is made very large. In that case, $A + \tfrac{1}{2} s(t)$ is much larger than $\tfrac{1}{2} \hat{s}(t)$, and the output of the envelope detector can be approximated by

$$B = \sqrt{[A + \tfrac{1}{2} s(t)]^2 + [\tfrac{1}{2} \hat{s}(t)]^2} \approx A + \tfrac{1}{2} s(t)$$ (4.77)

from which $s(t)$ is easily extracted. A receiver to accomplish this demodulation would differ only slightly from the standard (DSB) AM receiver. The catch is that the system is horribly inefficient. Large amounts of power are wasted in the sending of this large carrier term, and efficiencies of well under 5% are to be expected.

If power is a very minor consideration, and bandwidth and simple receiver design are major ones, the SSB system with large transmitter carrier and envelope demodulation would be used (television does something like this, as we shall see in the next section). If both bandwidth and power efficiency are serious considerations, but constant and sensitive monitoring of the local oscillator can be accomplished, SSB with synchronous demodulation may be used (we assume that the listener can tell when the signal sounds "right." This could rarely be done in data communications). This system is used in standard Ham radio and for overseas voice communications. For example, overseas news reports on radio probably come via this type of communications. Even though the stations use expensive crystal frequency control, you can occasionally hear the voice garbled due to a slight difference in transmitter and receiver oscillator frequencies.

For all of the above reasons, SSB does not find much application in common broadcast communication systems.

4.11   VESTIGIAL SIDEBAND TRANSMISSION

The only advantage of SSB over DSB is the economy of frequency usage. That is, SSB uses half the corresponding bandwidth required for DSB transmission of the same information signal. The primary disadvantage of SSB is the difficulty in building

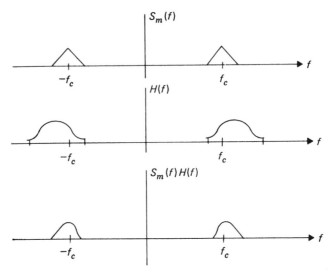

FIGURE 4.74 Vestigial sideband generated by non-ideal bandpass filtering of double sideband.

a transmitter or an effective receiver. If we could eliminate one of these disadvantages, SSB would become more attractive than it now appears.

Vestigial sideband posesses approximately the same frequency bandwidth advantage of SSB without the disadvantage of difficulty in building a modulator.

As the name implies, vestigial sideband (VSB) is a sloppy form of SSB where a vestige, or trace, of the second sideband remains.

We said earlier that one way of generating SSB is by perfect bandpass filtering of a double sideband modulated waveform. Suppose that we bandpass filtered the DSB wave, but used a non-ideal bandpass filter to do this. The result might resemble that shown in Fig. 4.74.

From the figure, we can see that the band of frequencies occupied by this VSB wave is not much larger than that occupied by the corresponding SSB waveform. If we can show that the demodulation of this wave is essentially the same as that for SSB, we have accomplished our objective. That is, instead of the complicated modulator required for SSB, we can use a modulator that is only slightly more complex than the standard DSB modulator.

The mathematical argument parallels that used to show that synchronous demodulation could be used for single sideband detection. We call $S_v(f)$ the Fourier Transform of the VSB wave, $v(t)$. $S_m(f)$ is the transform of the double sideband wave, and $H(f)$ is the system function of the bandpass filter. We therefore have (*see* Fig. 4.74)

$$S_v(f) = S_m(f)H(f) = \tfrac{1}{2}\left[S(f+f_c) + S(f-f_c)\right]H(f) \qquad (4.78)$$

We would now like to find those conditions under which $s_v(t)$, the VSB waveform, can be demodulated by a synchronous detector. If $s_v(t)$ is the input to a synchronous demodulator, the output will have the transform $Y(f)$, where

$$Y(f) = \begin{cases} \frac{1}{2}[S_v(f+f_c) + S_v(f-f_c)] & |f| < f_m \\ 0 & |f| > f_m \end{cases} \qquad (4.79)$$

In the above, we have taken the synchronous detector's multiplication by a sinusoid and lowpass filtering into account. Substituting Eq. (4.78) for $S_v(f)$, we see

$$Y(f) = \frac{1}{4}[S(f)][H(f+f_c) + H(f-f_c)] \qquad (4.80)$$

The multiplying factor, $[H(f+f_c) + H(f-f_c)]$ is sketched in Fig. 4.75 for a typical bandpass filter transfer function. We see that $y(t)$, the output of the synchronous

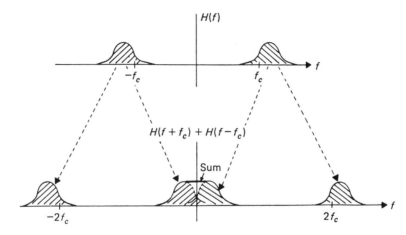

FIGURE 4.75 $H(f+f_c) + H(f-f_c)$ for a bandpass filter.

detector, will be proportional to $s(t)$ as long as $H(f+f_c) + H(f-f_c)$ is equal to a constant in the range of frequencies occupied by $s(t)$. In this case, the combination will resemble an ideal lowpass filter. In intuitive terms, the tail of the filter characteristic must be asymmetric about $f=f_c$. That is, the outer half of the tail must fold over and fill in any difference between the inner half values of the tail and the values for an ideal filter characteristic curve. This is not a very stringent restriction, and a sloppy common practical filter approximately accomplishes this.

With the addition of a strong carrier term, a transmitted carrier VSB waveform results:

$$s_v(t) + A \cos 2\pi f_c t \qquad (4.81)$$

If "A" is large, the VSB signal can be approximately demodulated using an envelope detector, just as in the SSB case. The proof of this will not be presented since it requires that we assume some specific form for $H(f)$.

In addition to simplified transmitter design, VSB possesses some additional attractive features when compared to SSB. First, since VSB is somewhere in between DSB and SSB, the size of a carrier term (A in Eq. (4.83)) required to permit envelope demodulation is smaller than that required by SSB. Therefore, although efficiency is not as great as that achieved in DSB, it is nevertheless higher than the SSB value. The second attractive feature of VSB arises out of a practical consideration. The filter required to separate one station from another (i.f. filter in the superheterodyne receiver case) need not be as perfect as that required for SSB station separation. The Fourier Transform of the VSB signal does not cut off sharply at the frequency limits, and therefore a less ideal filter can be used without appreciable distortion.

4.12   TELEVISION

Public television had its beginnings in England in 1927. In the United States, it started three years later, in 1930. These early forms used *mechanical* scanning of the picture to be transmitted. That is, a picture was changed into an electrical signal by scanning the entire image along a spiral starting at the center. This scanning was accomplished by means of a rapidly rotating wheel with holes cut in it. As the wheel rotated, light from various parts of the total picture passed through the holes.

During this early period, broadcasts did not follow any regular schedule. Such regular scheduling did not begin until 1939, during the opening of the New York World's Fair.

The concepts of television and picture transmission spread into many exciting areas; facsimile transmission, pictures from outer space, video telephone, videotext and cable TV represent a few examples. A cable TV revolution is occurring; two-way communication links will become common, and the TV will replace the newspaper, supermarket, baby-sitter, theater, and perhaps (heaven forbid!) the university campus. The theory about to be presented, although geared toward broadcast TV, is applicable to most forms of picture transmission, including video games and video computer terminals.

A Picture Is Worth a Thousand Words?

Nonsense! A picture is equivalent to far more than a thousand words. In fact, as a highly oversimplified example, we could take a picture of 1,001 words of text, thereby having a picture worth more than 1,000 words. But how much information does a picture contain? This is a serious technical question that must be answered before we can talk about specific ways of implementing transmission from one point to another.

While the information content of a message can be rigorously analyzed using concepts from the science of information theory we will make a simple adaptation for our present applications. We divide a picture into squares where each square is a certain shade (this will later be extended to color TV). The number of squares in any

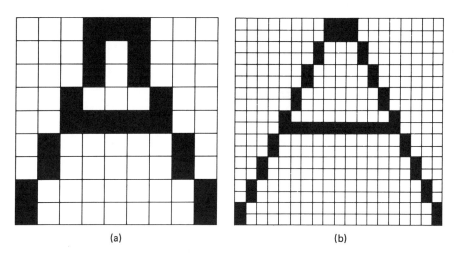

(a)   (b)

FIGURE 4.76   The letter "*A*" with two different resolutions.

given area determines the *resolution*. For example, Fig. 4.76(a) shows a picture of the letter "*A*" where 81 squares have been used to define the picture. In Fig. 4.76(b), the same letter is shown where the number of squares has increased to 342. The result is a "clearer," more highly resolved picture of the letter "A."

The number of squares needed depends upon the type of picture being considered. The less detail in a picture, the fewer squares are needed to adequately describe it.

The number of squares used in U.S. television is prescribed by law. The Federal Communications Commission (FCC) sets the number of boxes (known as *picture elements*) of U.S. TV at 211,000. This is divided into 426 elements in each horizontal line and 495 visible horizontal lines in each picture. All that television transmitters must do is to step through each of these 211,000 picture elements and send an intensity value for each one. The receiver interprets these transmissions and reconstructs the picture from the 211,000 intensity levels.

Now to the details. A TV receiver is not much different from a laboratory oscilloscope. A beam of electrons is shot toward a screen and bent by use of deflection plates. When a negative charge is placed on a plate, the electron beam is repelled. In the oscilloscope, we apply a sawtooth waveform to the horizontal deflection plates to sweep the beam from the left to the right edge of the screen (and then, more rapidly, back to the left). This traces a line. In TV receivers, we add a second dimension to this sweep. While the beam is rapidly sweeping from left to right on the screen, it is less rapidly sweeping from top to bottom on the screen. The net result is a series of (almost) horizontal lines on the screen, as sketched in Fig. 4.77. This is known as the TV *raster*.

The screen has 495 of these horizontal lines in order to comply with the FCC regulation. The beam then returns to the top of the screen, taking the equivalent time

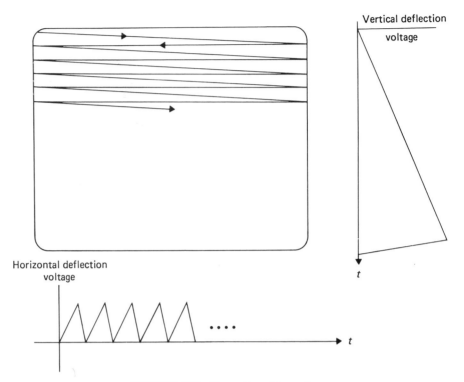

FIGURE 4.77 Generation of the raster.

of an additional 30 lines to do so. We can therefore think of the picture as having 525 lines, 30 of which occur during the *vertical retrace* time.

No times have yet been assigned to this process. The human eye requires a certain picture rate (number of times the entire screen is traced each second) to avoid seeing flicker, as in early motion pictures. This number is somewhere near 40 per second. U.S. TV decided to use 60 frames per second. This matches the frequency of household current and is chosen that way to minimize the effects of the video equivalent of 60-Hz hum.

The peaks of the 60-Hz power signal occur at intervals of $\frac{1}{60}$ sec. If the beam is tracing down the screen at this same rate, the effect of the power ripple will be stationary. If this were not the case, the effect would be a shaded bar rolling vertically, and the eye would be able to notice it much more readily than when the bar is stationary. At 525 lines/frame and 60 frames/sec, we take the product to get 31,500 lines/sec. The reciprocal of this gives the time per line as 31.75 μsec/line. Of this 31.75 μsec, 5.1 μsec is used for horizontal retrace from right to left. This leaves 26.65 μsec for the visible part of each horizontal line. Dividing this by the 426 elements in a line finally yields the time per element of 0.0625 μsec/element, or 16 million elements/sec. The system would therefore have to be capable of transmitting 16 million different shades (black, white, gray, etc.) per second. In the worst case, one may wish to display a perfect checkerboard design of alternating black and white squares. The system would then jump from the darkest to lightest and back again 16 million times a second.

If we now think of this light intensity information as a signal, we see from Fig. 4.78 that it has a fundamental frequency of 8 MHz. Thus, no matter what scheme we choose by which to transmit this, at least 8 MHz of bandwidth would be required.

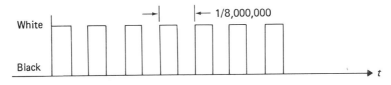

FIGURE 4.78   Intensity signal for checkerboard.

Here, the law steps in again, and the FCC says that the maximum bandwidth that the video signal can have is 4.1 MHz. Alas, how can these seemingly contradictory specifications (60 frames/sec to avoid flicker, 211,000 picture elements for proper resolutions, 4.1-MHz maximum bandwidth) be met?

Engineers had observed many decades of the development of motion pictures in which a similar predicament occurs. In standard motion pictures, 24 different pictures are shown each second. But 24 flashes/sec on a screen would appear to flicker considerably. Contemporary motion picture projectors flash each image twice. Thus, the frame rate is 48/sec, though only half of the frames represent new picture information.

Television's founders decided to play a similar trick. They cut the signal frequency in half by cutting the number of lines per second in half, from 525 to 262.5 lines/sec. However, it was necessary to still fill the entire screen each $\frac{1}{60}$ sec to prevent flicker. The technique for cutting the line frequency in half without changing the frame frequency is known as *interlaced scanning*. In the first $\frac{1}{60}$ sec, all of the odd-numbered lines are traced (ending with a $\frac{1}{2}$ line). The beam then returns to the top center of the screen to trace the even-numbered lines in the next $\frac{1}{60}$ sec. Thus, while it now takes $\frac{1}{30}$ sec to send all 211,000 picture elements, during this $\frac{1}{30}$ sec the entire screen is covered twice. The eye fills in the missing rows and detects no flicker. There is, or course, some loss of resolution on very fast-moving objects, but television was never intended to follow such high speeds anyway.

Signal Design and Transmission

We now translate this information into an electrical signal. If we plot light intensity as a function of time, a staircase function results. Figure 4.79 shows an example of the letter "T" in dark black followed by the punctuation "period" in light gray. The associated signal is shown where, for simplicity, interlaced scanning has not been shown and the number of lines has been drastically reduced.

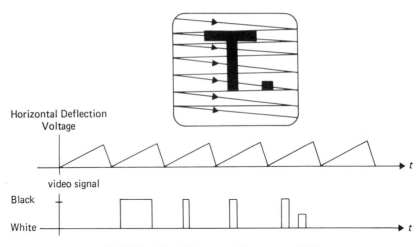

Horizontal Deflection
Voltage

video signal

Black

White

FIGURE 4.79　Video signal for message "*T*."

For broadcast TV, the information (video) signal would be a similar staircase (smoothed to reduce bandwidth) function where the minimum step width is 0.125 μsec. In the actual video signal, the voltage corresponds to the light intensity since, in the receiver, this voltage is used to control the electron gun. The higher the voltage applied to a grid placed between the gun and the screen, the fewer electrons hit the screen and the darker the spot on the screen. As an additional modification, the staircase function is smoothed. The eye cannot tell the difference between a smooth or rapid transition in the signal during 0.125 μsec. After all, this corresponds to only $\frac{1}{426}$ of the width of the TV screen.

There is one additional aspect to this electrical video signal known as *blanking*. While the beam on the cathode ray tube is retracing (either horizontally or vertically), it is desirable that the beam of electrons be "turned off" so that this retrace is not seen as a line on the screen.

Taking all of this into account, the video signal corresponding to the picture shown in Fig. 4.79 is redrawn as Fig. 4.80.

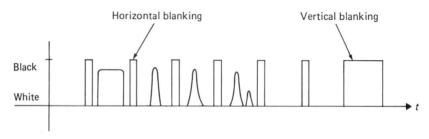

Horizontal blanking

Vertical blanking

Black

White

FIGURE 4.80　Figure 4.79 modified to include smoothing and blanking.

Synchronization

The transmitter at the TV studio is tracing rapidly from left to right and from top to bottom many times a second. It is sending a record of light intensity as a function of time. The receiver must be sure it is placing the transmitted intensity in the same spot on the screen as intended. If the beam in the receiver doesn't start a scan at the same instant that the transmitter does (corrected for transmission time), the picture will appear split at best and totally scrambled at worst. A means is thus required for synchronizing the two sweeping operations. This is done by means of synchronization pulses added to the video signal. The pulses are added during the blanking intervals, therefore not affecting what is seen on the screen. Figure 4.81 shows the same signal as Fig. 4.80 modified with the addition of synchronizing pulses.

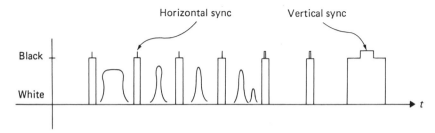

FIGURE 4.81   Video signal including sync pulses.

Two types of synchronizing pulses are shown in Fig. 4.81. The narrow pulses are horizontal synchronizing pulses, and the wide pulses are vertical synchronizing pulses.

The receiver separates these pulses from the remaining signal by means of a threshold circuit. The horizontal and vertical pulses are then separated by means of a simple RC integrator circuit. The integral of the wider vertical sync pulses is higher than that of the narrower horizontal pulses. The separated pulses are then used to synchronize (trigger) the horizontal and vertical oscillators.

Modulation Techniques

As discussed previously, the video signal has a maximum frequency of about 4 MHz. The FCC allocates 6 MHz of bandwidth to each television channel, and this space must contain both the video and audio sections of the transmitted signal. Obviously the use of double sideband AM must be rejected since this would require over 8 MHz of bandwidth for each channel.

Single sideband transmission is also rejected for the following reason. Recall that generation of SSB requires very sharp filtering of a double sideband signal to remove one of the sidebands. As a filter is designed to closely approach the desired amplitude characteristic, it becomes increasingly difficult to control the phase characteristic, which, ideally, should be linear with frequency. That is, in designing practical filters one can approach either the phase or amplitude characteristics as closely as desired, but to achieve both simultaneously is extremely difficult. In audio applications, phase deviations from the ideal characteristic are not very serious. They represent varying delays of the frequency components of the message, and the human ear is not sensitive to such variations. In a video signal, these varying delays would be manifested as position shifts on the screen. These are commonly referred to as *ghost images* and are highly undesirable. Thus, SSB is not used for television transmission.

The video portion of the TV signal is sent using vestigial sideband (VSB). A strong carrier is added to enable the use of the envelope detector for demodulation. The entire upper sideband and a portion of the lower sideband are sent. Figure 4.82 shows the frequency composition of a TV signal. Note that the audio and video are frequency multiplexed, and their carriers are separated by 4.5 MHz. The audio is sent via wideband FM, which is described in the following chapter.

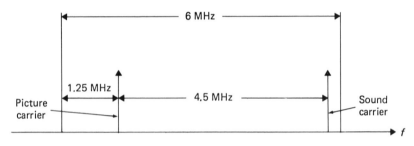

FIGURE 4.82   Frequency composition of TV signal.

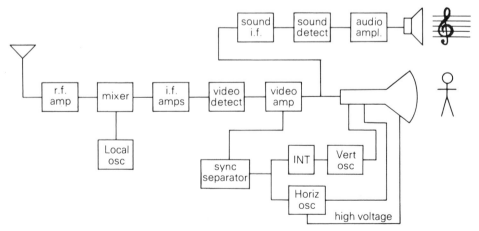

FIGURE 4.83   TV receiver block diagram.

Functional Block Diagram of TV Receiver

We are now in a position to examine the overall block diagram of a black and white (monochrome) TV receiver. A representative block diagram is shown in Fig. 4.83.

The composite (video plus audio) signal is received by the antenna and is amplified by an r.f. amplifier. It then enters the tuner which is a mixer and i.f. filter just as in the case of the superheterodyne radio receiver. The frequency band allocated to commercial television is shown in Table 4.1.

TABLE 4.1

Channel	Frequency Range (MHz)	Comments
2	54–60	
3	60–66	
4	66–72	
5	76–82	Note gap between channels 4 and 5
6	82–88	
7	174–180	Between 7 and 6 there is the FM radio band, aircraft radio, government, railroad, and police
8	180–186	
9	186–192	
10	192–198	
11	198–204	
12	204–210	
13	210–216	
14–83	470–890	

Most large cities with many TV stations use channels 2, 4, 5, 7, 9, 11, and 13. Examination of the table of frequency allocations shows that no two of these channels are adjacent to each other in frequency. This fact slightly eases the design requirements on the i.f. filters.

The i.f. frequency used for TV is 40 MHz. After i.f. amplification and filtering, the signal enters the video detector, which is simply an envelope detector.

Since the sound carrier is 4.5 MHz above the picture carrier frequency, a filter (trap) is used to separate the sound signal from the video signal. The sound is FM and is demodulated in a way to be described in the next chapter.

The synchronizing pulses are separated from the video signal by means of a threshold circuit (clipper) known as the *sync separator*. The vertical sync pulses are then distinguished from the horizontal by an integrator. Both sets of pulses are used to trigger the corresponding sweep oscillators. As an added bonus, the retrace portion of the horizontal deflection voltage is used to generate the very high (over 20 kV) voltage required on the CRT anode to pull the electrons in a straight line toward the face of the tube. This high voltage is generated by differentiating the retrace ramp of the voltage using what is known as a *flyback transformer*.

Color Television

Color TV takes advantage of the principle of primary colors. All colors, including black and white, can be formed as combinations of the three primary colors red, blue, and yellow. TV modifies this slightly by using red, blue, and green (a combination of blue and yellow) since phosphors that glow with these colors when excited with an electron beam are commonly available.

The color TV receiver is essentially three separate TVs, one generating red, one generating blue, and one generating green. By mixing these in varying strengths, any color can be formed.

Traditional (this is rapidly changing) color CRTs actually contain three separate electron guns, one for each color. The transmitted signal must therefore generate three separate video signals to control each of these guns.

But, you ask, where can these two additional signals be squeezed? We are already using all of the bandwidth allowed for TV. We need a system to transmit all three video signals without increasing the 6-MHz bandwidth *and*, at the same time, allowing for compatible transmission—a black and white TV receiver should be capable of receiving and reproducing a color signal (in black and white, of course).

Again, the engineers designing TV systems were ingenious. In fact, they used the dual of interlaced scanning.

If we were to examine the Fourier Transform of a black and white video signal, we would find it resembles a train of impulses. An actual signal for one trace of a picture contains 262.5 horizontal traces. Since most pictures contain some form of vertical continuity (that is, one horizontal line is very close to the next horizontal line in content), the video signal is almost periodic with fundamental frequency equal to the line frequency (15,750 lines/sec).

If the video signal were exactly periodic, its Fourier Transform would be a train of impulses at multiples of the fundamental frequency. Since the signal is almost periodic, the Fourier Transform consists of pulses (not impulses) centered around multiples of the line frequency.

Some caution is required. The above argument applies to most real life pictures. The argument fails if TV is pushed to its limits, and a high-resolution detailed picture is sent.

Since the Fourier Transform is essentially zero between multiples of the line frequency, additional information can be placed in these spaces by use of a form of frequency multiplexing or frequency interlacing. The additional information needed for color transmission modulates a carrier which is midway between two multiples of the line frequency. This is assured by using a carrier whose frequency is an odd multiple of half of the line frequency. The figure used is 3.579545 MHz, which is the 455th harmonic of half of the line frequency.

To make the signal compatible, the three signals are not the three colors. The three parameters sent are brightness, *hue* (position in the color spectrum), and *saturation* (how close is the color to a pure single frequency?). Instead of the word brightness, we use *luminance*. The required bandwidths of these three parameters are

4, 0.4, and 1.3 MHz, respectively. The color subcarrier is amplitude-modulated with the saturation signal and phase-modulated (see the next chapter) with the hue signal. The receiver demodulates this subcarrier using synchronous demodulation. The frequencies are matched at the transmitter and receiver by sending a segment of the exact sinusoidal subcarrier (the *color burst*) during each blanking interval.

4.13   AM STEREO

In this section, we shall very briefly introduce the basic concept of AM stereo. We will have to defer discussion of specific systems until the next chapter, since analysis of these systems requires a knowledge of phase modulation.

The concept of AM stereo is to send two independent audio signals within the 10-kHz bandwidth allocated by the FCC to each commercial broadcast station. Additionally, the FCC requires compatibility with existing monaural receivers. Thus, if the two signals represent the left and right channels, a monaural receiver must recover the sum of these two signals.

Quadrature modulation is one technique for sending two independent signals within the bandwidth for a single AM channel. If the two signals are designated $s_L(t)$ and $s_R(t)$, the quadrature modulation signal is given by

$$q(t) = s_L(t) \cos 2\pi f_c t + s_R(t) \sin 2\pi f_c t \qquad (4.82)$$

Sine and cosine are at 90° to each other and are therefore in phase quadrature. If $s_L(t)$ and $s_R(t)$ are audio signals with a maximum frequency of 5 kHz, then each of the modulation terms covers a band between $f_c - 5$ kHz and $f_c + 5$ kHz. When the two modulated signals are added together, the bandwidth of the sum is the same as the bandwidth of each individual term (The Fourier Transform is a linear operation.)

The composite signal can be rewritten as Eq. (4.83).

$$q(t) = \sqrt{s_L^2(t) + s_R^2(t)} \cos\left(2\pi f_c t - \tan^{-1}\frac{s_R(t)}{s_L(t)}\right) \qquad (4.83)$$

An envelope detector would receive the square root term in Eq. (4.83), which is a distorted version of the sum. This is unacceptable, and we shall see how it is modified in the next chapter.

In the meantime, Fig. 4.84 presents a block diagram of the stereo modulator and demodulator. The dashed block in the demodulator contains a phase lock loop (PLL). We shall discuss these loops in the next chapter. For now, we state that the output of this PLL with the input of Eq. (4.83) would be $\cos(2\pi f_c t - 45°)$. The various time functions as marked on the diagram are given below.

$$s_1(t) = \cos(2\pi f_c t - 45°)$$

$$s_2(t) = \cos 2\pi f_c t$$

$$s_3(t) = \sin 2\pi f_c t$$

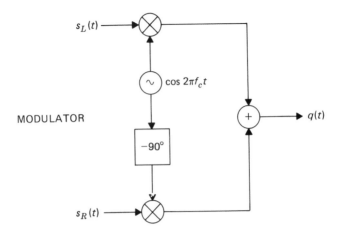

MODULATOR

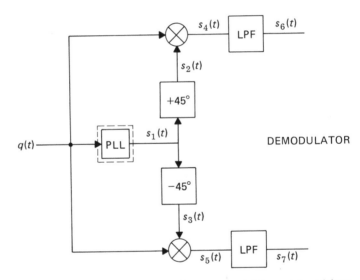

DEMODULATOR

FIGURE 4.84 Quadrature modulation stereo modulator and demodulator.

$$s_4(t) = s_L(t) \cos^2 2\pi f_c t + s_R(t) \sin 2\pi f_c t \cos 2\pi f_c t$$
$$s_5(t) = s_L(t) \sin 2\pi f_c t \cos 2\pi f_c t + s_R(t) \sin^2 2\pi f_c t$$
$$s_6(t) = \frac{1}{2} s_L(t)$$
$$s_7(t) = \frac{1}{2} s_R(t)$$

4.14   NOISE IN AM SYSTEMS

In the process of communication of a signal $s(t)$ via AM techniques, noise arises in various ways. The information signal $s(t)$ is corrupted by some noise before it even reaches the modulator in the transmitter. This noise occurs because of the presence of electronic devices (e.g., audio amplifiers) and also because of electromagnetic radiation and pickup in the transmitter wires, which act as small antennas. Thus, the signal at the output of the modulator would be of the form

$$[s(t) + n_1(t)] \cos 2\pi f_c t$$

where $n_1(t)$ is the noise. There is additional noise due to the fact that the carrier sinusoid is not a pure cosine wave. In fact, it contains harmonic distortion.

The modulated signal experiences multiplicative noise in the process of being transmitted from transmitter to receiver. This type of noise is due to both turbulence in the air and reflection of the signal. The turbulence causes the characteristics of the transmission medium, $H(f)$, to change constantly with time. The reflections, when recombined with the main path signal, either enhance the signal strength or subtract from it. Taking the multiplicative noise into account, the signal at the receiving antenna would be of the form

$$A\{1 + n_2(t)\} s_m(t)$$

where $n_2(t)$ is the multiplicative factor of noise.

The AM signal also experiences additive noise during transmission. This noise is generated by a multitude of sources, as mentioned in Chapter 3. For example, passing automobiles, static electricity, lightning, power transmission lines, and sunspots contribute to the overall noise effect. If one could listen to this noise, it would sound like static heard on radio transmission with occasional crackling sounds added. If the transmitted signal is $s_m(t)$, the received signal would be of the form

$$A s_m(t) + n_3(t)$$

where $n_3(t)$ is the additive noise.

An additional source of noise occurs in the receiver. Electronic devices and components are present, thus giving rise to thermal and shot noise. In addition, the wires in the receiver act as small antennas, thereby picking up some transmission noise. For purposes of analysis, this receiver noise can be treated as additive noise and included in $n_3(t)$.

Of the various types of noise introduced, the additive transmission noise is the most annoying type of noise. It normally contains the most power of the types discussed above. This is not to imply that other types of noise are not critical. Multiplicative transmission noise (turbulence) can become a significant factor at certain high frequencies. For example, if visible light frequencies are used for transmission, rain and clouds have a substantial impact upon the ability to receive a signal. To a certain extent, UHF television signals can experience multiplicative noise, which is more detrimental to reception than is additive transmission noise.

Synchronous Demodulation

We shall first analyze the case where coherent detection is employed. That is, the exact frequency and phase of the received carrier are known and can be reconstructed at the receiver. Synchronous demodulation can be used for either single or double sideband waveforms. It can also be used for transmitted or suppressed carrier transmission. Each of these cases will be analyzed individually.

Suppose first that a suppressed carrier double sideband AM waveform, $s(t) \cos 2\pi f_c t$, is transmitted, where $s(t)$ is the information signal. In the process of propagating from the transmitter to the receiver, the wave encounters attenuation, time delay, and additive noise. We will omit the time delay from the following analysis, since it is assumed to be precisely known (i.e., if you wish, substitute $t - t_0$ for t in the following equations). The received waveform at the antenna is therefore of the form

$$r(t) = Ks(t) \cos 2\pi f_c t + n(t)$$

where K is a constant which accounts for attenuation during transmission. The average signal power at the receiver is $\frac{1}{2}K^2 P_s$, where P_s is the power of $s(t)$. The additive noise, $n(t)$, is assumed to be white and has a power spectral density of $N_0/2$ watts/Hz for all values of f.

The synchronous demodulator is shown as Fig. 4.85. The bandpass filter preceding the multiplier is redundant, since the final lowpass filter would perform the

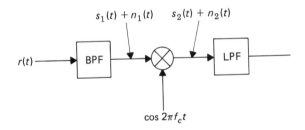

FIGURE 4.85   The synchronous demodulator.

same function. It is included to make the analysis simpler, since we can easily deal with bandlimited noise at the input to the multiplier. Its presence will not change the final result. While the above statement is true theoretically, it may not be true for the practical system. The signal at the output of the bandpass filter is smaller than that at the input. Signals outside the pass band of the filter have been rejected, as has any noise outside this band (theoretically, white noise has infinite power outside this band, so while the input to the filter has infinite power, the output power is finite). Practical multipliers can be overloaded by signals which are too large. Thus, the pre-multiplication filtering may have practical importance.

The signal portion at the output of this system would be $\frac{1}{2}Ks(t)$. In order to find the noise at the output, we start by observing that the noise at the input of the multiplier can be expanded into quadrature components. Thus, $n_1(t)$ is given by

$$n_1(t) = x(t) \cos 2\pi f_c t - y(t) \sin 2\pi f_c t$$

The power spectral densities of $x(t)$ and $y(t)$ were calculated in Illustrative Example 3.19 and are sketched in Fig. 4.86. (Note that there is a difference of a factor of 2 in

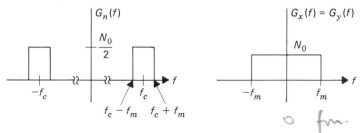

FIGURE 4.86 Power spectral density of quadrature noise components for double sideband AM.

the statement of the problem when compared with Illustrative Example 3.19.) The noise at the output of the multiplier, $n_2(t)$, is given by

$$n_2(t) = [x(t) \cos 2\pi f_c t - y(t) \sin 2\pi f_c t] \cos 2\pi f_c t$$

$$= \frac{1}{2} x(t) + \frac{1}{2} x(t) \cos 4\pi f_c t - \frac{1}{2} y(t) \sin 4\pi f_c t \qquad (4.84)$$

The only term in Eq. (4.84) which is not rejected by the lowpass filter is the first term, $\frac{1}{2}x(t)$. The other two terms have frequencies centered about $2f_c$. Thus, the noise waveform at the output of the detector is given by $\frac{1}{2}x(t)$.

In determining the effect of the noise, a quantity often calculated is the signal to noise ratio, S/N. This does not imply that, for example, in an audio communication system, the ear is sensitive only to noise powers. For example, a noise signal with extremely low power might be characterized by a series of clicks, which could be quite annoying in listening to a music signal. Nonetheless, in the case of filtered white noise, total noise power relative to the signal power is a good measure of the annoyance caused in an audio system.

The output noise power is the power of $\frac{1}{2}x(t)$, which is one fourth of the power of $x(t)$. This is given by

$$\overline{\frac{1}{4} x^2(t)} = \frac{2}{4} \int_0^\infty G_x(f)\,df$$

$$= \frac{1}{2} \int_0^{f_m} N_0\,df = \frac{1}{2} N_0\,f_m$$

Since the signal part of the output is $\frac{1}{2}Ks(t)$, the power is $\frac{1}{4}K^2P_s$, where P_s is the average power of $s(t)$. The output signal to noise ratio is then given by

$$S/N = \frac{K^2 P_s}{2 f_m N_0} \qquad (4.85)$$

Single Sideband Suppressed Carrier

We now use single sideband in place of double sideband in the above analysis. Assuming that the upper sideband is used, the received waveform will be

$$r(t) = \frac{1}{\sqrt{2}} K s(t) \cos 2\pi f_c t - \frac{1}{\sqrt{2}} K \hat{s}(t) \sin 2\pi f_c t + n(t)$$

The reason for introducing the constant of $1/\sqrt{2}$ will become evident when the received signal power is found. The signal power is given by

$$P = \frac{K^2}{2} \left[\overline{s^2(t) \cos^2 2\pi f_c t} + \overline{\hat{s}^2(t) \sin^2 2\pi f_c t} - \overline{2 s(t)\hat{s}(t) \cos 2\pi f_c t \sin 2\pi f_c t} \right]$$

where the bar over the function represents time average. The third term is equal to zero, since cosine and sine are at 90°. Looked at another way, $2 \cos 2\pi f_c t \sin 2\pi f_c t = \sin 4\pi f_c t$. This high frequency term averages to zero. On the other hand, $\cos^2 2\pi f_c t$ and $\sin^2 2\pi f_c t$ have DC terms of $\frac{1}{2}$. Thus,

$$P = \frac{K^2}{2} \left[\frac{1}{2} P_s + \frac{1}{2} P_{\hat{s}} \right]$$

where P_s is the power of $s(t)$ and $P_{\hat{s}}$ is the power of $\hat{s}(t)$. Now $\hat{s}(t)$ is the Hilbert Transform of $s(t)$, and results from putting $s(t)$ through a linear system with $H(f) = -j \, sgn(f)$. Thus, $|H^2(f)| = 1$ and the power spectral density of $\hat{s}(t)$ is identical to the power spectral density of $s(t)$. Thus, $P_{\hat{s}} = P_s$. The total signal power at the receiver is then given by

$$P = \frac{K^2}{2} \left[\frac{1}{2} P_{\hat{s}} + \frac{1}{2} P_s \right] = \frac{1}{2} K^2 P_s$$

This is the same as the received signal power for the double sideband case. Thus, the constant of $1/\sqrt{2}$ was chosen to match received signal power for the double sideband case. This is necessary to make a fair comparison between single sideband and double sideband. It would not mean much for double sideband to have, for example, twice the output signal to noise ratio as single sideband if more signal power were being received using double sideband.

The demodulator is the same as that shown in Fig. 4.85 except that the bandpass filter preceding the multiplier now covers the band of the upper sideband signal, f_c to $f_c + f_m$. The output signal would be

$$s_0(t) = \frac{1}{2\sqrt{2}} K s(t)$$

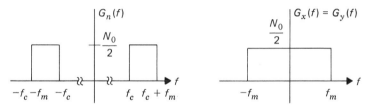

FIGURE 4.87   Power spectral density of quadrature noise components for single sideband AM.

Thus the output signal power is

$$P = \frac{K^2 P_s}{8}$$

The noise at $n_1(t)$ can again be expressed in quadrature form.

$$n_1(t) = x(t) \cos 2\pi f_c t - y(t) \sin 2\pi f_c t$$

The spectral densities are as derived in Illustrative Example 3.19 and are shown in Fig. 4.87. After the multiplier, the noise is given by

$$n_2(t) = \frac{1}{2} x(t) + \frac{1}{2} x(t) \cos 4\pi f_c t - \frac{1}{2} y(t) \sin 4\pi f_c t$$

and the output noise is again $\frac{1}{2} x(t)$. The output noise power is

$$P_n = \frac{2}{4} \int_0^\infty G_x(f)\, df = \frac{1}{2} \int_0^{f_m} \frac{N_0}{2}\, df = \frac{N_0 f_m}{4}$$

Finally, the signal to noise ratio is given by

$$\text{SNR} = \frac{K^2 P_s}{2 N_0 f_m}$$

This is *exactly the same* as the ratio found for double sideband suppressed carrier.

Illustrative Example 4.12

An information signal, $s(t) = 5 \cos 2{,}000\, \pi t$, is transmitted using DSBSC and demodulated using a synchronous demodulator. Noise with power spectral density $G_n(f) = 10^{-4}$ is added to the signal prior to reception. Find the signal to noise ratio at the output of the receiver.

Solution

We will assume that the signal is received without any attenuation. If this is not true, the attenuation factor must simply be included.

The signal to noise ratio can be found by direct application of the formula derived in this section.

$$\text{SNR} = \frac{K^2 P_s}{2N_0 f_m}$$

Since we assumed no attenuation, $K = 1$. The signal power, P_s, is equal to 12.5. $N_0/2$ is the height of the power spectral density of the noise, so $N_0 = 2 \times 10^{-4}$. The maximum frequency of the information signal, f_m, is given by 1000 Hz. Plugging these values into the formula yields

$$\text{SNR} = \frac{12.5}{4{,}000 \times 10^{-4}} = 31.25 = 14.95 \text{ dB}$$

Double Sideband Transmitted Carrier

Synchronous demodulation is used with AMTC if the carrier is not large enough to permit other forms of detection. The received waveform is given by

$$r(t) = K \sqrt{\frac{P_s}{P_s + A^2}} \, [s(t) + A] \cos 2\pi f_c t + n(t)$$

The constant is again chosen to match the received signal power in the suppressed carrier case. The received signal power is given by

$$P = \frac{P_s}{P_s + A^2} \, \overline{K^2 [s(t) + A]^2 \cos^2 2\pi f_c t}$$

$$= \frac{P_s}{P_s + A^2} \, \frac{1}{2} K^2 \overline{[s^2(t) + 2As(t) + A^2]}$$

$$= \frac{P_s}{P_s + A^2} \, \frac{1}{2} K^2 [P_s + A^2] = \frac{K^2 P_s}{2}$$

This is the same as the received signal power for the suppressed carrier case. In this derivation, we assumed that the average value of $s(t)$ is zero, as is true in most cases of interest.

The noise at the output of the synchronous demodulator is identical to that found in the double sideband suppressed carrier case. The output signal is given by

$$\frac{K}{2} [s(t) + A] \sqrt{\frac{P_s}{P_s + A^2}}$$

and the power of this signal is

$$P = \frac{K^2}{4} P_s$$

If the output signal is defined to be $A + s(t)$, the signal to noise ratio is identical to that of suppressed carrier. However, if the constant, A, is rejected, the signal part of the output is

$$\frac{Ks(t)}{2}\sqrt{\frac{P_s}{P_s + A^2}}$$

and the output signal power is

$$P = \frac{K^2}{4}\frac{P_s^2}{P_s + A^2}$$

This yields a signal to noise ratio of

$$\text{SNR} = \frac{K^2 P_s^2}{2(P_s + A^2)N_0 f_m} \tag{4.86}$$

As A increases, this signal to noise ratio decreases.

Incoherent Detection

If the frequency and phase of the carrier cannot be reconstructed at the receiver, we must use incoherent detection techniques. This includes the envelope detector, the square law detector, and the rectifier detector.

Envelope Detector

We shall first examine the envelope detector. Because of the non-linear operation, the analysis will be much more complex than that of the synchronous detector. In fact, we will find it advantageous to make certain approximations.

The signal at the receiver is again

$$r(t) = K\sqrt{\frac{P_s}{P_s + A^2}}\,[s(t) + A]\cos 2\pi f_c t + n(t)$$

where the constant has been chosen to keep received signal power at the same value as that used for the coherent detectors. We will pass this signal through a bandpass filter centered at the carrier frequency and with a bandwidth of $2f_m$. This will not only separate the desired signal from any other frequency multiplexed waveforms, but will also reduce the noise entering the envelope detector. As in the previous analyses, we can now expand the noise into quadrature components to get the following as the input to the envelope detector.

$$\left[K\sqrt{\frac{P_s}{P_s + A^2}}\,[A + s(t)] + x(t)\right]\cos 2\pi f_c t - y(t)\sin 2\pi f_c t$$

We will define a new constant in order to simplify the following equations.

$$K_1 \overset{\Delta}{=} K \sqrt{\frac{P_s}{P_s + A^2}}$$

The input to the envelope detector can be rewritten as a single sinusoid as

$$B(t) \cos\left(2\pi f_c t + \theta(t)\right) \qquad (4.87a)$$

where

$$B(t) = \sqrt{\left[K_1[A + s(t)] + x(t)\right]^2 + y^2(t)} \qquad (4.87b)$$

$$\theta(t) = \tan^{-1} \frac{y(t)}{K_1[A + s(t)] + x(t)}$$

Since the envelope detector, in its ideal form, is insensitive to phase variations, the output of the detector will be $B(t)$. Unfortunately, $B(t)$ contains non-linear operations, and therefore has powers of the noise components and cross products between the signal and noise. It will be helpful to examine the sinusoid at the input to the envelope detector by using phasors. Figure 4.88 is a phasor diagram of this sinusoid. In

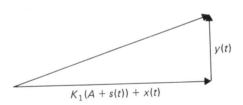

$y(t)$

$K_1(A + s(t)) + x(t)$

FIGURE 4.88   Phasor diagram of sinusoid at input to envelope detector with signal larger than noise.

sketching this, we have assumed that $K_1[A + s(t)]$ is larger than $y(t)$. Otherwise the angle between the two vectors would be greater than $45°$. This case will be analyzed later.

If we now assume that the signal to noise ratio is large, that is,

$$y(t) \ll K_1[A + s(t)]$$

then the amplitude can be approximated by

$$B(t) \approx K_1[A + s(t)] + x(t)$$

The advantage of this approximation is that the signal and noise appear explicitly as linear terms in this result. The signal power is $K_1^2 P_s$, while the noise power is the power of $x(t)$, which we previously found to be $2N_0 f_m$. The signal to noise ratio is then equal to

$$SNR = \frac{K_1^2 P_s}{2N_0 f_m} = \frac{K^2 P_s^2}{2N_0 f_m (P_s + A^2)}$$

This result is identical to that found for the synchronous demodulator. (See Eq. (4.86).)

Now we will assume the opposite extreme, that is, an extremely low signal to noise ratio.

$$x(t) \gg K_1[A + s(t)]$$

In order to see intuitively what type of result to expect, we shall redraw the phasor diagram, but this time referenced to the noise vector. This is shown in Fig. 4.89. The

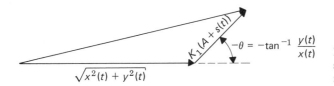

FIGURE 4.89   Phasor diagram of sinusoid at input to envelop detector with noise larger than signal.

length of the noise vector is $\sqrt{x^2(t) + y^2(t)}$. Note that the angle between the signal vector and the noise vector is

$$\theta(t) = \tan^{-1}\left(\frac{y(t)}{x(t)}\right)$$

The resultant vector is approximately given by

$$\sqrt{x^2(t) + y^2(t)} + K_1[A + s(t)]\cos\theta(t)$$

The only place where the signal appears is in the last term, and it is multiplied by a random noise term, $\cos\theta(t)$. It can be shown that $\theta(t)$ is uniformly distributed between $0°$ and $360°$. Thus, it is virtually impossible to recover the signal from this waveform.

We shall not analyze the intermediate signal to noise ratio case in detail. We have seen that at high signal to noise ratio, the envelope detector performance is similar to that of the synchronous demodulator, and at low signal to noise ratio, performance is a disaster. At some point in between these signal to noise ratio extremes, a threshold occurs below which performance degrades significantly. For purposes of approximation, we will assume that this threshold occurs when the signal amplitude is equal to the noise amplitude. If, for simplicity, we let A be much larger than $s(t)$, the threshold would occur when $n(t) = K_1 A$. Since $n(t)$ is random, we need talk about probabilities. In order to avoid this significant degradation, one often uses the benchmark that the following inequality holds 99% of the time:

$$n(t) < K_1 A$$

If we assume $n(t)$ is Gaussian, an examination of error function tables would indicate that to achieve this value, the standard deviation must be approximately equal to $K_1 A / 2.35$. The standard deviation of the bandlimited white noise is the square root of the power (or variance). Thus,

$$\sigma^2 = 2N_0f_m$$

Putting this all together yields

$$K_1A > 2.35\sqrt{2N_0f_m}$$

$$\frac{K_1^2A^2}{N_0f_m} > 11.045 = 10.4\,\text{dB}$$

Therefore, when the signal to noise ratio gets below about 10 dB, there is significant performance degradation. This degradation is often called *noise capture*. The result does not imply that above this threshold value one experiences good reception. In reality, the signal to noise ratio would have to be well above this value to get acceptable performance.

Square Law Detector

The square law detector is illustrated in Fig. 4.90. We have again included a pre-detection bandpass filter in order to separate the desired signal from other

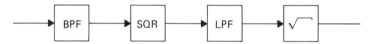

FIGURE 4.90   The square law detector.

multiplexed waveforms, and to reduce the noise entering the squarer. The output of the squarer is given by

$$r^2(t) = [K_1[A + s(t)] + x(t)]^2 \cos^2 2\pi f_c t + y^2(t) \sin^2 2\pi f_c t$$

where we again define the constant K_1 as

$$K_1 \overset{\Delta}{=} K\sqrt{\frac{P_s}{P_s + A^2}}$$

Both the square of the sine and the square of the cosine contain a DC term and a term at the double frequency. The output of the lowpass filter is then given by

$$[K_1[A + s(t)] + x(t)]^2 + y^2(t)$$

The final output is the square root of this quantity, or

$$\sqrt{[K_1[A + s(t)] + x(t)]^2 + y^2(t)}$$

This is seen to be exactly the same result as that obtained in Eq. (4.87b) for the envelope detector. Thus, the performance of the square law detector is identical to that of the envelope detector.

If we now assume a large carrier to signal ratio (i.e., A much larger than $s(t)$) and eliminate the square root operation, the signal at the output is approximately equal to $2K_1^2As(t)$. The noise at the output consists of two components. The first is $2K_1Ax(t)$ while the second results from passing $y^2(t)$ through the lowpass filter. If the power spectral density of $y^2(t)$ were known, it would be a simple matter to find the power at the output of the lowpass filter. It is, in general, quite difficult to find the power spectral density of the square of a random process. The interested student is referred to the references for details.

Rectifier Detector

The rectifier detector is shown in Fig. 4.91. The input to the rectifier is given by

$$B(t) \cos\left(2\pi f_c t + \theta(t)\right)$$

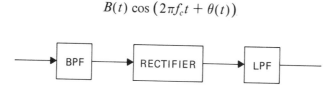

FIGURE 4.91   The rectifier detector.

as in Eq. (4.87a). When this quantity is rectified, the result is

$$|B(t)| \ |\cos\left(2\pi f_c t + \theta(t)\right)|$$

The rectified cosine is not strictly periodic because of the presence of the time varying phase. Nonetheless, it has a DC value of $\frac{1}{2}$ and the next component is centered at a frequency of $2f_c$ (or at f_c for the half wave rectifier). Thus, the lowpass filter output is given by

$$\tfrac{1}{2}|B(t)| = \tfrac{1}{2}B(t)$$

The absolute value sign has been removed in the above expression since $B(t)$ is non-negative. This result is again the same as that of the envelope detector except for the factor of $\frac{1}{2}$. Since this factor affects both the signal and noise, the performance of the rectifier detector is identical to that of the envelope detector.

PROBLEMS

4.1.   Give two reasons for using modulation rather than simply sending an information signal through the air.

4.2.   What is the reason for using AM transmitted carrier instead of AM suppressed carrier? Why might suppressed carrier be preferred over transmitted carrier?

4.3.   It is decided to amplitude-modulate a carrier of frequency 10^6 Hz with the signal shown below.

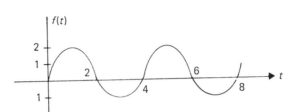

(a) If this modulation is performed as double sideband suppressed carrier (DSBSC), sketch the modulated waveform.

(b) The modulated wave of part (a) is the input to an envelope detector. Sketch the output of the envelope detector.

(c) If a carrier term is now added to the modulated wave to form DSBTC, what is the minimum amplitude of the carrier term such that an envelope detector could be used to recover $r(t)$ from the modulated signal?

(d) For the modulated signal of part (c), sketch the output of an envelope detector.

(e) Draw a block diagram of a synchronous detector that could be used to recover $r(t)$ from the modulated wave of part (a).

(f) Sketch the output of this synchronous detector if the input is the modulated waveform of part (c).

4.4. You are given the voltage signals, $r(t)$ and $\cos 2\pi f_c t$, and you wish to produce the AM wave, $r(t) \cos 2\pi f_c t$. Discuss two practical methods of generating this AM wave. Block diagrams would be useful.

4.5. The signal, $r(t) = (2 \sin t)/t$ is used to amplitude-modulate a carrier of frequency $f_c = 100/2\pi$ Hz. The modulated signal is sent through the air. At the same time, a strong signal, $r_s(t) = (\sin 99.5t)/t$ is being fed into a nearby antenna (without modulation). This adds to the desired modulated signal and both are received by the receiver. The receiver contains a synchronous demodulator which is perfectly adjusted. What is the output of the synchronous demodulator? (Assume any amplitudes you wish.)

4.6. Prove that the half wave rectifier circuit is a non-linear system.

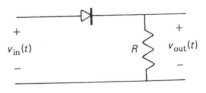

4.7. The waveform, $v_{in}(t)$, shown below is the input to an envelope detector. Sketch the output waveform.

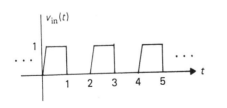
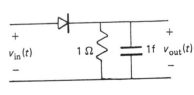

4.8. You are given the system shown below. An AMTC waveform is at the input. $h(t)$ is the periodic function shown, and $R(f)$ is as sketched. $R(f)$, $G(f)$, $S(f)$, and $Y(f)$ are the Fourier Transforms of $r(t)$, $g(t)$; $s(t)$, and $y(t)$ respectively. Assume $f_c \gg f_m$. Further assume that the presence of the impulse in $R(f)$ assures that $r(t) \geq 0$.

(a) *Sketch* $|G(f)|$.
(b) *Sketch* $|S(f)|$.
(c) *Sketch* $|Y(f)|$.

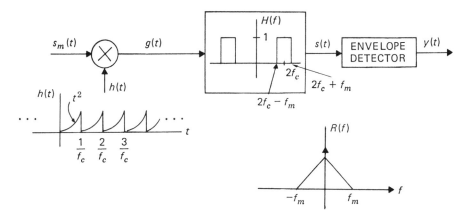

4.9. Given that the input to an envelope detector is

$$r(t) \cos 2\pi f_c t$$

where $r(t)$ is always greater than zero.

(a) What is the output of the envelope detector?
(b) What is the input average power in terms of the average power of $r(t)$?
(c) What is the average power at the output?
(d) Describe any apparent discrepancies.
(Hint: If you have trouble, do this problem for the special case of $r(t) = A$, and refer back to Problem 2.27.)

4.10. Given an information signal, $r(t)$, with $R(f)$ complex,

$$R(f) = A(f)e^{j\theta(f)}$$

Find the Fourier Transform of

$$r(t) \cos 2\pi f_c t$$

Also find the Fourier Transform of

$$r(t) \cos \left(2\pi f_c t + \tfrac{1}{4}\pi\right)$$

Note that the text has sketched only the magnitudes of these transforms.

4.11. You are given four different signals, each with $f_m = 10$ kHz. You wish to amplitude modulate and then frequency multiplex the four signals. The FCC has allocated the band between 100 kHz and 200 kHz to you. Sketch a block diagram of one possible transmitter scheme.

4.12. Figure 4.21 illustrated a non-linear device. Indicate the form of the output if the input is $x(t) = \cos 2\pi f_c t$. Try to use this result to incorporate the non-linear device in a frequency multiplexing system (similar to that in Problem 4.11). The non-linear device should enable you to build the system using only one carrier oscillator.

4.13. An AMTC signal, $s_m(t)$,

$$s_m(t) = [A + s(t)] \cos(2\pi f_c t + \theta)$$

is applied to both systems shown below. The maximum frequency of $s(t)$ is f_m, which is also the cutoff frequency of the lowpass filters. Show that the two systems will yield the same output. Also comment upon whether the two lowpass filters of system (b) can be replaced by a single filter following the square root operation or preceding the square root operation.

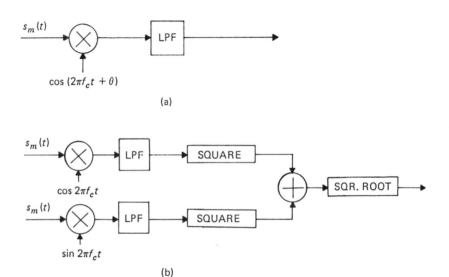

(a)

(b)

4.14. A synchronous demodulator is used for detection of an AMSC double sideband transmission. In designing the detector, the frequency is matched perfectly, but the phase differs from that of the received carrier by θ, as shown in the figure below.

    The phase difference is random and Gaussian distributed with a mean of zero and a variance of σ^2. If $\theta = 0$, the output is $\frac{1}{2}r(t)$.

    What must the variance of the phase error be so that the output amplitude is at least 50% of this "optimum" value 99% of the time?

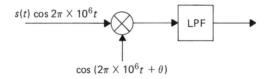

4.15. Consider the carrier injection system depicted in Fig. 4.37. The information signal is given by

$$r(t) = \cos 4\pi t + \cos 20\pi t + \cos 40\pi t + \cos 60\pi t$$

Suppose the bandpass filter in the receiver is tuned to the carrier frequency with a bandwidth that passes 3 Hz on either side of the carrier frequency. Thus, the reconstructed carrier is perturbed by the 2 Hz signal.

Find the output of the system. Now define the error as the difference between this output and an appropriately scaled version of $r(t)$. Find the percentage error (i.e., the ratio of the error power to the signal power).

4.16. Show that the system of Fig. 4.43 can be used as a demodulator. What is the maximum frequency that the lowpass filter should pass (f_m)?

Will this system still work if $T = 2/f_c$? What if $T = 3/f_c$?

4.17. Design an envelope detector to demodulate the following AM signal.

$$\left(5 + 10\frac{\sin 1{,}000t}{t}\right)\cos 2\pi \times 10^6 t$$

4.18. Investigate what type of non-linear device can be used as a demodulator for AMTC. That is, assume the system is as shown below.

The non-linearity can be expressed by

$$y(t) = \sum_{n=0}^{\infty} c_n x^n(t)$$

Specifically, what powers can be present for this system to work as a demodulator?

4.19. Calculate the envelope of the following waveform.

$$s(t) = \cos 10t + 17 \cos 30t \cos 1{,}000t$$

4.20. Find the Fourier Transform of the following waveform in terms of the transform of $r(t)$.

$$s(t) \cos (2\pi f_c t + \theta)$$

Use this result to find the Fourier Transform of

$$s(t) \sin 2\pi f_c t$$

4.21. A VSB signal is formed by amplitude modulating a carrier with the $s(t)$ shown below. A pure carrier term of amplitude 2 is added to the result. The double sideband signal is then filtered with the system function shown. Assume that f_m is large enough to pass all significant harmonics of the upper sideband.

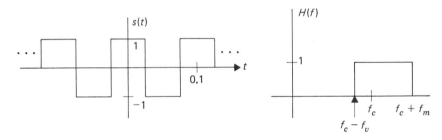

An envelope detector is used for demodulation. Find the minimum value of f_v such that the error is less than 10%. Define the error as the difference between the demodulated signal and $s(t)$. The percent error is the ratio of error power to signal power.

4.22. The system illustrated below is used in some simple scrambling operations. It essentially reverses frequencies. (That is, DC is switched to the highest frequency while the highest is switched to DC. Frequencies near f_m are flipped to be near 0.)

(a) Sketch the Fourier Transform of the output signal, $Y(f)$.
(b) Find the output time signal if

$$r(t) = 5 \cos 100t + 10 \cos 200t + 3 \cos 1{,}000t$$

(c) Design a system which would recover the original $r(t)$ from the scrambled output of the system.

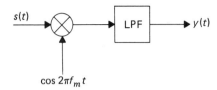

4.23. Prove that if $s(t)$ is even, then the Hilbert Transform of $s(t)$ is odd. Also prove the reverse. That is, if $s(t)$ is odd, the Hilbert Transform is even.

4.24. Starting with the transform of a lower sideband SSB wave,

$$s_{\text{lsb}}(f) = \tfrac{1}{2} H(f)[S(f - f_c) + S(f + f_c)]$$

where $H(f)$ is shown below, prove that a synchronous demodulator can be used to recover $r(t)$ from $s_{\text{lsb}}(t)$.

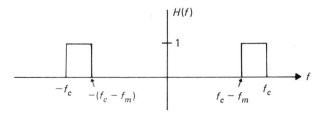

4.25. Discuss the trade-off decisions required if it were necessary to increase the vertical resolution of commercial TV by 10%.

4.26. Refer to the block diagram of Fig. 4.78. Pinpoint the problem block(s) which could result in the following symptoms:

(a) Horizontal bright line in the middle of the TV screen. Remainder of screen is dark. Sound OK.
(b) Picture is rolling vertically.
(c) Vertical line in the center of the screen. Remainder of screen is dark.
(d) No light on screen. Sound OK.
(e) Light on screen filling entire area but no sound or picture.

4.27. An information signal $s(t) = 5 \cos 1000t$ is transmitted using SSBSC and demodulated using a synchronous demodulator. Noise with power spectral density $G_n(f) = 10^{-4}$ is added to the signal during transmission. Find the S/N at the output of the receiver and express this in decibels.

4.28. A signal $s(t) = 20 \cos 1,000 \pi t + 10 \cos 2,000 \pi t$ is transmitted via SSB. Noise of power spectral density $G_n(f) = 10^{-3}$ is added during transmission.

(a) Sketch a block diagram of the required receiver.
(b) Find the S/N at the output of the receiver.
(c) If a band pass filter is added to the output of the receiver with $H(f) = 1$ for $400 < |f| < 1,100$, find the signal to noise improvement of this filter.

4.29. A signal $s(t) = 4 \sin (200t + 10°)$ is transmitted via DSBTC, DSBSC, and SSB. Noise of power spectral density $G_n(f) = 10^{-2}$ is added during transmission. Find the S/N at the output of the appropriate receiver for each case (assume that an envelope detector is used for the DSBTC case).

4.30. A signal, $s(t)$, is transmitted using single sideband AM. The power spectral density of $s(t)$ is shown below.

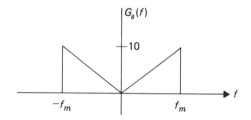

White noise of spectral density $N_0/2$ is added during transmission. Find the signal to noise ratio at the output of a synchronous demodulator.

4.31. A double sideband suppressed carrier waveform with $f_m = 5$ kHz and $f_c = 1$ MHz is transmitted. Non-white noise with power spectral density as shown below adds to the signal prior to detection with a synchronous demodulator.

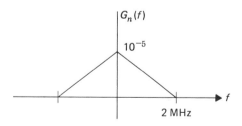

Find the signal to noise ratio at the output of the detector assuming that the power of the signal, $s(t)$, is 1 watt.

4.32. Derive an expression for the noise at the output of the second system shown in Problem 4.13. Compare this to the noise at the output of the first system (the usual synchronous demodulator). Comment on the advantages and disadvantages of each system.

Chapter 5

Angle Modulation

As in the type of modulation previously discussed, we start ·with an unmodulated carrier wave,

$$s_c(t) = A \cos(2\pi f_c t + \theta) \tag{5.1}$$

If f_c is varied in accordance with the information we wish to transmit, the carrier is said to be *frequency modulated*. However, when f_c is varied with time, $s_c(t)$ is no longer a sinusoid, and the definition of frequency which we have used all of our lives must be modified accordingly.

Let us examine three time functions.

$$s_1(t) = A \cos 6\pi t \tag{5.2a}$$

$$s_2(t) = A \cos(6\pi t + 5) \tag{5.2b}$$

$$s_3(t) = A \cos(2\pi t e^{-t}) \tag{5.2c}$$

The frequencies of $s_1(t)$ and $s_2(t)$ are clearly 3Hz. The frequency of $s_3(t)$ is, at present, undefined. Our old definition of frequency does not apply to this type of waveform. Some thought would indicate that the usual intuitive definition of frequency can be stated as follows: "If $s(t)$ can be put into the form $A \cos(2\pi f t + \theta)$, where A, f, and θ are constants, then the frequency is *defined* as f Hz." $s_3(t)$ cannot be put into this form.

As is usually done to get out of a predicament such as this, a new quantity will be defined (recall envelope discussion). This new quantity is *instantaneous frequency*. We shall define instantaneous frequency in such a way that, in those cases where the old intuitive definition of frequency can be applied, the two definitions will yield the same value.

Given $s(t) = A \cos \theta(t)$ where A is a constant, the instantaneous frequency of $s(t)$ is defined as the time derivative of $\theta(t)$. Note that at any time, t_0, $\theta(t_0)$ is the argument of the cosine function, and therefore can be thought of as its phase.

Before continuing, we note that this definition is not as restrictive as it might appear. Any time function, $s(t)$, can be put into the form, $s(t) = A \cos \theta(t)$. One does this by setting $\theta(t) = \cos^{-1}[s(t)/A]$, where A is chosen large enough so that the inverse cosine is defined. That is,

$$\left| \frac{s(t)}{A} \right| \leq 1 \qquad \text{for all } t$$

The symbol $f_i(t)$ will be used for the instantaneous frequency.

$$2\pi f_i(t) \overset{\Delta}{=} \frac{d\theta}{dt} \text{ (rps)} \tag{5.3}$$

For example, for $s_2(t)$ above, $\theta(t) = 6\pi t + 5$ and $(\frac{1}{2\pi})d\theta/dt$ is equal to 3 Hz as we found using our old definition. For $s_3(t)$, $\theta(t) = 2\pi t e^{-t}$, and $f_i(t) = e^{-t} - te^{-t}$. We have no intuitive answer with which to compare this value, so we must accept it as the definition of the frequency of $s_3(t)$.

Illustrative Example 5.1

Find the instantaneous frequency of the following waveform,

$$s(t) = \begin{cases} \cos 2\pi t & t < 1 \\ \cos 4\pi t & 1 \leq t \leq 2 \\ \cos 6\pi t & 2 < t \end{cases}$$

Solution

This wave is of the form

$$s(t) = \cos[tg(t)2\pi] \tag{5.4}$$

where $g(t)$ is as sketched in Fig. 5.1. Therefore,

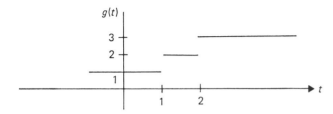

FIGURE 5.1 $g(t)$ for Illustrative Example 5.1.

$$f_i(t) = \frac{d}{dt}[tg(t)] = g(t) + t\frac{dg}{dt}$$

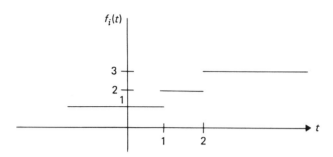

FIGURE 5.2 $f_i(t)$ for Illustrative Example 5.1.

This instantaneous frequency is sketched in Fig. 5.2. Note that the frequency is undefined at $t = 1$ and $t = 2$.

Illustrative Example 5.2

Find the instantaneous frequency of the following function:

$$s(t) = 10 \cos 2\pi[1{,}000t + \sin 10\pi t] \tag{5.5}$$

Solution

For this waveform,

$$\frac{1}{2\pi} \theta(t) = 1{,}000t + \sin 10\pi t$$

and

$$f_i(t) = \frac{1}{2\pi} \frac{d\theta}{dt} = 1{,}000 + 10\pi \cos 10\pi t$$

This instantaneous frequency is sketched in Fig. 5.3.

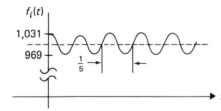

FIGURE 5.3 $f_i(t)$ for Illustrative Example 5.2.

In frequency modulation, we modulate (vary) $f_i(t)$ with the signal, $s(t)$, just as in amplitude modulation we modulated A with the signal. Since, intuitively, we wish to shift the frequencies of $s(t)$ up to the vicinity of f_c for efficient transmission, we will immediately add the constant, f_c, to $f_i(t)$. This is done instead of first trying $f_i(t)$ of the form of $s(t)$. What we are saying here is that, in FM, there is no analogy to the suppressed carrier case in AM. Therefore,

$$f_i(t) = f_c + k_f s(t) \qquad (5.6)$$

where f_c and k_f are constants. Given this $f_i(t)$, the total transmitted waveform with this instantaneous frequency is given by

$$\lambda_{fm}(t) = A \cos \theta(t) \qquad (5.7)$$

where

$$\theta(t) = 2\pi \int_0^t f_i(\tau)\, d\tau = 2\pi \left[f_c t + k_f \int_0^t s(\tau)\, d\tau \right] \qquad (5.8)$$

We have assumed $\theta(0) = 0$.

Finally, the modulated waveform is given by

$$\lambda_{fm}(t) = A \cos 2\pi \left[f_c t + k_f \int_0^t s(\tau)\, d\tau \right] \qquad (5.9)$$

where the constant of integration has been set equal to zero since it represents an arbitrary time delay which for audio signals is not a form of distortion. Note that $\lambda_{fm}(t)$ is a pure carrier wave if $s(t) = 0$. This would not be true if we hadn't added the constant, f_c, in Eq. (5.6).

Illustrative Example 5.3

Sketch the FM and AMSC modulated waveforms for the information signals of Fig. 5.4.

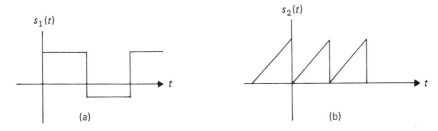

$s_1(t)$

$s_2(t)$

(a)

(b)

FIGURE 5.4   Waveforms for Illustrative Example 5.3.

Solution

The AMSC and FM waveforms corresponding to the information signals shown in Fig. 5.4 are illustrated in Fig. 5.5. You should stare at these results until you are convinced that you have a feel for what is taking place. The concept of a frequency being controlled by the

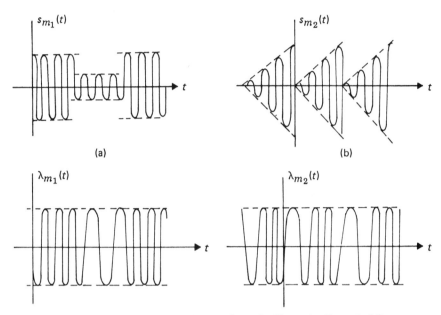

FIGURE 5.5   AMSC and FM waveforms for Illustrative Example 5.3.

amplitude of the information signal is a difficult one to grasp. Note that as $s(t)$ increases, it is the *frequency* of the FM wave which increases, not the amplitude, which remains constant.

When FM was first investigated, it was reasoned that the frequency of $\lambda_{fm}(t)$ varied from $f_c + k_f[\min_t s(t)]$ to $f_c + k_f[\max_t s(t)]$. Therefore, by making k_f arbitrarily small, the frequency of $\lambda_{fm}(t)$ can be kept arbitrarily close to f_c (i.e., small variations). This would result in a great bandwidth savings without resorting to SSB or other techniques. It was quickly realized that this reasoning is completely fallacious. That is, given a signal whose instantaneous frequency varies from f_1 to f_2, we shall show that the Fourier Transform of the signal is certainly *not* confined to the range of frequencies between f_1 and f_2. Clearly, one must be careful not to confuse the concept of instantaneous frequency with that of frequency in the Fourier Transform representation.

Proceeding in the usual manner which we have established for modulation systems' treatment, we must now show that this form of transmitted wave satisfies the three criteria for a modulating scheme. That is, it must be capable of effecting efficient transmission through the air; be in a form that allows multiplexing; and $s(t)$ must be uniquely recoverable from the modulated waveform. The demonstration of these properties is easier said than done. We will be forced to make many approximations. In the end, we will be satisfied with a far less detailed analysis than was possible in the AM wave case.

In order to show that FM is efficient and can be adapted to frequency multiplexing, the Fourier Transform of the FM wave must be found. We shall find things much simpler if we first divide FM into two classes depending upon the size of the constant, k_f. A relatively simple approximation is possible for very small values of k_f.

For reasons that will become obvious, the case where $k_f \max |s(t)|$ is small is known as *narrowband FM*. If $k_f \max |s(t)|$ is not very small, *wideband FM* results.

5.1   NARROWBAND FM

If $k_f \max |s(t)| \ll 1$, the approximate Fourier Transform of $\lambda_{fm}(t)$ can be found. We start with the general form of the FM wave,

$$\lambda_{fm}(t) = A \cos 2\pi \left[f_c t + k_f \int_0^t s(\tau)\, d\tau \right] \qquad (5.10)$$

In order to avoid having to rewrite the integral many times, we define

$$g(t) \overset{\Delta}{=} \int_0^t s(\tau)\, d\tau$$

Therefore,

$$\lambda_{fm}(t) = A \cos 2\pi [f_c t + k_f g(t)] \qquad (5.11)$$

Using the trigonometric cosine sum formula, $\lambda_{fm}(t)$ can be rewritten as follows:

$$\lambda_{fm}(t) = A \cos 2\pi f_c t \cos 2\pi k_f g(t) - A \sin 2\pi f_c t \sin 2\pi k_f g(t) \qquad (5.12)$$

If k_f is small enough such that $k_f g(t)$ is always much less than one, the cosine of $2\pi k_f g(t)$ is approximately equal to unity, and the sine approximately equal to the radian argument (i.e., we are taking the first term in a Taylor series expansion for the sine and cosine functions). With these approximations, $\lambda_{fm}(t)$ becomes

$$\lambda_{fm}(t) = A \cos 2\pi f_c t \cos 2\pi k_f g(t) - A \sin 2\pi f_c t \sin 2\pi k_f g(t)$$
$$\approx A \cos 2\pi f_c t - 2\pi A g(t) k_f \sin 2\pi f_c t \qquad (5.13)$$

We could easily find the transform of this expression if we only knew the transform of $g(t)$. Recall that $g(t)$ is the integral of $s(t)$. By referring to the properties of the Fourier Transform, $G(f)$ can be expressed in terms of $S(f)$.

$$G(f) = \frac{S(f)}{j2\pi f}$$

That is, integration in the time domain corresponds to division by $j2\pi f$ in the frequency domain (compare this to LaPlace Transforms, where integration corresponds to division of the transform by s).

Therefore, if we make the usual assumption that $s(t)$ is bandlimited to frequencies below f_m, $g(t)$ must also be bandlimited to these frequencies. This is true since, for all values of f for which $S(f)$ is equal to zero, $S(f)/(j2\pi f)$ must also be zero. We can now write the transform of $\lambda_{fm}(t)$ in terms of $S(f)$. We start by writing $\sin 2\pi f_c t$ in terms of its complex exponential expansion.

$$\lambda_{fm}(t) = A \cos 2\pi f_c t - 2\pi A k_f g(t) \left[\frac{e^{j2\pi f_c t} - e^{-j2\pi f_c t}}{2j} \right]$$

$$\Lambda_{fm}(f) = \frac{A}{2} \left[\delta(f - f_c) + \delta(f + f_c) \right] - \frac{2\pi A k_f}{2j} \left[G(f - f_c) - G(f + f_c) \right]$$

$$= \frac{A}{2} \left[\delta(f - f_c) + \delta(f + f_c) \right] + \frac{2\pi A k_f}{4\pi} \left[\frac{S(f - f_c)}{f - f_c} - \frac{S(f + f_c)}{f + f_c} \right]$$

$$(5.14)$$

The magnitude of this transform is sketched in Fig. 5.6.

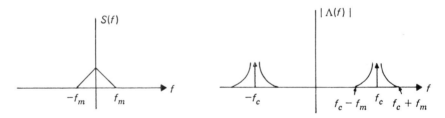

FIGURE 5.6   Magnitude transform of a narrowband FM waveform.

The fact that the transform, $G(f)$, apparently approaches infinity at $f = 0$ is not too troublesome. The important consideration is whether the corresponding time function is bounded. Indeed, the impulse also goes to infinity although the inverse transform is bounded.

In real life situations, $S(f)$ can usually be assumed to go to zero as f approaches zero. For example, standard audio signals are essentially zero below about 15 Hz. Therefore, $G(f)$, which is equal to $S(f)/2\pi jf$ would also be zero for frequencies below 15 Hz.

Figure 5.6 illustrates that, as far as narrowband FM is concerned, the first two objectives of modulating are automatically achieved. That is, the frequencies present can be made as high as is necessary for efficient transmission by raising f_c to any desired value. By using different carrier frequencies separated by at least $2f_m$, many signals can be transmitted simultaneously on the same channel. As for the third requirement, being able to recover $s(t)$ from the modulated waveform, we shall defer that consideration to our later study. We do this since the demodulation process does not depend upon the value of k_f. The same detector will be used for both small and large k_f.

We now digress for a moment to present two vector plots. In later work these will prove useful in comparing AM and FM with respect to noise rejection.

In the AM transmitted carrier case, for a sinusoidal information signal,

$$s_m(t) = A \cos 2\pi f_c t + \cos 2\pi f_m t \cos 2\pi f_c t \qquad (5.15)$$

where we have set $s(t) = \cos 2\pi f_m t$

This can be written as

$$s_m(t) = Re\{e^{j2\pi f_c t}[A + \tfrac{1}{2}e^{j2\pi f_m t} + \tfrac{1}{2}e^{-j2\pi f_m t}]\} \qquad (5.16)$$

where "Re" stands for "real part of."

If we plotted the phasor of this sinusoidal signal (the quantity in the braces in Eq. (5.16)), it would have a large angular term due to the $2\pi f_c t$ phase factor. Recall that $f_c \gg f_m$. We will use a standard technique of assuming that a stroboscopic picture of this phasor is taken every $1/f_c$ seconds. That is, we will only plot the phasor of the inner bracket in Eq. (5.16). This is shown as Fig. 5.7.

Note that the resultant is in phase with the carrier term, A, as expected since there is no phase variation in AM.

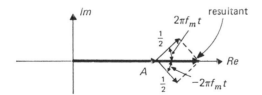

FIGURE 5.7 "Stroboscopic" phasor plot for AM waveform.

For narrowband FM, with the same information signal we have

$$f_i(t) = f_c + k_f \cos 2\pi f_m t \qquad (5.17)$$

and, from Eq. (5.13),

$$\lambda_{fm}(t) = A \cos 2\pi f_c t - \frac{k_f}{f_m} A \sin 2\pi f_m t \sin 2\pi f_c t \qquad (5.18)$$

This can be rewritten as

$$\lambda_{fm}(t) = Re\left\{ e^{j2\pi f_c t}\left[A - \frac{Ak_f}{2jf_m} e^{-j2\pi f_m t} + \frac{Ak_f}{2jf_m} e^{+j2\pi f_m t} \right] \right\} \qquad (5.19)$$

The quantity in the inner brackets of Eq. (5.19) is plotted in Fig. 5.8.

Note that the resultant has almost the same amplitude as A. The reason that the amplitudes are not exactly the same is that Eq. (5.13) is an approximation. If we had

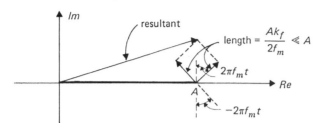

FIGURE 5.8 "Stroboscopic" phasor plot for narrowband FM waveform.

not used an approximation, the amplitude must be constant and only the frequency varies.

We shall defer the discussion of narrowband FM modulators to Section 5.3.

5.2   WIDEBAND FM

If k_f is not small enough to make the approximations for the sine and cosine valid, we have what is known as *wideband FM*. The transmitted signal is of the form found earlier. That is,

$$\lambda_{fm}(t) = A \cos 2\pi[f_c t + k_f g(t)] \qquad (5.20)$$

If $g(t)$ were given, the Fourier Transform of this FM waveform could be evaluated. We would simply calculate $\Lambda_{fm}(f)$ from the defining integral for the transform. (A difficult computational task!) However, we do not wish to restrict ourselves to a particular $s(t)$. We only know that $s(t)$ is bandlimited to frequencies below f_m. Therefore, the only information given about $g(t)$ is that it is bandlimited to frequencies below f_m. With this information only, it is not possible to find the Fourier Transform of the FM wave. That is, in the AM case the transform of the modulated waveform was very simply related to the transform of the information signal. In the FM case, this is not true. Since we cannot find the Fourier Transform of the FM wave in the general case, let's see how much we can possibly say about the modulated signal.

In order to show that FM can be efficiently transmitted and that several channels can be multiplexed, we must only gain some feeling for what range of frequencies the modulated wave occupies (note that we are now discussing wideband FM since all of these questions have already been answered for the narrowband FM signal). That is, if we knew that $\Lambda_{fm}(f)$ occupied the band of frequencies between f_1 and f_2 without knowing the exact form of the Fourier Transform, we would be satisfied. This will, indeed, be the only information we will be able to obtain in general about the transform of the FM wave. The modulation and demodulation processes will be analyzed entirely in the time domain. Let us emphasize this fact. We will only use the Fourier Transform domain to get a rough idea of the range of frequencies occupied by the modulated waveform. This is an extremely significant point.

Even finding this range of frequencies is impossible in general, so we shall begin by restricting ourselves to a specific type of information, or modulating signal. $s(t)$ will be assumed to be a pure sinusoid. Very few people are interested in transmitting a pure sinusoid. It would not constitute a very pleasing piece of music since it would be a flat sounding single tone for all time. We shall, however, make one of our frequent generalizations after analysis of this special case is completed.

Assuming that $s(t) = a \cos 2\pi f_m t$, where "a" is a constant amplitude,

$$f_i(t) = f_c + k_f s(t)$$
$$= f_c + ak_f \cos 2\pi f_m t \qquad (5.21)$$

and

$$\lambda_{fm}(t) = A \cos \left[2\pi f_c t + \frac{ak_f}{f_m} \sin 2\pi f_m t \right] \qquad (5.22)$$

Rather than be forced to use trigonometric sum-difference relationships at this time, we shall use the complex exponential notation, and later take the real part of the result.

$$\lambda_{fm}(t) = Re \left\{ A \exp \left(j2\pi f_c t + \frac{jak_f}{f_m} \sin 2\pi f_m t \right) \right\} \qquad (5.23)$$

In order to find the Fourier Transform of this, we recognize that the second part of the exponent,

$$\frac{ak_f}{f_m} \sin 2\pi f_m t$$

is a periodic time function. Therefore,

$$\exp \left(\frac{jak_f}{f_m} \sin 2\pi f_m t \right)$$

is also a periodic function. That is, the only place where "t" appears is in the term $\sin 2\pi f_m t$, which is itself periodic. We shall let

$$\beta \triangleq \frac{ak_f}{f_m} \qquad (5.24)$$

in order to avoid having to write this term many times. Therefore,

$$e^{j\beta \sin 2\pi f_m t}$$

is periodic with fundamental frequency, f_m. It can be expanded in a Fourier Series to yield

$$e^{j\beta \sin 2\pi f_m t} = \sum_{n=-\infty}^{\infty} c_n e^{jn2\pi f_m t} \qquad (5.25)$$

where the Fourier coefficients are given by

$$c_n = \frac{1}{T} \int_{-T/2}^{T/2} e^{j\beta \sin 2\pi f_m t} e^{-jn2\pi f_m t} \, dt \tag{5.26}$$

and

$$T = \frac{1}{f_m}$$

This integral cannot be evaluated in closed form. It does however converge to some real value (*see* Problem 5.3). Thus the fact that it cannot be expressed in closed form should not bother anybody. One can always approximate the value of the integral to any desired degree of accuracy using numerical techniques and a digital computer. As a last resort, one can build a trough in the shape of the real part of the integral, fill it with water, and see how much it holds. This situation is not basically different from the case of the integral of cos t, which also cannot be expressed in closed form. This latter case arises often enough that it has been tabulated under the name "sin t."

The expression in Eq. (5.26) turns out to be a function of n and β (*see* Problem 5.4). That is, given n and β, one can find a number for the integral evaluation. These numbers are given the name, *Bessel function of the first kind*, and are tabulated under the symbol, $J_n(\beta)$. Given any value of n and β, we simply look up the corresponding value of $J_n(\beta)$.

Bessel Functions

In order to appreciate the properties of the Fourier Transform of the wideband FM waveform, it is necessary to gain some familiarity with Bessel functions.

The Bessel function of the first kind is generated as solutions of the following differential equation:

$$x^2 \frac{d^2 y}{dx^2} + x \frac{dy}{dx} + (x^2 - n^2) y(x) = 0 \tag{5.27}$$

Although the Bessel function is defined for all values of n, we will only be concerned with the positive and negative real integers. It can be shown that, for integer values of n,

$$J_{-n}(x) = (-1)^n J_n(x) \tag{5.28}$$

Figure 5.9 shows $J_0(x)$, $J_1(x)$, and $J_2(x)$ as a function of x. Note that for very small x, $J_0(x)$ approaches unity while $J_1(x)$ and $J_2(x)$ approach zero.

It will be necessary for us to observe the behavior of the Bessel function as n gets large. Figure 5.10 is a plot of $J_n(10)$ as a function of n. The function of Fig. 5.10 appears to be in underdamped oscillation for negative n, but again note that we are only concerned with integer values of n. For these values, the symmetry of Eq. (5.28) holds, and we can focus attention upon the positive n axis. The important observation

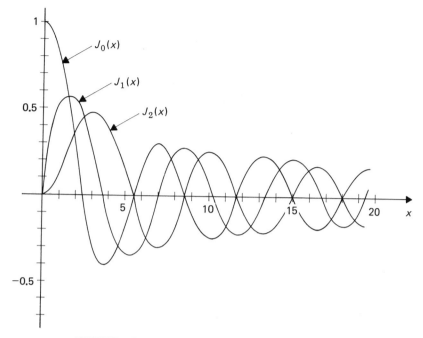

FIGURE 5.9 $J_0(x)$, $J_1(x)$, and $J_2(x)$ as a function of x.

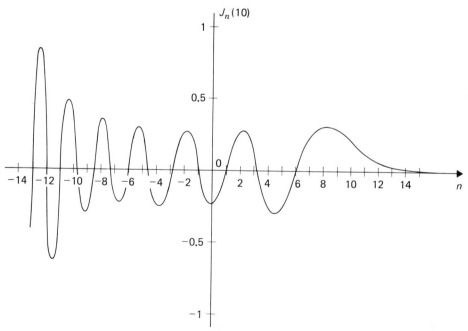

FIGURE 5.10 $J_n(10)$ as a function of n.

to make is that for n greater than about 9, the Bessel function asymptotically approaches zero. In fact, for fixed n and large x, the Bessel function can be approximated by

$$J_n(x) \approx \frac{(\frac{1}{2}x)^n}{\Gamma(n+1)}$$

where $\Gamma(n+1)$ is the *Gamma function*. Gamma functions approach infinity for arguments greater than 2. For example, the value of the Gamma function for arguments of 2, 3, 4, 5, and 6 is 1, 2, 6, 24, and 120 respectively. Since the Gamma function is in the denominator, it can be seen that the Bessel function gets small rapidly with increasing n. This will become an important observation in finding the bandwidth of the FM waveform.

Returning now to Eq. (5.26), we see that this can be rewritten, using the Bessel function, as

$$c_n = J_n(\beta)$$

Equation (5.23) becomes

$$\lambda_{fm}(t) = Re\left\{e^{j2\pi f_c t}\sum_{n=-\infty}^{\infty} J_n(\beta)e^{jn2\pi f_m t}\right\} \tag{5.29}$$

Since $e^{j2\pi f_c t}$ is not a function of n, we can bring it under the summation sign to get

$$\lambda_{fm}(t) = Re\left\{A\sum_{n=-\infty}^{\infty} j_n(\beta)e^{jt2\pi(nf_m+f_c)}\right\}$$

Finally, taking the real part of the expression, we obtain

$$\lambda_{fm}(t) = A\sum_{n=-\infty}^{\infty} J_n(\beta)\cos 2\pi(nf_m+f_c)t \tag{5.30}$$

The FM waveform, in this case, is simply a sum of cosines. Its transform will be a train of impulses in the frequency domain. Taking the transform of the FM wave term by term, we get $\Lambda_{fm}(f)$ as sketched in Fig. 5.11.

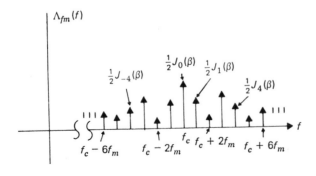

FIGURE 5.11   Transform of wideband FM wave with sinusoidal modulating signal (positive f shown).

It appears that, unless $J_n(\beta)$ is equal to zero for "n" above a certain value, the Fourier Transform of the FM wave has frequency components spaced by f_m and extending over all frequencies. Unfortunately, the Bessel functions are not equal to zero for large values of n. That is, one FM signal seems to occupy the entire band of frequencies. This is certainly disastrous if we wish to frequency multiplex.

Our previous discussion of Bessel functions shows that, for fixed β, the functions $J_n(\beta)$ eventually approach zero as n increases. That is, if we decide to approximate as zero all $J_n(\beta)$ that are less than some specified value, the FM wave will be essentially bandlimited. We can find some N for any given β such that, for $n > N$, $J_n(\beta)$ is less than any specified amount.

Note that the Bessel functions do not necessarily decrease monotonically. In fact, by choosing β to be any of the zeros of $J_0(x)$ in Fig. 5.9, $J_0(\beta) = 0$ and the carrier term is eliminated from the FM waveform. In the AM case, such an accomplishment would increase power efficiency. In FM, elimination of the carrier does not buy anything since the total power remains constant.

In determining the bandwidth of the FM wave, the question becomes one of how many significant sidebands exist in Fig. 5.11. We can see from Fig. 5.9 that if β is below about 0.5, $J_2(\beta)$ is below 0.03. Thus, the second harmonic at a frequency of $f_c + 2f_m$ is about 3% of the carrier. The first harmonic is about 24% of the carrier amplitude at $\beta = 0.5$. Examination of a table of Bessel functions would show that all higher harmonics are considerably below these values and, in fact, decrease approximately inversely to the Gamma function, as discussed earlier. Thus, for small values of β, the only significant sidebands are the ones at $f_c + f_m$ in frequency. (If 3% is not small enough to be neglected, the interpretation of "small β" must be narrowed to smaller values than 0.5.) For large values of β, we can use the result inferred from Fig. 5.10. Thus, for example, for $\beta = 10$, the significant terms are $J_n(10)$ for n between about -10 and $+10$. Thus, the significant sidebands are those between $f_c - 10f_m$ and $f_c + 10f_m$.

The preceding observations indicate that, for small β, the bandwidth of the FM wave is approximately $2f_m$, while for large β the bandwidth is approximately $2\beta f_m$. Note that for small β, the result corresponds to that of narrowband FM as expected, since small $k_f \max|s(t)|$ implies small β.

Behavior of the Bessel function between the two extremes requires detailed analysis and approximations. One of the first people to do this analysis was John Carson, who in 1922 proposed the following rule of thumb which has been widely adopted. The bandwidth of the FM wave is approximately given by

$$BW \approx 2(\beta f_m + f_m) \tag{5.31}$$

We see that for large β, this is approximately equal to $2\beta f_m$, and for small β, $2f_m$ as desired.

Recalling that $\beta = ak_f/f_m$, we see that Eq. (5.31) becomes

$$BW \approx 2(ak_f + f_m) \tag{5.32}$$

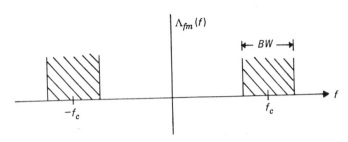

FIGURE 5.12 Band of frequencies occupied by FM wave.

We have accomplished our objective for the sinusoidal information signal case. That is, given a, k_f, and f_m, we can find the approximate bandwidth, and we therefore know that the Fourier Transform occupies that portion of the frequency axis which is shaded in Fig. 5.12.

Illustrative Example 5.4

Find the approximate band of frequencies occupied by an FM wave with carrier frequency 5,000 Hz, $k_f = 10$ Hz/v, and
(a) $s(t) = 10 \cos 10\pi t$ volts
(b) $s(t) = 5 \cos 20\pi t$ volts
(c) $s(t) = 100 \cos 2{,}000\pi t$ volts

Solution

The bandwidth in each case is approximately given by
(a) $\text{BW} \approx 2(ak_f + f_m) = 2[10(10) + 5] = 210$ Hz
(b) $\text{BW} \approx 2(ak_f + f_m) = 2[5(10) + 10] = 120$ Hz
(c) $\text{BW} \approx 2(ak_f + f_m) = 2[100(10) + 1{,}000] = 4{,}000$ Hz

The band of frequencies occupied is therefore
(a) 4,895 Hz to 5,105 Hz
(b) 4,940 Hz to 5,060 Hz
(c) 3,000 Hz to 7,000 Hz

Unfortunately, Eq. (5.32) does not tell us too much about the general case when the modulating signal is not a pure sine wave. You might be tempted to say that, since any $s(t)$ can be approximated by a sum of sinusoids, we need only sum the individual sinusoidal transforms. This is not true, since

$$\cos 2\pi[f_c t + k_f(g_1(t) + g_2(t))]$$

is *not* equal to

$$\cos 2\pi[f_c t + k_f g_1(t)] + \cos 2\pi[f_c t + k_f g_2(t)]$$

That is, the FM wave due to the sum of two information signals is not the sum of the two corresponding FM waveforms. People sometimes call this property "non-linear modulation." (Recall the discussion concerning AM and linear time varying systems.)

As promised, it is now time for a rather bold generalization. What we desire is an equation similar to Eq. (5.32) that applies to a general modulating information signal. We therefore ask what each term in Eq. (5.32) represents in terms of the information signal. f_m is clearly the highest frequency component present. In the single sinusoid case, it is the only frequency present. Indeed, one would expect the bandwidth of the modulated waveform to be dependent upon the highest frequency component of the information signal, $s(t)$. The term, ak_f, represents the maximum amount that $f_i(t)$ deviates from f_c. That is, if we sketch $f_i(t)$, we can define a maximum frequency deviation as the maximum value by which $f_i(t)$ deviates from its "*dc*" value of f_c. (*See* Fig. 5.13.)

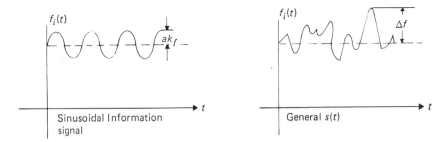

FIGURE 5.13   Definition of maximum frequency deviation.

Recall that for the sinusoidal $s(t)$ case, we found

$$f_i(t) = f_c + k_f s(t) = f_c + ak_f \cos 2\pi f_m t$$

The maximum frequency deviation is given the symbol Δf and is defined as

$$\Delta f \stackrel{\Delta}{=} \max_t [f_i(t) - f_c] = \max_t [f_c + k_f s(t) - f_c] = \max_t [k_f s(t)] \tag{5.33}$$

which is clearly defined for any general signal, $s(t)$.

We can now make an educated guess that Eq. (5.32) can be modified to read

$$\text{BW} \approx 2(\Delta f + f_m) \tag{5.34}$$

for any general $s(t)$ information signal.

One can give an extremely intuitive interpretation to Eq. (5.34). If Δf is much larger than f_m (wideband FM), the frequency of the carrier is varying by a large amount, but at a relatively slow rate. That is, the instantaneous frequency of the carrier is going from $f_c - \Delta f$ to $f_c + \Delta f$ very slowly. It therefore approximates a pure sine wave over any reasonable length of time. We can almost think of it as a sum of many sine waves with frequencies between these two limits. The transform is therefore approximately a superposition of the transform of each of these many sinusoids, all lying between the frequency limits. It is therefore reasonable to assume that its bandwidth is approximately the width of this frequency interval, or $2\Delta f$. On the other hand, for very small Δf, we have a carrier which is varying over a very small range of frequencies, but doing this relatively rapidly. One can almost think of this as being caused by two oscillators, one at a frequency of $f_c - \Delta f$, and the other at $f_c + \Delta f$, each being on for half of the total time, and alternating with the other. Thus each can be considered as being multiplied by a gating function at a frequency of Δf. (*See* Fig. 5.14.)

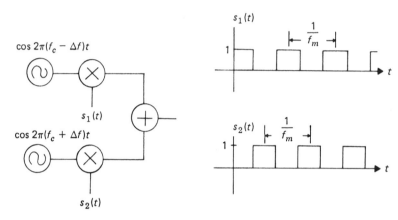

FIGURE 5.14   Visualization of narrowband FM.

From an analysis similar to that of the gated modulator studied in Chapter 4, we would find that the band of frequencies occupied is approximately that from $f_c - \Delta f - f_m$ to $f_c + \Delta f + f_m$. For small Δf, this gives a bandwidth of $2f_m$, thus verifying Eq. (5.34). The details of this derivation are left to Problem 5.10.

The previous analysis was highly intuitive. It is doubtful that one could suggest such plausibility arguments if the desired result were not known in advance.

It should be clear that the bandwidth of an FM wave increases with increasing k_f. Therefore at this point there appears to be no reason to use other than very small values of k_f (narrowband FM) and thus require the minimum bandwidth. We state without elaboration at this time that the advantages of FM over AM lie in its ability to reject unwanted noise. This ability to reject noise increases with increasing k_f, and therefore, with increasing bandwidth.

Illustrative Example 5.5

A 10-MHz carrier is frequency modulated by a sinusoidal signal of 5-kHz frequency such that the maximum frequency deviation of the FM wave is 500 kHz. Find the approximate band of frequencies occupied by the FM waveform.

Solution

We must first find the approximate bandwidth. This is given by

$$BW \approx 2(\Delta f + f_m)$$

For this case, we are told that Δf is 500 kHz, and the maximum frequency component of the information is 5 kHz. In this case this is the *only* frequency component of the information, and therefore, certainly is the maximum. Therefore the bandwidth is approximately

$$BW \approx 2(500 \text{ kHz} + 5 \text{ kHz}) = 1{,}010 \text{ kHz}. \tag{5.35}$$

Thus the band of frequencies occupied is centered around the carrier frequency, and ranges from 9,495 kHz to 10,505 kHz. The FM signal of this example represents a wideband FM signal. If it were narrowband, the bandwidth would only be 10 kHz as compared with 1,010 kHz found above.

Illustrative Example 5.6

A 100-MHz carrier is frequency modulated by a sinusoidal signal of one volt amplitude. k_f is set at 100 Hz/volt. Find the approximate bandwidth of the FM waveform if the modulating signal has a frequency of 10 kHz.

Solution

Again we use the only formula which is available. That is,

$$BW \approx 2(\Delta f + f_m)$$

Since the information, $s(t)$, has unit amplitude, the maximum frequency deviation, Δf, is given by k_f, or 100 Hz. f_m is simply equal to 10 kHz. Therefore,

$$BW \approx 2(100 + 10 \text{ kHz}) = 20{,}200 \text{ Hz} \tag{5.36}$$

Since f_m is much greater than Δf in this case, the above signal would be a narrowband signal. The bandwidth necessary to transmit the same information waveform using double sideband AM techniques would be 20 kHz. This is almost the same amount required in this example.

Illustrative Example 5.7

An FM waveform is described by

$$\lambda_{fm}(t) = 10 \cos[2 \times 10^7 \pi t + 20 \cos 1{,}000 \pi t] \tag{5.37}$$

Find the approximate bandwidth of this waveform.

Solution

f_m is equal to 500 Hz. In order to compute Δf, we first find the instantaneous frequency, $f_i(t)$.

$$f_i(t) = \frac{d}{dt} (2 \times 10^7 \pi t + 20 \cos 1{,}000 \pi t)/2\pi$$

$$= 1 \times 10^7 - 10{,}000 \sin 1{,}000 \pi t \tag{5.38}$$

The maximum frequency deviation is the maximum value of $10{,}000 \sin 1{,}000 \pi t$, which is simply

$$\Delta f = 10{,}000 \text{ Hz}$$

The approximate bandwidth is therefore given by

$$BW \approx 2(10,000 + 500) = 21,000 \text{ Hz} \qquad (5.39)$$

This is clearly a wideband waveform since Δf is much greater than f_m.

5.3   MODULATORS

We have shown that in the case of narrowband or wideband FM waveforms the transform of the modulated waveform is bandlimited to some range of frequencies around f_c, the carrier frequency. The first two criteria of a useful modulation system are therefore satisfied. We can transmit efficiently by choosing f_c high enough. We can frequency multiplex many separate signals by making sure that the adjacent carrier frequencies are separated by a sufficient amount such that the transforms of the FM waveforms do not overlap in frequency.

We need now only show a few ways to generate an FM wave and to recover the information signal from the received waveform.

In the narrowband case, a simple trigonometric identity allows us to rewrite the FM waveform in the following form (we repeat Eq. (5.13)),

$$\lambda_{fm}(t) = A \cos 2\pi f_c t - 2\pi A g(t) k_f \sin 2\pi f_c t \qquad (5.40)$$

where $g(t)$ was the integral of the information signal. This immediately leads to the block diagram of the narrowband FM modulator shown in Fig. 5.15.

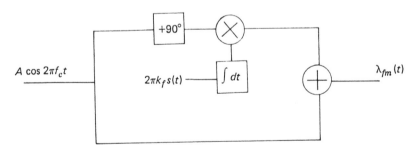

FIGURE 5.15   Narrowband FM modulator.

We now make the general observation that if the narrowband FM output of this modulator is put through a frequency multiplier, it can be made into a wideband waveform. A frequency multiplier is a device which multiplies all frequency components of its input by a constant.

We start with

$$\lambda_{fm}(t) = \cos 2\pi[f_c t + k_f g(t)] \qquad (5.41)$$

which has an instantaneous frequency given by

$$f_i(t) = f_c + k_f s(t)$$

If we multiply all frequencies by a constant, C, we get a new FM wave with an instantaneous frequency of $f_i(t)$.

$$f_i(t) = Cf_c + Ck_f s(t) \qquad (5.42)$$

The maximum frequency deviation of this new waveform is C times the maximum frequency deviation of the old waveform. The maximum frequency of the information signal (which can be thought of as being $Cs(t)$ for the modified waveform) is still f_m. That is, although the frequency is covering a much wider band, it is changing just as rapidly as it did for the narrowband case. Therefore, if C is chosen large enough, the frequency deviation may end up being large enough to cause the resulting wave to be labelled *wideband*.

This can be seen quite easily in the frequency domain. If the original wave is narrowband, it occupies a bandwidth of approximately $2f_m$ around f_c as shown in Fig. 5.16. If we now multiply all frequencies by C, the new waveform occupies

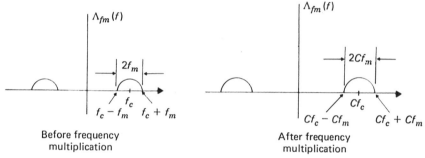

FIGURE 5.16   Frequency multiplication to change narrowband FM into wideband FM.

frequencies from $C(f_c - f_m)$ to $C(f_c + f_m)$ as shown. Its bandwidth has increased accordingly. If the bandwidth is much larger than $2f_m$ (i.e., C is much larger than 1), then the new wave must be classified as wideband. Thus we have a technique for generating the wideband waveform from the narrowband signal. This yields the FM modulator shown in Fig. 5.17. This type of system is known as an *Indirect FM Modulator*.

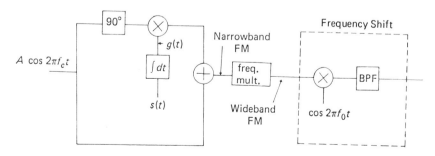

FIGURE 5.17   Indirect FM modulator for wideband FM.

In practice, a squarer can be used as a frequency doubler. This is possible due to the trigonometric identity,

$$\cos^2 x = \tfrac{1}{2}(1 + \cos 2x). \qquad (5.43)$$

A diode acts as a squarer over a limited range of inputs. If a frequency multiplication factor of 4 is desired, two squarers can be used.

We note that while the frequency multiplier increases the bandwidth of the FM wave, it also increases the carrier, or center frequency. If for some reason we were not happy with the new carrier frequency, Cf_c, we could easily use the standard technique of multiplying by a cosine wave and then bandpass filtering in order to shift the entire transform to any desired section of the frequency axis. This shift would not change the bandwidth of the wave which would therefore remain wideband. The shifting process is shown in the dotted section of Fig. 5.17. The essential difference between shifting all frequencies via heterodyning and multiplying of all frequency components by a constant should be clarified in the reader's mind before he continues.

There are also more direct techniques for generating a wideband FM wave. An electronic oscillator has an output frequency which depends upon the resonant frequency of a tuned L-C circuit. If either L or C of the tuned circuit is a function of time, the output of the oscillator will be a sinusoid with time-varying instantaneous frequency.

Figure 5.18 shows a highly simplified block diagram of a feedback oscillator.

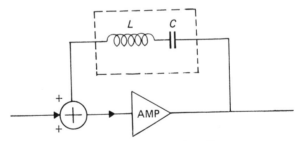

FIGURE 5.18   A feedback oscillator.

The output frequency would be

$$f_i(t) = \frac{1}{2\pi\sqrt{LC}}$$

If either L or C is varied, the output frequency will also vary. That is,

$$f_i(t) = \frac{1}{2\pi\sqrt{L(t)C(t)}} \qquad (5.44)$$

If FM is our goal, the trick becomes varying L or C such that

$$f_i(t) = f_c + k_f s(t)$$

Due to the presence of the square root relationship, achieving this particular $f_i(t)$ requires a complicated control over L or C. However, since the required variations are relatively small (i.e., $\Delta f \ll f_c$), the square root can be approximated by a linear term. Thus if C or L varies with $s(t)$ such that

$$C = C_0 + k_1 s(t) \tag{5.45a}$$

or

$$L = L_0 + k_2 s(t) \tag{5.45b}$$

and the constants, k_1 or k_2 are small, the output will be an FM wave.

The question now is whether or not we can vary a capacitance (or inductance) in this manner. It certainly isn't practical to physically move the knob of a variable capacitor back and forth.

Electronic devices do exist which exhibit a capacitance which is dependent upon an input voltage. We could feed $f(t)$ into one of these devices (e.g., from a microphone) and immediately get the desired results.

Two such devices are the *reactance tube* and the *varactor diode*. The resulting circuit is known as a *voltage-controlled oscillator (VCO)*.

5.4   DEMODULATORS

The problem of FM demodulation can be stated as follows. Given $\lambda_{fm}(t)$ in the form

$$\lambda_{fm}(t) = A \cos 2\pi \left[f_c t + k_f \int_0^t s(\tau) \, d\tau \right]$$

recover the information signal, $s(t)$. We note that the entire analysis to follow will be performed in the time domain since we do not know the Fourier Transform of the FM wave. We only have some feeling for the range of frequencies occupied by the transform.

Demodulators fall into one of two broad classifications. The first of these employ discriminators, which are devices which discriminate one frequency from another by transforming frequency changes to amplitude changes. The amplitude changes are detected just as was done in AM. The second category of demodulator uses a phase locked loop to match a local oscillator to the modulated carrier frequency.

Discriminator Detectors

Since differentiation of a sinusoid is a process which multiplies the sinusoid by its instantaneous frequency, we start by investigating the derivative of the FM waveform.

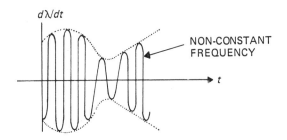

NON-CONSTANT
FREQUENCY

FIGURE 5.19   Derivative of FM
waveform.

$$\frac{d\lambda}{dt} = -2\pi A[f_c + k_f s(t)]\sin 2\pi \left[f_c t + k_f \int_0^t s(\tau)\, d\tau \right] \qquad (5.46)$$

This derivative is sketched in Fig. 5.19 for a typical $s(t)$.

If we now assume that the instantaneous frequency of the sinusoidal part in Eq. (5.46) is always much greater than f_m (a fair assumption in real life), this carrier term fills in the area between the amplitude and its mirror image. That is, we have exaggerated the carrier in sketching Fig. 5.19. Actually, the area between the upper and lower outline should be shaded in due to the extremely high carrier frequency. Thus even though the carrier frequency is not constant, the envelope of the waveform is still clearly defined by

$$2\pi |A[f_c + k_f s(t)]| \qquad (5.47)$$

The slight variation in the frequency of the carrier would not even be noticed by an envelope detector.

Illustrative Example 5.8

Use the concept of pre-envelope to prove that the envelope of

$$r(t) = -2\pi A[f_c + k_f s(t)]\sin 2\pi \left[f_c t + k_f \int_0^t s(\tau)\, d\tau \right] \qquad (5.48)$$

is given by

$$2\pi |A[f_c + k_f s(t)]|$$

Solution

Using techniques similar to those of Illustrative Example 4.11 we find that the pre-envelope of $r(t)$ is

$$z(t) = -2\pi A[f_c + k_f s(t)]\sin 2\pi \left[f_c t + k_f \int_0^t s(\tau)\, d\tau \right]$$
$$\qquad - 2\pi j A[f_c + k_f s(t)]\cos 2\pi \left[f_c t + k_f \int_0^t s(\tau)\, d\tau \right] \qquad (5.49)$$

The envelope of $r(t)$ is the magnitude of $z(t)$,

$$|z(t)| = 2\pi\sqrt{A^2[f_c + k_f s(t)]^2 \left\{ \sin^2 2\pi(f_c t + k_f \int_0^t s(\tau)\,d\tau) + \cos^2 2\pi(f_c t + k_f \int_0^t s(\tau)\,d\tau) \right\}} = 2\pi|A[f_c + k_f s(t)]| \quad (5.50)$$

A differentiator followed by an envelope detector can therefore be used to recover $2\pi A[f_c + k_f s(t)]$ from $\lambda_{fm}(t)$. This is shown in Fig. 5.20. Since f_c is always much greater than $k_f s(t)$, we have removed the absolute value sign.

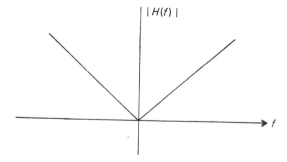

FIGURE 5.20   An FM demodulator.

We note that the above analysis did not assume anything about the size of k_f. The demodulator of Fig. 5.20 therefore works well for either wideband or narrowband FM signals.

The occurrence of the envelope detector should give one clue that AM is somehow rearing its head again. Indeed this is the case as the following analysis will reveal.

If we examine the transfer function of a differentiator, the operation of the FM demodulator will become immediately clear. The system function of a differentiator is

$$H(f) = 2\pi j f \quad (5.51)$$

with magnitude characteristic sketched in Fig. 5.21.

FIGURE 5.21   Magnitude characteristic of differentiator.

The magnitude of the output of the differentiator is linearly related to the frequency of its input. The differentiator therefore changes FM into AM!! When a differentiator is used in this manner, it is often called a *discriminator*.

In practice, the differentiator need not be constructed. Any system which has a system function that is approximately linear with frequency in the range of frequencies of interest will change the FM into AM. Thus, even a sloppy bandpass filter will work as a discriminator if we operate on the "up slope," as shown in Fig. 5.22 (or the "down slope" as a similar argument would show). The filter would have to be followed by an envelope detector.

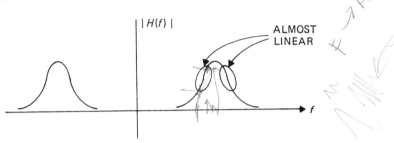

FIGURE 5.22   A bandpass filter as a discriminator.

If quality of reproduction is not a critical factor, one can actually use an AM table radio to receive FM signals. A heterodyner would first have to be constructed to shift the desired FM signal into the range of frequencies receivable on the AM receiver. The AM radio is then tuned to a frequency slightly (several kHz) above or below the shifted FM carrier frequency. Since the image rejection filter in the AM receiver is not ideal, it will change the FM signal to an AM signal (i.e., discriminate). The receiver's envelope detector will then complete the demodulation. This technique is actually used in inexpensive converters in order to listen to police and emergency calls. These are often transmitted via narrowband FM.

Most of the non-linearity in the slope of the bandpass filter characteristic of Fig. 5.22 can be eliminated if the characteristic is subtracted from a shifted version of

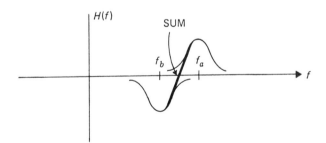

FIGURE 5.23   Improving the linearity of the discriminator.

itself as shown in Fig. 5.23. That is, we take the difference between the output of two bandpass filters whose center frequencies are separated as shown.

Figure 5.24 shows a circuit diagram of a demodulator which uses the principle shown in Fig. 5.23. The tuned circuit consisting of the upper half of the output winding of the transformer and C_1 is tuned to f_a (see Fig. 5.23) and the tuned circuit consisting of the other half of the output winding and C_2 is tuned to f_b.

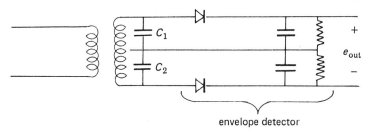

FIGURE 5.24 Slope demodulator.

One more possible form of the discriminator is called the "phase shift" detector. This is shown in block diagram form as Fig. 5.25. If t_0 is made small enough, the delay and subtraction forms an approximation to the derivative operation.

$$s(t) - s(t - t_0) \approx t_0 \frac{ds}{dt} \tag{5.52}$$

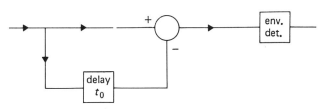

FIGURE 5.25   Phase shift demodulator.

Phase Locked Loop[1]

The *phase locked loop* (PLL) is a feedback circuit which can be used to demodulate angle modulated waveforms. Feedback circuits are often used to reduce an error term toward zero. In the case of the PLL, the error term is the phase difference between the input signal and a reference sinusoid.

   The phase locked loop incorporates a *voltage-controlled oscillator* (VCO) in the feedback loop. A voltage-controlled oscillator is an oscillator whose output frequency is a function of its input voltage. As we saw in the previous section, an FM modulator is a VCO. A typical loop configuration is shown in Fig. 5.26. The loop compares the phase of the input signal to the phase of the signal at the output of the VCO. If the phase difference between these two signals is anything other than zero, the output frequency of the VCO is adjusted in a manner which forces this difference down to zero.

---

[1]Engineering is not as dull as some would argue. As evidence of this, we point to an ongoing debate within the Communication Society of IEEE regarding a semantic issue. The phase locked loop is sometimes referred to as a phase "lock" loop. We shall use the former nomenclature, since it is the loop and not the phase which is in the locked condition.

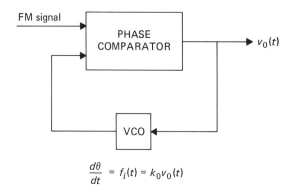

FM signal

$$\frac{d\theta}{dt} = f_i(t) = k_0 v_0(t)$$

FIGURE 5.26   The phase locked loop.

The output of the phase comparator, $v_0(t)$, forms the input to the VCO. The output of the VCO is an FM waveform with an instantaneous frequency, $f_i(t)$, that is proportional to the phase difference between the input and the VCO output. If, for example, the input frequency is higher than the VCO frequency, the phase difference between the two will increase linearly with time causing the VCO frequency to increase until it matches that of the input. Thus, the VCO output attempts to follow the input frequency. Since the frequency of the VCO output is proportional to the voltage at its input, it is this input voltage which also tries to follow the frequency of the loop input signal. Thus, monitoring the input to the VCO yields a demodulated version of the FM waveform.

The simplest method of phase comparison consists of a product operation followed by a lowpass filter, as shown in Fig. 5.27. Suppose that the two inputs to the

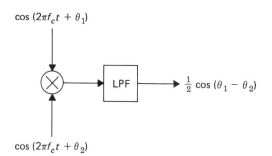

$\cos (2\pi f_c t + \theta_1)$

$\frac{1}{2} \cos (\theta_1 - \theta_2)$

$\cos (2\pi f_c t + \theta_2)$

FIGURE 5.27   A multiplier used as a phase comparator.

comparator are $\cos(2\pi f_c t + \theta_1)$ and $\cos(2\pi f_c t + \theta_2)$. The output of the lowpass filter will then be $\frac{1}{2} \cos(\theta_1 - \theta_2)$. This output will be a maximum if the two phases are equal. It is preferable to have minimum rather than maximum output when phases are equal since feedback loops normally drive a particular parameter toward zero. To achieve a minimum output at the desired condition, one of the inputs is changed from cosine to sine. The output of the lowpass filter will then be $\frac{1}{2} \sin(\theta_1 - \theta_2)$. This output will be zero if the phases are equal. That is the desired lock condition.

In some cases, it is desirable to have a comparator output which varies *linearly* with the phase difference rather than sinusoidally. If the input cosine and sine signals are severely clipped, they can be thought of as square waveforms. A representative situation is illustrated in Fig. 5.28. The product of the two square waves is now averaged by the lowpass filter. This average is proportional to the fraction of time that the square waves are equal. Therefore, the output amplitude is linearly related to the phase difference.

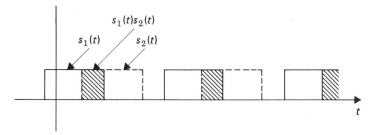

FIGURE 5.28   Severe clipping results in a linear phase comparator.

Using the multiplier as a phase comparator, we can redraw the PLL with FM input and with a general "loop filter" in place of the lowpass filter. This system is shown in Fig. 5.29. The output of the VCO is an FM waveform with instantaneous frequency

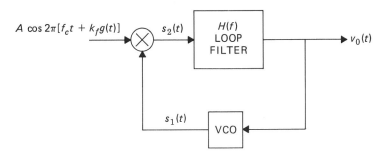

FIGURE 5.29   The phase locked loop using a multiplier as the phase comparator.

$$f_i(t) = f_c + k_0 v_0(t) \qquad (5.53)$$

Note that with a zero input voltage, the output frequency is the same as the carrier frequency. k_0 is a constant that determines the sensitivity of the VCO. The signal, $s_1(t)$ in Fig. 5.29, is given by

$$s_1(t) = 2B \sin 2\pi \left[f_c t + k_0 \int_0^t v_0(t)\, dt \right]$$

The multiplying factor of 2 is included to eliminate factors of $\frac{1}{2}$ in the following equations. $2B$ is the amplitude associated with the VCO. The signal, $s_2(t)$, is the product of two sinusoids. We will expand this product and drop the higher frequency term, since the loop filter will usually reject this high frequency term.

$$s_2(t) = 2AB \cos 2\pi[f_c t + k_f g(t)] \sin 2\pi \left[f_c t + k_0 \int_0^t v_0(t) \, dt \right]$$

$$= AB \sin 2\pi \left[k_f g(t) - k_0 \int_0^t v_0(t) \, dt \right]$$

Two phase factors are defined as follows.

$$\theta_{fm}(t) = 2\pi k_f g(t)$$

$$\theta_0(t) = 2\pi k_0 \int_0^t v_0(t) \, dt$$

The phase locked loop of Fig. 5.29 can be redrawn as Fig. 5.30 where the input is now the phase of the FM waveform as defined above. The *phase error* is

$$\theta_e(t) = \theta_{fm}(t) - \theta_0(t)$$

The presence of the block labelled "sin( )" makes this a non-linear feedback system. If we used a linear phase comparator as described earlier, the loop would be linear. When $\theta_e(t) = 0$, the loop is in phase lock.

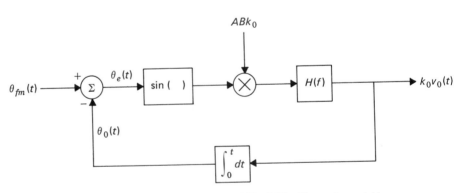

FIGURE 5.30   Phase locked loop of Fig. 5.29 with angular variables.

For simplicity of analysis, we will assume that the loop operates near phase lock. Thus, $\theta_e(t)$ is small, and sin $\theta_e(t)$ can be approximated by $\theta_e(t)$. The resulting linearized loop is shown in Fig. 5.31.

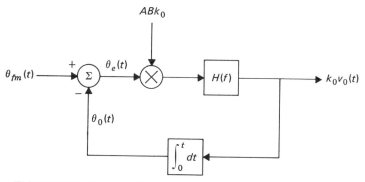

FIGURE 5.31   Linearized approximation to phase locked loop of Fig. 5.30.

First Order Loop

If the loop filter is an ideal lowpass filter, the PLL is known as *first order*. Since we have already accounted for the bandlimiting operation of the filter by eliminating high frequency terms, we can set $H(f) = 1$. The loop equations can be written directly from Fig. 5.31.

$$v_0(t) = AB[\theta_{fm}(t) - \theta_0(t)] \tag{5.54}$$

$$= AB\left[\theta_{fm}(t) - k_0 \int_0^t v_0(t)\, dt\right]$$

Taking the derivative of both sides of the equation yields $[dv_0(t)]/dt$

$$\frac{d\theta_{fm}(t)}{dt} + ABk_0 v_0(t) = AB = ABk_f s(t)$$

The overall system function of the loop is

$$H_L(f) = \frac{V_0(f)}{S(f)} = \frac{ABk_f}{jf + ABk_0} \tag{5.55}$$

The magnitude of this system function is shown in Fig. 5.32. The half-power frequency is $f = ABk_0$. This gives a measure of the frequency range of information signals which the loop can respond to properly.

An indicator of how quickly the loop can respond to changes in $s(t)$ is given by the step response. This is the integral of the impulse response, $h(t)$, which is found from $H(f)$. It can also be evaluated directly from the differential equation

$$a(t) = \frac{k_f}{k_0}[1 - \exp(-2\pi ABk_0 t)]u(t)$$

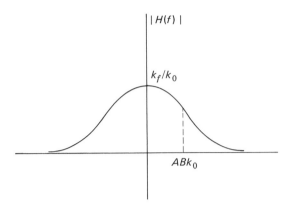

FIGURE 5.32 Magnitude of the loop transfer function.

5.5   PHASE MODULATION

There is no basic difference between phase modulation and frequency modulation. In spite of this fact, we will begin our study of phase modulation by treating it as an independent technique. We will continue this approach until a point is reached where the similarity between the two techniques cannot be overlooked.

We start with the unmodulated carrier signal

$$s_c(t) = A \cos 2\pi(f_c t + \theta)$$

If θ is varied in accordance with the information we wish to transmit, the carrier is said to be *phase modulated*. We will therefore let θ vary in the following manner.

$$\theta = \theta(t) = k_p s(t)$$

where k_p is a constant factor associated with phase modulation. We use the subscript "p" to distinguish this constant from k_f used in FM (if for no other reason than the fact that the two constants have different units). The total phase modulated waveform will therefore be of the form

$$\lambda_{pm}(t) = A \cos 2\pi[f_c t + k_p s(t)] \tag{5.56}$$

with corresponding instantaneous frequency

$$f_i(t) = f_c + k_p \frac{ds}{dt} \tag{5.57}$$

This form does not look basically different from that of the FM signal. If the information signal used to frequency modulate a carrier were ds/dt, the resulting FM waveform would look exactly the same as that of a carrier phase modulated with $s(t)$. If a signal which is either a phase or frequency modulated carrier is received, there is no way of telling whether it is an FM or PM wave without knowing $s(t)$ beforehand.

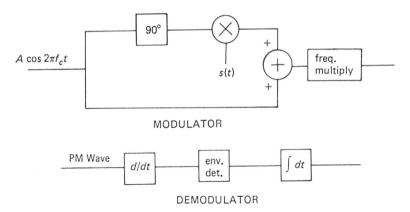

FIGURE 5.33   A phase modulator and demodulator.

Of course knowledge of $s(t)$ is absurd. If it were known, why waste time transmitting it?

The techniques of generating and receiving phase modulated carriers are the same as those for FM with the addition of an integrator or differentiator in the proper places. Figure 5.33 shows a modulator and a demodulator. The student should compare these with the block diagrams of the corresponding FM systems.

There is a definition applied to phase modulation which is analogous to that of maximum frequency deviation for FM. That is, the *maximum phase deviation*, $\Delta\theta$, is defined as the maximum amount by which the phase deviates from that of an unmodulated carrier wave. It is therefore equal to the maximum value of $k_p s(t)$.

The bandwidth of the phase modulated waveform is found by forgetting the fact that it is not an FM wave. We write the form of the modulated carrier, find f_m and Δf, and plug these values into the "rule of thumb" formula previously developed.

Illustrative Example 5.9

The following waveform is received,

$$g(t) = 5 \cos(2\pi \times 10^6 t + 200 \sin 2\pi \times 500t)$$

(a) What is the approximate bandwidth of the waveform?
(b) If this represents an FM wave, what is $s(t)$, the information signal?
(c) If this represents a phase modulated wave, what is $s(t)$, the information signal?

Solution

(a) The approximate bandwidth of this signal is given by

$$BW \approx 2(\Delta f + f_m)$$

The instantaneous frequency of the waveform is given by

$$f_i(t) = 10^6 + \frac{d}{dt}(200 \sin 2\pi \times 500t)/2\pi$$

$$= 10^6 + 10^5 \cos 2\pi \times 500t$$

f_m is the maximum frequency component of $\cos 2\pi \times 500t$, which is simply 500 Hz. Δf is the maximum value of $10^5 \cos 2\pi \times 500t$, which is 10^5 Hz. The bandwidth is therefore given by

$$BW \approx 2(10^5 + 500) \approx 2 \times 10^5 \text{ Hz.}$$

Since the bandwidth for the narrowband case would be only 1,000 Hz, this is clearly a very wideband waveform. We have not yet needed to know whether the wave is FM or PM.

(b) If this is an FM wave, we would view the instantaneous frequency,

$$f_i(t) = f_c + k_f s(t)$$
$$= 10^6 + 10^5 \cos 2\pi \times 500t \tag{5.58}$$

The information signal is easily identified as

$$s(t) = \cos 2\pi \times 500t$$

We note that the assumption was made that $k_f = 10^5$. Our result for $s(t)$ might therefore be wrong by a constant if 10^5 is not the actual value of k_f used. Since a constant multiplier does not represent distortion of the waveform, it does not concern us here.

(c) If this were a PM wave, we would view the instantaneous phase,

$$\theta(t) = 2\pi f_c t + k_p s(t) = 2\pi \times 10^6 t + 200 \sin 2\pi \times 500t \tag{5.59}$$

It is again easy to identify the information signal as

$$s(t) = \sin 2\pi \times 500t \tag{5.60}$$

Here we have assumed that $k_p = 200$. The argument of part (b) concerning the constant multiplier applies again here.

While the difference between $\cos 2\pi \times 500t$ and $\sin 2\pi \times 500t$ may not appear significant, this will not be the case if $s(t)$ is a more complicated waveform. That is, while differentiation of a pure sinusoid does not alter its basic shape, differentiation of a general time function certainly represents a severe form of distortion. It is therefore important to know what form of modulation was used before one attempts demodulation of an incoming signal.

Since phase modulation is so similar to frequency modulation, a logical question at this time is, "why bother with phase modulation?" It would appear that there is no reason for using one form in preference to the other.

Actually one can see one practical difference by viewing the block diagrams of the modulators and demodulators (Figs. 5.17, 5.20, and 5.33). FM requires an integrator in the modulator. PM requires the integrator in the demodulator. If integrators are expensive, and many receivers are required while only one transmitter is used, FM would prove cheaper than PM.

We also mention at this time that integration of the information signal *before* transmission through the air may make it less susceptible to noise interference. This will be explored fully in Section 5.7.

5.6   BROADCAST FM AND STEREO

If you pick up any FM program guide, you will realize that adjacent stations are separated by 200 kHz. The FCC has assigned carrier frequencies of the type

$$101.1 \text{ MHz}, \quad 101.3 \text{ MHz}, \quad 101.5 \text{ MHz}, \quad \dots$$

to the various transmitting stations. Thus, while the bandwidth allocated to each station in AM was 10 kHz, it is an impressive 200 kHz for FM stations. This gives each broadcaster lots of room to work with.

The maximum frequency of the information signal is fixed at 15 kHz. Compare this with the 5-kHz figure for f_m in the AM broadcast case. It should now be obvious why quality music is reserved for FM.

To transmit this signal via narrowband FM would require a bandwidth of $2f_m$, or 30 kHz. Since 200 kHz is available, we see that broadcast FM can afford to be wideband.

Using the approximate formula developed for the bandwidth of an FM signal, we see that a maximum frequency deviation of about 85 kHz is possible. The actual figure used is a conservative 75 kHz.

FM Stereo

What about FM stereo? Where could we possibly squeeze the second channel?

We have already answered these questions. If narrowband FM had been used instead of wideband, we would have only required 30 kHz out of the 200 kHz allocated to one station. Sufficient frequency space therefore exists.

We begin with a system similar to that described in Illustrative Example 4.5. The two audio signals, $s_1(t)$ and $s_2(t)$ are frequency multiplexed. We choose a carrier frequency of 38 kHz in order to yield an 8-kHz safety band between the two transforms. The system, and the corresponding transforms are sketched in Fig. 5.34.

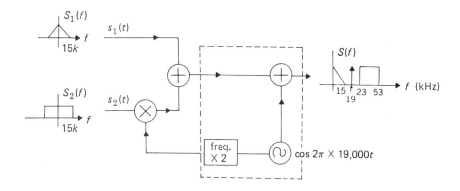

FIGURE 5.34 Multiplexing stereo signals.

Two different low frequency shapes have been chosen for $S_1(f)$ and $S_2(f)$ in order to make it easier to track these shapes through the analysis.

This composite signal represents a perfectly legitimate time function with an upper frequency of 53 kHz. It can therefore be transmitted via narrowband FM and be well within the allotted 200-kHz bandwidth. We would actually use only 106 kHz of the band.

At the receiver we would simply FM demodulate to recover the multiplexed signal. The remainder of the receiver is identical to that described in Illustrative Example 4.5 (*see* Section 4.4). We separate the two signals using a lowpass filter and then synchronously demodulate the upper channel using a 38-kHz sinusoid. Synchronous detection must be used, since the carrier frequency (38 kHz) is *not* much higher than the highest frequency component of the information signal (15 kHz). Therefore envelope detection could not be used, even if a carrier term were added. To avoid the problems inherent in synchronous demodulation, we play a trick. The original multiplexed signal generated by the system of Fig. 5.34 is added to a sinusoid of frequency 19 kHz before the frequency modulation (this is taking place at the transmitter). The 38-kHz carrier is formed by doubling the frequency (squaring) so that the two sinusoids are of the same phase and have frequencies which are exactly in the ratio of 2×1. This is illustrated in the dashed box in Fig. 5.34.

At the receiver, this 19-kHz sinusoid is separated after FM demodulation. This is easy to accomplish due to the safety band mentioned earlier. The sinusoid is then doubled in frequency and used in the synchronous demodulator. The frequency doubling is again performed by a squarer. As a bonus, it is this 19-kHz sinusoid that is used to light the "stereo" indicator light on the receiver panel. This *sub-carrier* is only present during a stereo broadcast. The entire receiver is illustrated in Fig. 5.35.

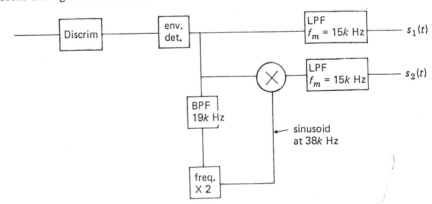

FIGURE 5.35   FM stereo receiver.

A monaural (single channel) receiver consists of discriminator, envelope detector, and lowpass filter. It would therefore recover the signal $s_1(t)$ when used in the above system. In order to make this system compatible, one sends sum $(L + R)$ and difference $(L - R)$ signals as $s_1(t)$ and $s_2(t)$. The monaural receiver only recovers the sum channel. The stereo receiver of Fig. 5.35 recovers the sum and difference. If they are added together in a simple resistive circuit, the result is the left channel signal. If the difference is taken, again in a resistive circuit, the result is the right channel. This sum and difference procedure is known as a *matrix* operation.

The unused portion of the allotted band between 53 kHz and 100 kHz deviation from f_c is sometimes used for so-called SCA (Subsidiary Communications Authorization) signals. These include the commercial-free music heard in some restaurants, transmitted on a subcarrier of 67 kHz.

If four-channel FM stereo becomes popular, the third and fourth channels will occupy this region. They will probably be transmitted over subcarriers of 72 kHz and 92 kHz with a maximum audio frequency of 8 kHz. The reduction from an f_m of 15 kHz is being considered since the two additional channels will only carry the "fill-in" rear sounds.

This discussion illustrates the flexibility of FM broadcast systems. This flexibility is due to the fact that the bandwidth of an FM signal depends upon more than just the bandwidth of the information (or audio in this case) signal.

AM Stereo Revisited

We discussed quadrature amplitude modulation stereo (QAM) in Chapter 4. Here we present several other forms of AM stereo. These techniques are not exclusively AM but employ a combination of AM and angle modulation. The reason they are still known as "AM stereo" is that they are designed to transmit two independent audio signals in the 10-kHz band allocated to each broadcast AM station.

One technique (proposed in two different forms by Belar and by Magnavox) angle modulates the carrier with one audio signal and amplitude modulates the resulting modulated carrier with the second signal. The angle modulation is narrowband.

Figure 5.36 presents one possible AM/FM configuration. This is a simplified version of the Belar AM stereo system. The frequency deviation of the FM is $\Delta f = 320$ Hz, which is much less than the maximum audio frequency of 5 kHz. This represents narrowband FM, so the resulting modulated signal can be kept within the 10-kHz band. The system is compatible, since an envelope detector will recover the sum signal, as required by the FCC. The limiter in the receiver removes the amplitude modulation to leave an FM wave with approximately constant amplitude.

The idealized system of Fig. 5.36 requires some modification to make it practical. Timing is important in a stereo system. If the $L - R$ and $L + R$ channels are not time synchronized, the original signals cannot be recovered without distortion. Since the paths traversed by the $L - R$ and $L + R$ signals in the transmitter of Fig. 5.36 do not take identical lengths of time, it is necessary to insert a time delay in the $L - R$ line so the two signals can be properly aligned at the output.

A more complex technique of stereo transmission again starts by amplitude modulating a frequency modulated carrier with the sum signal, $L + R$, as in the preceding system. However, the carrier is not angle modulated with the difference

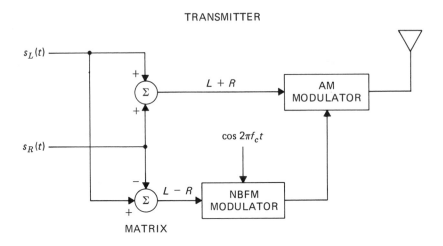

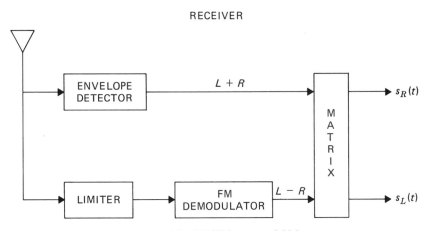

FIGURE 5.36   AM/FM system of AM stereo.

signal. Instead, the angle modulation is performed in such a way that the left channel information is carried on the lower sideband and the right channel on the upper sideband. To do so requires a relatively complex system involving a 90° phase shift between $L + R$ and $L - R$ signals (recall the discussion of SSB and the Hilbert Transform) and also employing automatic gain control (agc) circuitry. One advantage of this system is that it is possible to produce a stereo effect with two monaural receivers. If one of the receivers is tuned slightly above the carrier and the other slightly below, a (non-optimum) stereo effect is produced. (If you live in the Southwest of the United States, you may recall early experiments using this technique on XETRA, Tijuana, Mexico.) The details of the analysis are left to the problems at the end of this chapter.

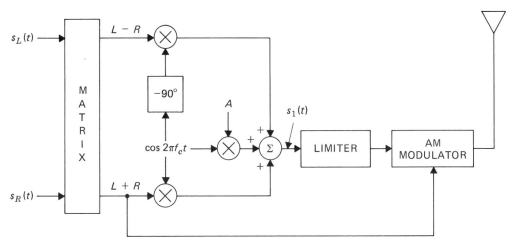

FIGURE 5.37   The compatible QAM (C-QAM) system for AM stereo.

Yet a third technique is a refinement of the quadrature AM stereo system discussed in Chapter 4. Recall that the problem with that system was one of compatibility. A monaural receiver would recover the square root of the sum of the squares of the left and right signals. This is a highly distorted version of the sum waveform.

A *compatible QAM (C-QAM)* system (a simplified version of the Motorola configuration) is shown in Fig. 5.37. The system begins with a QAM signal where the two signal components are the sum and difference waveforms. The signal at $s_1(t)$ is therefore given by

$$s_1(t) = \sqrt{[A + s_L(t) + s_R(t)]^2 + [s_L(t) - s_R(t)]^2} \cos \left(2\pi f_c t \right.$$

$$\left. - \tan^{-1} \left[\frac{s_L(t) - s_R(t)}{A + s_L(t) + s_R(t)} \right] \right) \qquad (5.61)$$

We now perform an operation to replace the square root multiplying factor by the sum signal. That is, we restore

$$s_2(t) = [A + s_L(t) + s_R(t)] \cos \left(2\pi f_c t - \tan^{-1} \left[\frac{s_L(t) - s_R(t)}{A + s_L(t) + s_R(t)} \right] \right)$$
$$\qquad (5.62)$$

This is done by limiting the amplitude of the QAM waveform [Eq. (5.61)] to yield a constant amplitude, and then by amplitude modulating with the sum signal.

The system of Fig. 5.37 produces a signal which is compatible, since an envelope detector recovers the sum signal. The problem is that the stereo receiver must perform a complex function. Given the waveform of Eq. (5.62), the receiver must first replace the $L + R$ multiplying term by the square root multiplying term of Eq. (5.61). The information to do this restoration of the QAM waveform is present because the sum signal exists as the envelope, and the difference signal can be found by detecting the phase and combining this phase with the sum signal waveform. The problem is one of performing these operations with relatively simple circuitry. The receiver uses a PLL to detect the phase and then remodulates the received waveform with this phase. The details are left to the exercises at the end of this chapter.

5.7   NOISE IN FM SYSTEMS

Since FM is a non-linear form of modulation, analysis of noisy systems will prove much more difficult than it was in the AM case. We will again have to be content with some approximations and generalizations.

Let us assume that a noise-free FM waveform is transmitted.

$$\lambda_{fm}(t) = A \cos(2\pi f_c t + k_f g(t))$$

where $g(t)$ is the time integral of the information signal. Noise is added to this FM signal along the path between the transmitter and receiver. The additive noise is assumed to be white with power spectral density $N_0/2$ for all f. The combination of noise and signal, $r(t)$, enters the receiver antenna as shown in Fig. 5.38. Although the figure indicates that the noise is added prior to the receiver, this model can also be used for the case where the receiver itself adds noise.

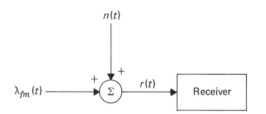

FIGURE 5.38   Sum of FM signal and noise enters receiver.

FM Receivers

An FM waveform carries its information in the form of the frequency of the time function. The amplitude of the FM wave remains constant. Thus, the signal information is contained in the zero crossings of the FM wave. The FM waveform can be clipped at an extremely low level (i.e., effectively changed from a continuous waveform to a square wave) without loss of signal information. Additive noise can significantly affect the amplitude of the FM wave, but it perturbs the zero crossings to a smaller degree. Receivers therefore often clip, or limit, the amplitude of the received waveform prior to frequency detection. This provides a constant amplitude waveform as input to the discriminator. One effect of clipping is to introduce higher harmonic terms (think of the relationship between a sine wave and square wave with the same

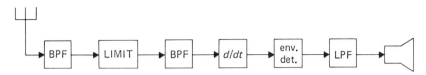

FIGURE 5.39   FM receiver.

period). A post-detection lowpass filter can be used to reject these higher harmonics.

A block diagram of the simplified FM receiver is shown in Fig. 5.39. In most cases, the receiver uses the heterodyne concept and a fixed i.f. filter. The simplified receiver of Fig. 5.39 will exhibit the same performance as that of the heterodyne receiver.

Signal to Noise Ratio

The qualitative analysis of noise in FM systems is facilitated by a phasor presentation. We used phasors earlier (*see* Fig. 5.8) in describing the FM waveform. Figure 5.40 illustrates a phasor diagram of the received FM signal plus additive

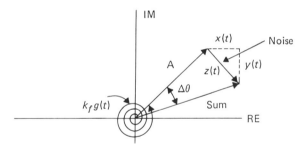

FIGURE 5.40   Phasor diagram for FM signal and noise. Note that the constant carrier frequency term has been removed.

noise. The signal phasor has length A and angle $k_f g(t)$. This phasor wiggles around according to $k_f g(t)$, while its amplitude remains constant. The noise phasor can be added to the signal when the noise is expanded in quadrature components.

$$n(t) = x(t)\cos 2\pi f_c t - y(t)\sin 2\pi f_c t$$

$x(t)$ and $y(t)$ are low frequency random processes. It is the bandpass filter in the receiver which allows us to use this bandlimited quadrature representation for the noise.

The angle of the resultant phasor contains all of the useful information. The length of the vector can be disregarded, since the limiter will eliminate any amplitude variations.

As long as the magnitude of the noise vector does not exceed A, the maximum angular difference between the resultant and the signal vector, $\Delta\theta$, is $90°$. The limiting value of $90°$ would only occur when the noise amplitude reaches A, and the vectors are colinear and $180°$ out of phase. As the noise gets smaller, the angular variations around the signal vector would be limited to decreasing limits less than $90°$.

The angle of the signal vector is $k_f g(t)$. For large k_f (i.e., wideband FM), this angle can be considerably larger than $360°$. Thus, the degradation caused by the noise can be made arbitrarily small by increasing k_f.

As a first step in quantifying the preceding observations, we shall analyze the special case of a large carrier to signal ratio. That is, assume A is much larger than $g(t)$. Thus, in the limit, a pure carrier is being transmitted. The signal plus noise entering the limiter is then given by

$$r(t) = A\cos 2\pi f_c t + x(t)\cos 2\pi f_c t - y(t)\sin 2\pi f_c t$$

This can be rewritten as a single sinusoid by using trigonometric identities.

$$r(t) = \sqrt{[A + x(t)]^2 + y^2(t)}\cos\left(2\pi f_c t + \tan^{-1}\frac{y(t)}{A + x(t)}\right) \qquad (5.63)$$

The output of the limiter has constant amplitude. The clipping operation creates higher harmonics, and the normalized limiter output can be approximated by

$$\cos\left(2\pi f_c t + \tan^{-1}\frac{y(t)}{A + x(t)}\right) + \text{higher harmonics} \qquad (5.64)$$

All receiver operations following the limiter are linear in the modulating signal. We therefore can neglect the higher order harmonics, since the output lowpass filter will reject them.

If we neglect higher order terms, the output of the discriminator is the derivative of the inverse tangent shown above. This is given by

$$\frac{d\theta}{dt} = \frac{\{x(t) + A\}\,dy/dt - y(t)\,dx/dt}{y^2(t) + \{x(t) + A\}^2} \qquad (5.65)$$

The signal of Eq. (5.65) must then be put through a lowpass filter, as shown in Fig. 5.36. In order to perform this operation, we must first find the autocorrelation function of the signal and then the power spectral density and, finally, arrive at the power spectral density of the output of the filter. Following all of these operations, we could evaluate the power of the output noise and thereby find the S/N at the output of the FM receiver. These mathematical operations are somewhat involved, and one usually makes an additional simplifying assumption at this point. We assume a high carrier to noise ratio (CNR). That is, we assume that

$$A \gg x(t) \qquad A \gg y(t)$$

With this assumption, Eq. (5.65) can be reduced to

$$\frac{d\theta}{dt} = \frac{dy/dt}{A} \qquad (5.66)$$

Note that in both Eq. (5.65) and Eq. (5.66) we omit the constant f_c, as is done in all FM receivers.

The goal is to find the noise power, so, at this point, we shall investigate the power spectral density of the output noise as given by Eq. (5.66). From Illustrative Example 3.19, we have that the power spectral density of $y(t)$ is N_0 for f between $-BW/2$ and $+BW/2$, where BW is the bandwidth of the receiver input filter. We can find the power spectral density of dy/dt as in Problem 2.18. That is, since a differentiator has transfer function $H(f) = 2\pi jf$, the power spectral density of $d\theta/dt$ is given by $4\pi^2 f^2/A^2$ multiplied by the density of $y(t)$. This density is shown in Fig. 5.41. Since the discriminator output is of interest only for frequencies between $-f_m$

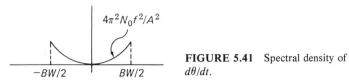

FIGURE 5.41   Spectral density of $d\theta/dt$.

and $+f_m$ (i.e., the audio is lowpass filtered before entering the speakers), the output noise power can be explicitly calculated as

$$P_N = 2 \int_0^{f_m} S_{N_0}(f)\, df = \frac{8\pi^2}{3} \frac{N_0}{A^2} f_m^3 \tag{5.67}$$

Since the output signal is $k_f s(t)$, the output signal power is $k_f^2 P_s$, where P_s is the power of the information signal. The output signal to noise ratio for high CNR is finally given by

$$S/N = \frac{3A^2 k_f^2 P_s}{8\pi^2 f_m^3 N_0} \tag{5.68}$$

Note from this result that as the carrier power increases, the output S/N increases. This is not the case in AM. Note also that as k_f increases, the S/N increases. Thus, as FM becomes more wideband, the noise becomes less disrupting.

Illustrative Example 5.10

Assume that $s(t) = a \cos 2\pi f_m t$. This information signal frequency modulates a large carrier and experiences additive white noise of spectral density $N_0/2$. Find the signal to noise ratio at the output of the receiver.

Solution

Since the problem states that a "large" carrier is modulated, we can make the assumption that the large carrier to signal ratio results can be used to approximate the solution. From Eq. (5.61), with $P_s = a^2/2$, we have

$$S/N = \frac{a^2 k_f^2 3A^2}{16 N_0 f_m^3 \pi^2} \tag{5.69}$$

We wish to make one observation from this result. Recall that for a sinusoidal information signal the FM bandwidth is given by $2(ak_f + f_m)$. If ak_f is much larger than f_m (wideband FM), this bandwidth can be approximated by $2ak_f$. Therefore, the signal to noise ratio of Eq. (5.69) can be rewritten as

$$S/N = \frac{3A^2 BW^2}{16 N_0 \pi^2 f_m^3} \tag{5.70}$$

Equation (5.70) clearly shows the dependence between the S/N and the bandwidth of the FM wave.

Pre-emphasis

Equation (5.68) indicates that the signal to noise ratio of an FM waveform increases with increasing k_f. Thus, a noise advantage is realized by increasing the bandwidth of the FM signal.

An even greater noise advantage can be realized by observing that the noise at the output of the FM receiver is not white even though the input noise is assumed to be white. Figure 5.41 illustrates that the output noise increases parabolically with increasing frequency. Therefore, the higher information frequencies are more severely affected by noise than the lower frequencies. A more efficient use of the band of frequencies would occur if these higher signal frequencies were emphasized and the lower frequencies were de-emphasized (so that the total power remains constant). The signal could therefore be shaped by a filter with $H(f)$ as shown in Fig. 5.42. This

FIGURE 5.42 $H(f)$ of pre-emphasis filter.

pre-shaping would be done prior to transmission. All signal frequencies would then be equally affected by the noise. The filter, of course, represents a signal distortion. The filtering operation must therefore be "undone" at the receiver by an inverse filter known as the *de-emphasis filter*. This filter would have transfer function $1/H(f)$. The composite filter transfer function of the two filters is unity.

The pre-emphasis filter characteristic of Fig. 5.42 changes the noise spectral density in order to make the noise white. This will assure equal noise disturbance over all signal frequencies. Many signals in real life drop off as frequency increases. It may therefore be desirable to "overemphasize" the signal with a filter characteristic which increases more rapidly with frequency than that shown.

Commercial broadcast FM uses pre-emphasis. In the interest of economy, the exact function shown in Fig. 5.42 is not used. A simple RC lowpass filter is used in the receiver for de-emphasis. The inverse filter is used at the transmitting station for pre-emphasis.

Threshold Effect

Let us return now to Fig. 5.40; we made the assumption that the signal vector magnitude, A, was larger than the noise vector magnitude, $|z|$. If the reverse is true, the diagram can be redrawn as Fig. 5.43. Alas, the receiver doesn't know the

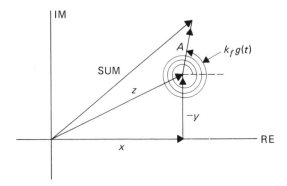

FIGURE 5.43 Phasor diagram of signal plus noise when noise is larger than signal (carrier removed).

difference between signal and noise. Here the signal vector causes perturbations upon the noise vector, and the resultant angle can differ from that of the *noise* by no more than 90°, regardless of how large $k_f g(t)$ becomes. *The receiver is locking onto the noise and suppressing the poor signal.* Their roles have reversed. This phenomenon is known as *noise capture.*

The signal amplitude remains constant at A. As the average length of the noise vector approaches A, we can expect this random quantity to be sometimes larger and sometimes smaller than A. Figure 5.44 shows a possible trajectory of the noise vector, and the resultant sum as the noise vector hovers around a magnitude of A. A counterclockwise rotation around the origin is illustrated. Also in the figure are the resulting angle and its derivative, the frequency. Note that the pulse is positive and has an area of 2π. If the encirclement of the origin had been in the clockwise direction, the pulse would have been negative. In cases where the origin is not encircled, the average value (total area) of the angular variations is zero. The spike in the frequency waveform would result in an audible *click* in the receiver audio output. Since the noise vector is sweeping around as the quadrature components, $x(t)$ and $y(t)$, change, we can expect occasional clicks as $|z|$ approaches A. The larger the average value of $|z|$ gets, the more frequent will be the clicks until noise capture takes effect. After this point, the audio output would simply be approximately that of the random noise—a rushing sound in the speaker.

Since the noise is random, the clicks cannot be characterized deterministically. We can only derive probability averages. One quantity of interest is the average number of clicks per second. We shall sketch the derivation of this parameter.

For simplicity, the phasor diagram will be rotated to align the horizontal axis with the signal vector. The new diagram is redrawn as Fig. 5.45. In order for the

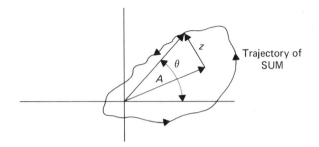

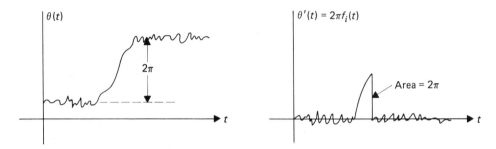

FIGURE 5.44   Trajectory of sum phasor and resulting phase and frequency as a function of time.

trajectory to cross the horizontal axis at point P, $y(t)$ must have a zero crossing and, at the same time, $x(t)$ must be less than $-A$. Since the noise quadrature components, $x(t)$ and $y(t)$, are independent of each other, the probability of this joint event is the product of the two individual probabilities. $x(t)$ and $y(t)$ are assumed to be Gaussian random processes. Thus, the probability that $x(t)$ is less than $-A$ is given by a complementary error function. The probability of a zero axis crossing can be found from the joint density of the function and its derivative. (E.g., if the function is

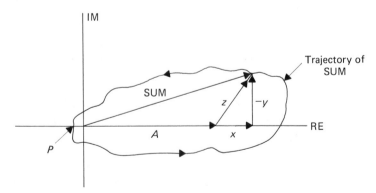

FIGURE 5.45   Realinged phasor diagram to illustrate conditions for circling of the origin.

negative at a particular point in time with a sufficiently positive derivative, a zero crossing results. Details appear in the references.) The probability of a trajectory circling the origin within a time interval Δt is then given by

$$Pr(\text{click}) = \frac{BW\Delta t}{2\sqrt{3}} \text{ erfc } \left(\sqrt{\frac{A^2}{2N_0 BW}} \right) \tag{5.71}$$

The average number of encirclements, or clicks, per second is then given by

$$CPS_{\text{AVG}} = \frac{BW}{2\sqrt{3}} \text{ erfc } \left(\sqrt{\frac{A^2}{2N_0 BW}} \right) \tag{5.72}$$

where BW is the bandwidth of the FM waveform. Note that the argument of the complementary error function is the square root of the signal to noise ratio. Thus if we define this signal to noise ratio as

$$\rho = \frac{A^2}{2N_0 BW}$$

the average number of clicks per second becomes

$$CPS_{\text{AVG}} = \frac{BW}{2\sqrt{3}} \text{ erfc}(\sqrt{\rho}) \tag{5.73}$$

As the signal to noise ratio, ρ, increases, the complimentary error function decreases and the number of clicks per second also decreases. As the bandwidth of the FM wave increases, the number of clicks per second increases, since additional noise is being admitted to the system.

The area under each pulse in Fig. 5.44 is 2π. We can therefore get an approximation to the output signal to noise ratio near threshold by adding the power of the clicks to the noise power found in the high CNR case. The power of the clicks is estimated in a manner similar to that used for impulse noise analysis. The final result is

$$\text{SNR} = \frac{3BW\rho k_f^2 P_s / f_m^3 4\pi^2}{1 + (BW)^3 \sqrt{3} \text{ erfc }(\sqrt{\rho})/16\pi^2 f_m^3} \tag{5.74}$$

As the average number of clicks per second approaches zero, the denominator of Eq. (5.74) approaches unity and the SNR becomes that of Eq. (5.68), the high CNR case.

Phase Locked Loops

We saw earlier in this chapter that a phase locked loop can be used to demodulate angle modulated waveforms. We also saw that a first order loop PLL has an associated time constant in its response to rapid changes in the information signal. This exponential response helps improve performance near threshold, since the spikes (clicks) are very narrow and will be attenuated by the loop.

It can be shown (*see* the references, in particular Taub and Schilling) that above threshold, the performance of the PLL is the same with or without the addition of a limiter. Below threshold, the performance improves if the limiter is *omitted*. The phase locked loop exhibits what is known as *threshold extension*. The input signal to noise ratio can be several dB lower than that of the discriminator detector before threshold occurs.

PROBLEMS

5.1. Find the instantaneous frequency of $s(t)$, where

$$s(t) = 10[\cos(10t)\cos(30t^2) - \sin(10t)\sin(30t^2)]$$

5.2. Find the instantaneous frequency, $f_i(t)$, of $s(t)$, where

$$s(t) = 2e^{-t}U(t)$$

5.3. In Eq. (5.26), c_n was found to be

$$c_n = \frac{1}{T} \int_{-T/2}^{T/2} e^{j\beta \sin 2\pi f_m t} e^{-jn2\pi f_m t} \, dt$$

where $T = 1/f_m$. Show that the imaginary part of c_n is zero. That is, show that c_n is real.

5.4. Show that c_n as given in Eq. (5.26) (*see* Problem 5.3) is not a function of f_m or T. That is, show that c_n only depends upon β and n. (Hint: Let $x = 2\pi f_m t$ and make a change of variables.)

5.5. Find the approximate Fourier Transform of $s(t)$, where

$$s(t) = \cos\left[\frac{ak_f}{f_m}\sin 2\pi f_m t\right]$$

Do this by expanding cos (x) in a Taylor series expansion and retaining the first three terms.

5.6. Repeat Problem 5.5 for $s(t)$ given by

$$s(t) = \sin\left[\frac{ak_f}{f_m}\sin 2\pi f_m t\right]$$

5.7. Use the results of Problems 5.5 and 5.6 to find the approximate Fourier Transform of the FM waveform

$$\lambda_{fm}(t) = \cos\left[2\pi f_c t + \frac{ak_f}{f_m}\sin 2\pi f_m t\right]$$

5.8.   An information signal is as shown below.

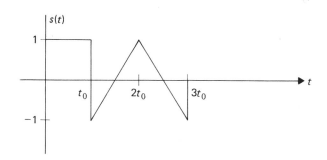

The carrier frequency is 1 MHz, $k_f = 10^4$, and $k_p = 20$. Sketch the FM and PM waveforms, and label significant frequencies and values.

5.9.   Find the approximate band of frequencies occupied by an FM waveform of carrier frequency 2 MHz, $k_f = 100$ Hz/volt, and
(a) $s(t) = 100 \cos 150t$ volts.
(b) $s(t) = 200 \cos 300t$ volts.

5.10.   Find the approximate band of frequencies occupied by an FM waveform of carrier frequency 2 MHz, $k_f = 100$ Hz/volt, and

$$s(t) = 100 \cos 150t + 200 \cos 300t \text{ volts}$$

Compare this with your answers to Problem 5.9.

5.11.   Consider the system shown below where the output is alternately switched between a source at frequency $f_c - \Delta f$ and a source at frequency $f_c + \Delta f$. This switching is done at a frequency f_m. Find and sketch the Fourier Transform of the output, $y(t)$. You may assume that the switch takes zero time to go from one position to another. (Hint: *See* Fig. 5.14.)

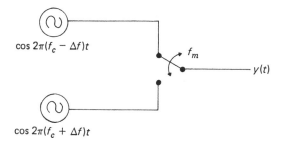

5.12.   A 10-MHz carrier is frequency modulated by a sinusoid of unit amplitude and $k_f = 10$ Hz/volt. Find the approximate bandwidth of the FM signal if the modulating signal has a frequency of 10 kHz.

5.13.   A 100-MHz carrier is frequency modulated by a sinusoid of frequency 75 kHz such that $\Delta f = 500$ kHz. Find the approximate band of frequencies occupied by the FM waveform.

5.14.   Find the approximate band of frequencies occupied by the waveform

$$\lambda(t) = 100 \cos[2\pi \times 10^5 t + 35 \cos 100\pi t]$$

5.15.   Find the approximate pre-envelope and envelope of

$$s(t)\cos\left[2\pi f_c t + k_f \int_0^t s(\sigma)\, d\sigma\right]$$

where the maximum frequency of $s(t)$ is much less than f_c.

5.16.   A phase locked loop is used to demodulate an FM waveform. The information signal is a voice waveform with maximum frequency of 5 kHz. The comparator has an output which varies linearly with input phase difference. An input phase difference of 90° causes a 25-volt change in the output. The VCO output varies linearly and changes by 10 kHz for each 1 volt input change. Will this phase locked loop act as an effective demodulator? (Hint: Observation of Eq. (5.54) shows that AB is the sensitivity of the comparator.)

5.17.   Show that the C-QAM signal of Eq. (5.62) can be processed to produce the QAM signal of Eq. (5.61). Do this by first detecting the phase (e.g., with a phase locked loop) and then multiplying this phase by $s_2(t)$.

5.18.   An angle modulated wave is described by

$$\lambda(t) = 50 \cos[2\pi \times 10^6 t + 0.001 \cos 2\pi \times 500t]$$

(a) What is $f_i(t)$, the instantaneous frequency of $\lambda(t)$?
(b) What is the approximate bandwidth of $\lambda(t)$?
(c) Is this a narrowband or wideband angle modulated signal?
(d) If this represents an FM signal, what is $s(t)$, the information signal?
(e) If this represents a PM signal, what is $s(t)$?

5.19.   You are given the sum $(L + R)$ and difference $(L - R)$ signals in a stereo receiver system. Design a simple resistive circuit that could be used to produce the individual left and right signals from the sum and difference waveforms.

5.20.   An information signal $s(t) = 5 \cos 2000\pi t$ frequency-modulates a carrier of frequency 10^6 Hz. $A = 100$ and $k_f = 100$ Hz/volt. Noise of power spectral density 10^{-3} is added during transmission.

(a) Find the S/N at the output of an FM receiver, and express this in decibels.
(b) If the output is put through a bandpass filter with $H(f) = 1$ for $900 < |f| < 1,100$, find the signal to noise improvement of the filter.
(c) Repeat parts (a) and (b) for AM DSBTC. Compare the results.

5.21.   Repeat Problem 5.20 for $s(t) = 5 \cos 2,000\pi t + 3 \cos 4000\pi t$.

5.22.   Repeat Problem 5.20 for $s(t) = (100 \sin 1,000t)/t$.

5.23.   Repeat the analysis of Problem 5.20 if PM is used instead of FM (give answers in terms of k_p). Does a range exist for k_p such that PM gives better performance than FM? If so, give that range.

5.24.   You are given the FM signal

$$\lambda(t) = 100 \cos(2\pi \times 10^5 t + 35 \cos 100\pi t)$$

This signal is received along with additive white noise of power spectral density $N_0/2 = 0.1$. Find the average number of clicks per second in the output of the detector.

5.25.   An FM signal is given by

$$\lambda(t) = 50 \cos(2\pi \times 10^6 t + 0.001 \cos 2\pi \times 500t)$$

Noise of power spectral density $N_0/2 = 0.05$ is added to the signal. The sum of signal plus noise enters the detector. Is this detector operating near threshold? If your answer is yes, find the average number of clicks per second at the output of the detector.

5.26.   An FM signal is given by

$$\lambda(t) = A \cos(2\pi \times 10^5 t + 35 \cos 100\pi t)$$

White noise with power of $N_0 = 0.01$ watts/Hz is added to this signal, and the sum forms the input to a detector.
(a) Find the value of A so that the average number of clicks per second is 5.
(b) For the value of A in part (a), find the signal to noise ratio at the detector output.

Chapter 6

Pulse Modulation

Prior to the present chapter we have been discussing analog continuous forms of communication. An analog signal can be used to represent a quantity which can take on a continuum of values. For example, the pressure wave emitted by a pair of vocal chords when a person speaks constitutes an analog continuous waveform.

A *discrete* signal is a signal which is not continuous in time. That is, it only occurs at disconnected points of the time axis. As an example, a teletype signal would be considered as a discrete signal.

6.1   SENDING ANALOG SIGNALS BY DISCRETE TECHNIQUES

The sampling theorem suggests one technique for changing a bandlimited analog signal, $s(t)$, into a discrete time signal. We need only sample the continuous signal at discrete points in time. For example, a list of numbers could be specified, $s(0)$, $s(T)$, $s(2T), \ldots$, where $T < \frac{1}{2} f_m$. In order to transmit this discrete sampled signal the list of numbers could be read over a telephone or written on a piece of paper and sent through the mail. Neither one of these two suggested techniques is very desirable for obvious reasons.

A more attractive technique would be to modulate some parameter of a carrier signal in accordance with the list of numbers. This modulated signal could then be transmitted over wires or, if the band of frequencies it occupies allows, through the air.

What type of carrier signal should be used in this case? Since the information is in discrete form, it would be sufficient to use a discrete carrier signal as opposed to the continuous sine wave used previously.

We therefore choose a periodic pulse train as a carrier. The parameters which can be varied are the height, width and the position of each pulse. Varying one of these three leads to pulse amplitude modulation (PAM), pulse width modulation (PWM), and pulse position modulation (PPM), respectively.

Pulse amplitude modulation is the most commonly used technique and is the only one of the three which is easy to analyze mathematically.

6.2   PAM—GENERAL DESCRIPTION

Figure 6.1 illustrates an unmodulated carrier wave, $s_c(t)$, a representative information signal, $s(t)$, and the resulting pulse modulated signal, $s_m(t)$.

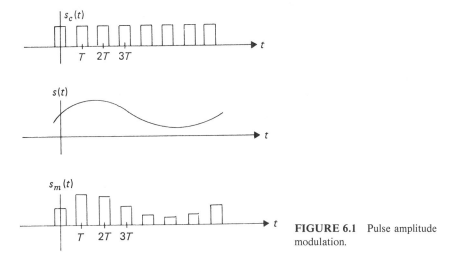

FIGURE 6.1   Pulse amplitude modulation.

In this form, PAM allows only the amplitude of the carrier pulses to vary. The pulse shapes remain unchanged. We therefore note that $s_m(t)$ is *not* the product of $s(t)$ with $s_c(t)$. If this were the case, the heights of the various pulses would not be constant, but would follow the curve of $s(t)$ as shown in Fig. 6.2.

$s_m(t)$ certainly contains sufficient information to enable recovery of $s(t)$. This is simply a restatement of the sampling theorem.

In order to assess transmission system requirements, we will now find the Fourier Transform of $s_m(t)$. A simple observation makes this calculation straightforward.

We observe that if $s(t)$ is first sampled with a train of impulses, the sampled wave consists of impulses whose areas (strengths) are equal to the sample values. If we now put the impulse sampled wave through a filter which changes each impulse into a rectangular pulse, the output of the filter is $s_m(t)$. This is illustrated in Fig. 6.3.

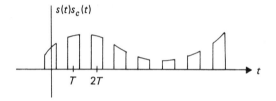

FIGURE 6.2   Product of $s(t)$ and $s_c(t)$ from Fig. 6.1.

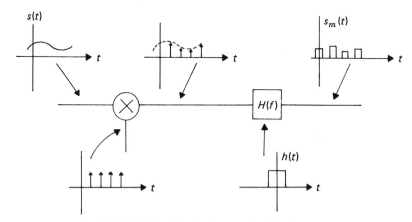

FIGURE 6.3   Idealized generation of $s_m(t)$.

The fact that the filter is non-causal will be of no consequence since we will never actually construct this system.

The input to this filter is

$$\sum_{n=-\infty}^{\infty} s(nT)\delta(t - nT) \tag{6.1}$$

The output of the filter is therefore a superposition of delayed impulse responses. This is given by

$$\sum_{n=-\infty}^{\infty} s(nT)h(t - nT) \tag{6.2}$$

which is seen to be $s_m(t)$. We note that if the discrete carrier, $s_c(t)$, were composed of anything other than rectangular pulses (e.g., triangular) we would simply have to modify $h(t)$ accordingly (*see* Illustrative Example 6.1).

In practice, one would certainly not use a system such as that shown in Fig. 6.3 to generate $s_m(t)$. If the impulse sampled waveform could be generated, it would be sent without modification. The reasons for this will become clear when we speak about multiplexing in a later section. The system of Fig. 6.3 will simply be used as an analysis tool in order to find the Fourier Transform of the PAM wave.

Proceeding to calculate this transform, we know that the transform of the ideal impulse sampled wave is given by

$$\sum_{n=-\infty}^{\infty} s(nT)\delta(t - nT) \leftrightarrow \frac{1}{T} \sum_{n=-\infty}^{\infty} S(f - nf_0) \qquad (6.3)$$

where

$$f_0 \triangleq \frac{1}{T}$$

This was derived in Section 1.10.

Therefore, the transform of the output of the system of Fig. 6.3 is given by

$$\frac{1}{T} H(f) \sum_{n=-\infty}^{\infty} S(f - nf_0) \qquad (6.4)$$

where $H(f)$ is the transform of $h(t)$ as sketched in Fig. 6.4. Figure 6.5 shows the resulting transform of the PAM waveform. We have sketched a general low frequency limited $S(f)$ as we have many times before.

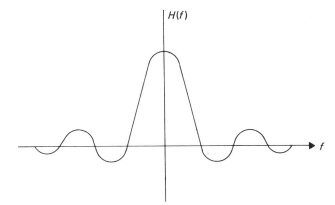

FIGURE 6.4   Transform of $h(t)$ given in Fig. 6.3.

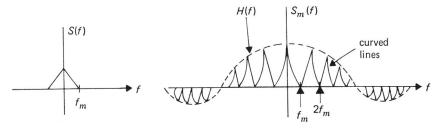

FIGURE 6.5   Transform of output of system in Fig. 6.3.

Illustrative Example 6.1

An information signal is of the form

$$s(t) = \frac{\sin \pi t}{\pi t} \tag{6.5}$$

It is transmitted using PAM. The carrier waveform is a periodic train of triangular pulses as shown in Fig. 6.6. Find the Fourier Transform of the modulated carrier.

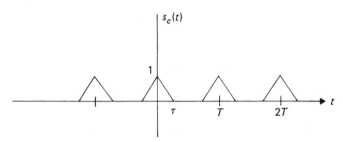

FIGURE 6.6   Carrier for Illustrative Example 6.1.

Solution

We again refer to the system shown in Fig. 6.3. The output of the ideal impulse sampler has a transform given by

$$S_s(f) = \frac{1}{T} \sum_{n=-\infty}^{\infty} S(f - nf_0) \tag{6.6}$$

where $S(f)$ would be the transform of $\sin \pi t/\pi t$. This transform is a pulse, and is shown in Fig. 6.7.

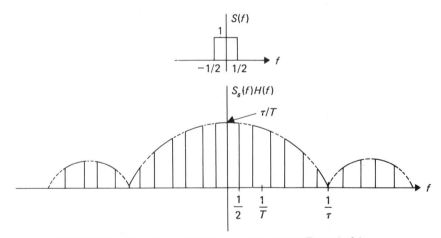

FIGURE 6.7   Transform of PAM wave of Illustrative Example 6.1.

In this case, the filter would have to change each impulse into a triangular pulse. Its impulse response is therefore a single triangular pulse which has as its transform (*see* Appendix III),

$$H(f) = \frac{4 \sin^2 (f\tau/4\pi)}{\tau f^2/16\pi^2} \tag{6.7}$$

Finally, the transform of the PAM waveform is given by the product of $S_s(f)$ with $H(f)$ as shown in Fig. 6.7.

The significant general observation to make about the transform of a PAM wave is that it occupies all frequencies from zero to infinity. It would therefore seem that neither efficient transmission through the air nor frequency multiplexing are possible.

Since the most significant part of the transform of the PAM wave lies around zero frequency, one often uses either AM or FM to send PAM waves. That is, we treat the PAM wave as the information signal and modulate a sine wave carrier with it. If PAM is used in this manner one must obviously invoke new arguments to justify its existence. Otherwise it would be logical to question why one doesn't just transmit the original information signal by AM or FM. The answer to such a question should be obvious if the original information is in discrete form. If it is in continuous analog form one can only justify the "double modulation" on the basis of processing advantages which depend upon the specific application.

After either amplitude or frequency modulation with the PAM wave the bandwidth is huge (actually it is infinity). For this reason pulse modulation combined with AM or FM is usually not transmitted in the same way as our other forms of modulated signals are. It is often sent over wires which are capable of transmitting a wide range of frequencies efficiently. Sometimes it is sent through the air at microwave frequencies. These are high enough so that the large bandwidth does not seem as overpowering relative to the carrier frequency.

Since frequency multiplexing is ruled out it would appear that only one signal can be sent at a time. However, we will see that a different form of multiplexing is ideally suited to pulse modulated waves. Before exploring this new type of multiplexing we shall discuss practical methods of modulation and demodulation of PAM waves.

Modulators

We have already seen one technique for simulating a PAM modulator. The gated modulator used for continuous AM generation (*see* Fig. 4.14) can be adapted to PAM use. It is simply necessary to eliminate the bandpass filter from the output. Figure 6.8 illustrates the modulator, and the diode bridge implementation of the block diagram.

The output waveform resembles the PAM wave. The only variation is that the heights of the pulses are not constant, but follow the shape of the information signal.

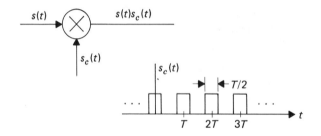

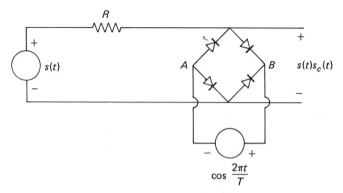

FIGURE 6.8   The gated PAM modulator.

As the pulse widths approach zero, the pulse heights approach a constant. Since the shape of this form of PAM depends upon the signal values over a range around each sampling point, we refer to this as *natural sampled PAM*.

The *flat top*, or *instantaneous sampled* PAM can be generated with a modified sample and hold circuit. The modulator is shown in Fig. 6.9. The only difference between this modulator and the *sample and hold* circuit is the addition of the discharge switch. This is needed because the sample value is not held until the next sample comes along. It is only held for the duration of the PAM pulse. Switch S_1 closes instantaneously at the sample instants. This charges the capacitor to the

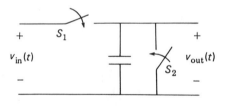

FIGURE 6.9   Idealized PAM modulator.

sample value. S_2 closes periodically at a fixed short interval after each sample point, thus discharging the capacitor and returning the output voltage to zero. With both switches open, the capacitor has no discharge path and remains charged at the previous sample value until S_2 closes.

In the practical realization of this system, S_1 would have to be preceded by some electronics in order to supply the energy to charge the capacitor very rapidly. The capacitor would have to be followed by electronics to prevent the following circuitry from providing a discharge path for the capacitor while both switches are open.

Demodulators

One type of natural sampled PAM demodulator is adapted from the sampling theorem. That is, since the natural sampled PAM modulator is simply a sampler, a lowpass filter can be used to recover the original waveform.

In the case of flat top PAM, the appropriate demodulator is indicated from the Fourier Transform shown in Fig. 6.7. From this figure, one can observe that the low frequency section of $S_m(f)$ is of the form $S(f)H(f)$. Therefore, $s(t)$ can be recovered from $s_m(t)$ by first injecting $s_m(t)$ into a filter with transfer function $1/H(f)$ and then passing the result through a lowpass filter. The first of these two filters is known as an *equalizer* since it cancels the effects of the pulse shape. These two series filters can be combined into a single filter with system function as shown in Fig. 6.10.

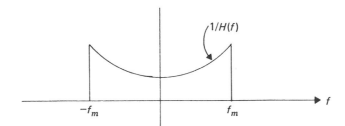

FIGURE 6.10   System function of PAM demodulator.

In order to arrive at a very simple second type of PAM demodulator, one often approximates the lowpass demodulator filter by a sample and hold circuit. The output of such a circuit consists of periodic samples of the input. For times between the sample points the output remains constant at the previous sample value. The output of the sample and hold circuit with PAM as the input is therefore a step approximation to the original information signal. This is illustrated in Fig. 6.11. One form of idealized sample and hold circuit consists of a switch that closes momentarily at the sample instants, and a capacitor which charges up and remains charged during the time that the switch is open. This is shown in Fig. 6.12 and is the same circuit as that of Fig. 6.9 except that the discharge switch has been eliminated. As before, this idealized circuit would need some associated electronics to cancel the effects of loading.

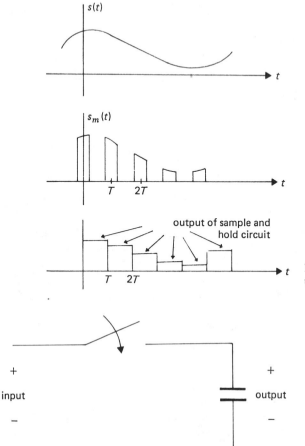

FIGURE 6.11   Sample and hold circuit used as PAM demodulator.

FIGURE 6.12   Simple form of a sample and hold circuit.

The sample and hold is an approximation to the demodulator of Fig. 6.10. The mathematical derivations to illustrate this concept are relegated to the problems at the back of this chapter. Caution must be observed in making the comparison since the sample and hold is not a linear system with respect to continuous input signals. This is true since the output depends only upon the sample values, and not upon functional values in between the sample points. However, with the added constraint of the sampling theorem, the input signal is uniquely characterized by its sample values. With this restriction, the sample and hold becomes an approximation to a continuous lowpass filter.

We shall see that the sample and hold circuit is easily adaptable to the type of multiplexing used in pulse systems. For this reason, it is often preferred to the lowpass filter as a demodulation system.

6.3   TIME DIVISION MULTIPLEXING

We have continually stressed that signals could be separated from each other if they are non-overlapping in either time or frequency. Since the bandwidth of the pulse modulated waveform is extremely high (theoretically, it is infinity), frequency separation is not practical. Fortunately, the ideal pulse modulated waveform possesses portions of the time axis where the waveform is identically zero. Therefore, time separation is possible. When signals are added together such that they do not overlap in time, this is known as *time division multiplexing (TDM)*.

We can easily illustrate time division multiplexing (TDM) by returning to our original example of transmitting a signal by reading the list of sample values over a telephone.

Suppose that the reader is capable of reading two numbers every second, but the maximum frequency of $s(t)$ is only $\frac{1}{2}$ Hz. The sampling theorem tells us that at least one sample must be sent every second. If one sample per second is sent, the reader is idle for one half of every second. During this half second, a sample of another signal could be sent. Alternating between samples of the two signals is called *time division multiplexing*. The person at the receiver will realize that every other number is a sample of the second signal. The two signals can therefore be easily separated.

This operation can be described schematically with two switches, one at the transmitter and one at the receiver. The switches alternate between each of two positions making sure to take no longer than one sampling period to complete the entire circle. (*See* Fig. 6.13.)

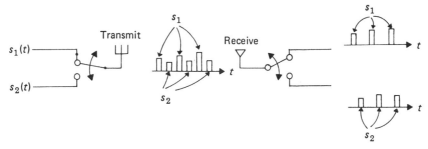

FIGURE 6.13   TDM of two signals.

If we now increase to 10 channels, we get a system as shown in Fig. 6.14.

The switch must make one complete rotation fast enough so that it arrives at channel # 1 in time for the second sample. The receiver switch must rotate in synchronism with that at the transmitter. In practice, this synchronization is not trivial to effect. If we knew exactly what was being sent as # 1, we could identify its samples at the receiver. Indeed, a common method of synchronization is to sacrifice one of the channels and send a known synchronizing signal in its place.

The only thing that limits how fast the switch can rotate, and therefore how many channels can be multiplexed, is the fraction of time required for each PAM

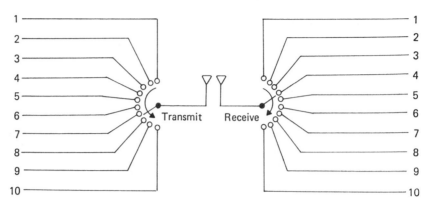

FIGURE 6.14   TDM of 10 signals.

signal. This fraction is the ratio of the width of each pulse to the spacing between adjacent samples of the sample channel. We can now finally appreciate the fact that ideal impulse samples would be beautiful. If this were possible, an infinite number of channels could be multiplexed.

It looks as if we are getting something for nothing. Apparently the only limit to how many channels can be time division multiplexed is placed by the width of each sample pulse.

The minimum width of each sample pulse, and therefore the maximum number of channels which can be multiplexed, is determined by the bandwidth of the communication channel. Examination of Illustrative Example 6.1 (Fig. 6.7) will show that the bandwidth of the PAM wave increases as the width of each pulse decreases. Note that we are assuming some practical definition of bandwidth is used since the ideal bandwidth is infinity.

Problem 6.8 at the end of this chapter indicates that the bandwidth of a multiplexed signal is independent of the number of channels multiplexed, provided that the width of each pulse remains unchanged. Therefore once the pulse width is chosen, system potential would be wasted unless the maximum possible number of non-overlapping signals is squeezed together.

6.4   PULSE WIDTH (DURATION) MODULATION

We again start with a carrier which is a periodic train of pulses as in Fig. 6.1. Figure 6.15 shows an unmodulated carrier, a representative information signal $s(t)$, and the resulting pulse width modulated waveform.

The width of each pulse has been varied in accordance with the instantaneous sample value of $s(t)$. The larger the sample value, the wider the corresponding pulse. Since the pulse width is not constant, the power of the waveform is also not constant. Thus, as the signal amplitudes increase, transmitted power also increases.

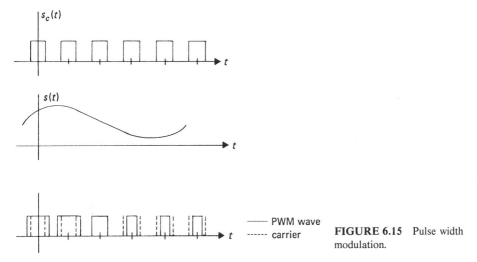

FIGURE 6.15   Pulse width modulation.

In attempting to find the Fourier Transform of this modulated waveform we run into severe difficulties. These difficulties are not unlike those which we encountered in trying to analyze frequency modulation.

Pulse width modulation is a non-linear form of modulation. A simple example will illustrate this. If the information signal is a constant, say $s(t) = 1$, the PWM wave will consist of equal width pulses. This is true since each sample value is equal to every other sample value. If we now transmit $s(t) = 2$ via PWM, we again get a pulse train of equal width pulses, but the pulses would be wider than those used to transmit $s(t) = 1$. The principle of linearity dictates that if the modulation is indeed linear, the second modulated waveform should be twice the first. This is not the case as is illustrated in Fig. 6.16.

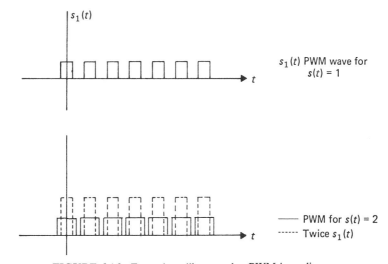

FIGURE 6.16   Example to illustrate that PWM is nonlinear.

If one assumes that the information signal is slowly varying, then adjacent pulses will have almost the same width. Under this condition an approximate analysis of the modulated waveform can be performed. In so doing, a PWM waveform can be expanded in an approximate Fourier Series type expansion. Each term in this series is actually an FM wave instead of being a pure sine wave. We will not carry out this analysis here.

In our study of FM, we found that the easiest way to demodulate a signal was to first convert it to AM. We shall take a similar approach to PWM. That is, we will try to show how one would go about converting one form of pulse modulation into another.

The significant difference between PAM and PWM is that PAM displays the sample values as the pulse amplitudes, and PWM displays these values as the "time amplitudes." We therefore search for a technique to convert back and forth between time and amplitude. We require something whose amplitude depends linearly upon time. A sawtooth waveform as sketched in Fig. 6.17 answers this need.

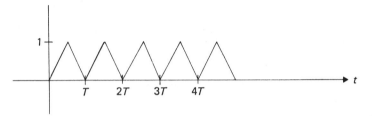

FIGURE 6.17   A sawtooth waveform.

We shall now describe one possible use of the sawtooth to generate PWM. This is not necessarily the simplest system.

We start with an information signal, $s(t)$. This is put through a sample and hold circuit. The sawtooth is then subtracted, and the resulting difference forms the input to a threshold circuit. The output of the threshold circuit is a narrow pulse each time the input goes through zero. Each of these zero crossings then triggers a bistable multivibrator.

The entire system is sketched in Fig. 6.18. Typical waveforms are shown in Fig. 6.19.

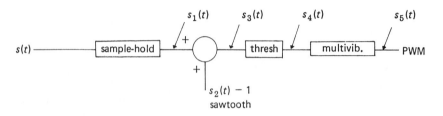

FIGURE 6.18   One way of generating PWM.

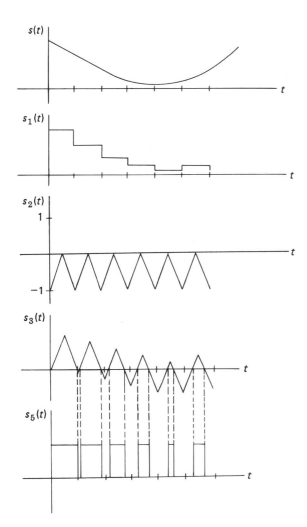

FIGURE 6.19 Typical normalized waveforms in the system of Fig. 6.18.

A technique similar to the ramp (sawtooth) can be used to change PWM to PAM, and, therefore, to demodulate PWM. If each pulse in a PWM waveform is integrated, the result is a ramp whose final value is the area of the pulse. Since the pulse height is constant, the area is proportional to the width.

Therefore, if a PWM waveform forms the input to an integrator which starts at zero at each sampling point, the output of the integrator is a series of ramps. The final value of each ramp is proportional to the previous sample value. Thus, if the integrator output is sampled at points just prior to the original sample points, the result is essentially a sampled version of the original waveform.

6.5   PULSE POSITION MODULATION

Pulse position modulation (PPM) possesses the noise advantage of PWM without the problem of a variable power which is a function of signal amplitude. An information signal, $s(t)$, and the corresponding PPM waveform are illustrated in Fig. 6.20. We see that the larger the sample value, the more the corresponding pulse will deviate from its unmodulated position.

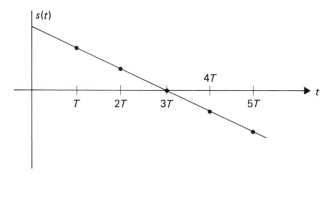

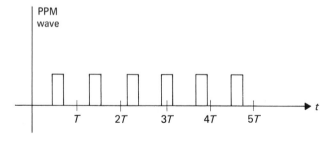

FIGURE 6.20   PPM—the unmodulated pulses would fall at multiples of T.

A PPM waveform can be simply derived from a PWM waveform. We first detect each "downslope" in the PWM wave. We can do this by first differentiating the PWM wave and then looking for the negative pulses in this derivative (ideally they will be impulses). If we were to now place a pulse at each of these points, the result would be a PPM wave. This process is illustrated in Fig. 6.21.

At this time it would be reasonable to ask what the purposes of the present section were, other than to confuse the student. We have probably said much less about this very complicated area of communication theory than we have about some of the much simpler areas. Part of the justification for this is that PWM and PPM are sufficiently complex that not a great deal can be said in general. As such, we have only defined the two types of modulation and have attempted to show possible ways to produce these waveforms. Beyond this, we will close by mentioning that PWM and

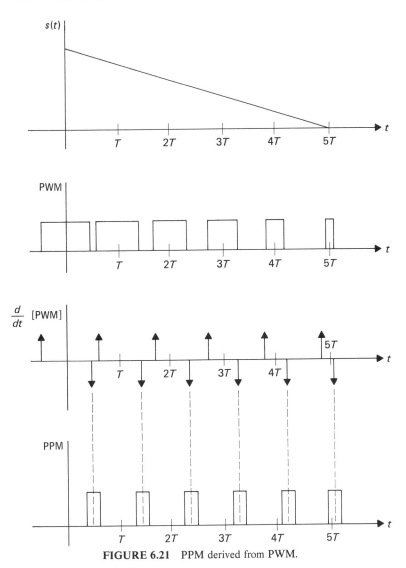

FIGURE 6.21   PPM derived from PWM.

PPM afford a noise advantage over PAM. This is true since the sample values are not contained in the pulse amplitudes. These amplitudes are easily distorted by noise. This noise advantage is similar to that afforded by FM when compared to AM. In both cases we see that going to the more complicated form of modulation leads to certain disadvantages. In FM we required more space on the frequency axis. In pulse modulation, the two non-linear forms will require more space on the time axis. One must be sure that adjacent sample pulses do not overlap. If pulses are free to shift around or to get wider, as they are in PPM and PWM, one cannot simply insert other pulses in the spaces (i.e., multiplex) and be confident that no interaction will occur.

Sufficient spacing must be maintained to allow for the largest possible sample value.

Besides taking more space, these non-linear forms of modulation are difficult, if not impossible, to analyze. The systems necessary to modulate and demodulate are more complicated than those required for linear forms of modulation.

In continuous forms of modulation, if a noise improvement was required, there was little choice but to turn to FM. In discrete pulse modulation, an alternative to PWM and PPM does exist as we shall see in the following chapter.

6.6   CROSS TALK

We now consider one source of error in a noise-free pulse modulation system. As an illustration, assume that phone calls were being transmitted via PAM and that 20 channels were being time division multiplexed. If we look at the transmitted pulse train, we can identify pulses numbered 1, 21, 41, 61, 81, . . . as being the samples of the first signal and pulses 2, 22, 42, 62, 82, . . . as samples of the second signal. If the demultiplexing system is not perfectly synchronized, a portion of pulses 2, 22, 42, . . . will be picked up by the receiver for channel 1. That is, pulse 1 will be corrupted by pulse 2, pulse 21 will be corrupted by pulse 22, etc. The person trying to listen to the first channel will hear part of the second channel in the background. This is an example of an error known as *cross talk*.

The above simplified example attributed cross talk to poor synchronization. In fact, this is not a major source of crosstalk. If a single pulse in the train in traced through the system from transmitter to receiver, we find that it will change shape from one point to another. As an idealized example, if the original pulse were a perfect square, it would have rounded corners after transmission. This is due to the fact that the system function of any transmission medium must approach zero as the frequency approaches infinity. The sharp corners of the pulse represent infinite frequencies and will therefore be rounded by the lowpass filter effect. Figure 6.22 shows a perfect

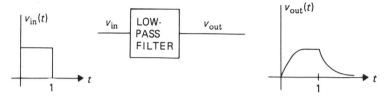

FIGURE 6.22   Effect of lowpass filter on pulse.

square pulse and then the result of passing this pulse through a simple lowpass filter comprised of a resistor and capacitor combination. It should be clear that the rounded pulse occupies a wider portion of the time axis than the original pulse. It is therefore possible for the received pulses in a train to overlap even though the transmitted pulses do not overlap. This pulse widening and overlap due to the transmission characteristics are the cause of cross talk. It has its exact dual in a frequency multiplexed system. For example, in broadcast AM radio, if one station were to spread beyond its allocated 5 KHz of bandwidth, a receiver tuned to an adjacent station would hear the undesired signal in the background. The frequency spreading would occur if the signal were gated in time, just as the time spreading in TDM pulse systems results from gating in frequency (the lowpass filter characteristic of the transmission system is a frequency gate).

We will see in the next chapter that in a digital communication system, the pulses used to transmit the sample values will be replaced by pulses used to transmit numbers. For that reason, the pulses are often referred to as symbols, and cross talk is the same as *intersymbol interference*.

There are several ways of reducing the effects of intersymbol interference. Pulse synchronization is a minor problem, and care should be exercised to design the system so that timing jitter (wiggling, or slight random variations in the timing of sampling points) is a minimum.

Pulse spreading is a major cause of cross talk. This spreading can be decreased by increasing the bandwidth of the transmission system. Unfortunately, this approach is a luxury which requires a flexibility we don't often have in a communication system. However, one parameter over which we do have control is the shape of the pulses used to transmit the sample values.

Suppose first that we use ideal impulses and that the channel can be modelled as an ideal lowpass filter. The pulses at the receiver will then be of the form of $(\sin t)/t$ as shown in Fig. 6.23. This result will apply for non-ideal impulses if the product of the

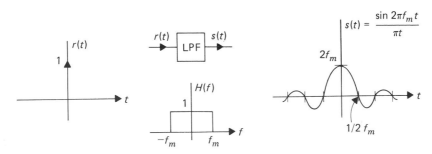

FIGURE 6.23   Pulse shaping with the ideal lowpass filter channel.

transmitted pulse transform with the system function of the channel is equal to the lowpass characteristic. That is, the transmitted pulse need not be an ideal impulse to obtain this result. The transmitted waveform can be changed as long as the channel characteristic is adjusted inversely with the Fourier Transform of the transmitted pulse.

The $(\sin t)/t$ waveform has the very nice property that it is zero at evenly spaced points away from the origin. In fact, if the highest frequency the channel passes is f_m, the zeros of the spread pulse are spaced by $1/f_m$. Thus, if the sampling rate is any integral multiple of this spacing, the cross talk can be completely eliminated by

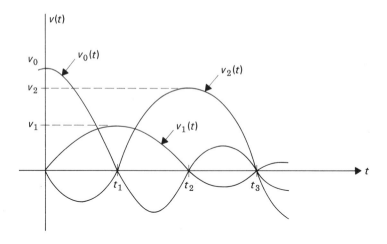

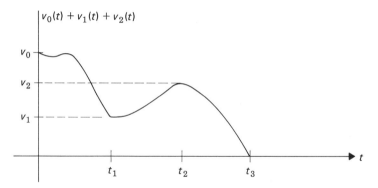

FIGURE 6.24   Ideal lowpass filter shaping does not change sample values.

sampling at the midpoints of each interval. Figure 6.24 shows a composite of three sequential pulses and the resulting sum. The sample values of the sum are the correct original sample values.

Figure 6.24 is really no more than a restatement of the sampling theorem in reverse. If the channel passes frequencies up to $1/2T_s$, where T_s is the sampling period, then the individual sample values are independent. Any single sample value can be chosen independently of adjacent sample values.

The ideal bandlimited pulse shape is difficult to achieve because of the sharp corners on the frequency spectrum. A desirable compromise is the *raised cosine* characteristic. The Fourier Transform of this pulse is similar to the square transform of the ideal lowpass filter, except that the transition from maximum to minimum follows a sinusoidal curve. This is shown in Fig. 6.25. The value of the constant K determines the width of the constant portion of the transform. If $K = 0$, the transform

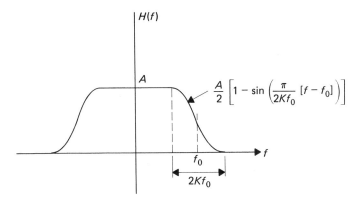

FIGURE 6.25   Raised cosine frequency characteristic.

is that of the ideal lowpass pulse of Fig. 6.23. If $K = 1$, the flat portion is reduced to a point at the origin. The time function corresponding to this Fourier Transform is

$$h(t) = A \frac{\sin 2\pi f_0 t}{2\pi f_0 t} \frac{\cos 2\pi K t f_0}{1 - (4K f_0 t)^2} \qquad (6.8)$$

This time function is sketched for several representative values of K in Fig. 6.26. Note that for $K = 0$, the time function is of the form $(\sin t)/t$. This goes to zero at multiples of $1/2f_0$. For $K = 1$, the response not only goes to zero at these points, but also at points midway between these values (except for the first set of points around the origin). Note that for $K = 1$, the transform goes up to $2f_0$. Therefore, to compare this with the ideal lowpass case, we must set $f_m = 2f_0$. Doing this, we see that it is possible to communicate at the same sampling rate for the ideal lowpass and for the raised cosine system. However, with the raised cosine system, intersymbol interference does occur for one pulse on either side of the desired pulse. Of course, by using the raised cosine with twice the bandwidth, no cross talk occurs. Even with the same bandwidth, there are relatively simple ways to eliminate this limited form of interference (*Partial response signalling*—see References).

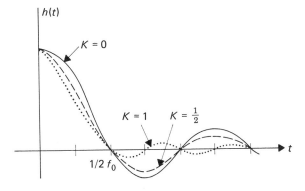

FIGURE 6.26   Time function corresponding to transform of Fig. 6.25.

PROBLEMS

6.1. Sketch the schematic for a circuit that will produce the product $s(t)s_c(t)$ where $s(t)$ and $s_c(t)$ are as shown in Fig. 6.1.

6.2. Given the product, $s(t)s_c(t)$ as in Problem 6.1, sketch a system that could be used to recover $s(t)$. Verify that this system works.

6.3. You are given a low frequency bandlimited signal, $s(t)$. This signal is multiplied by the pulse train, $s_c(t)$ as shown below. Find the transform of the product, $s(t)s_c(t)$. What restrictions must be imposed so that $s(t)$ can be uniquely recovered from the product waveform?

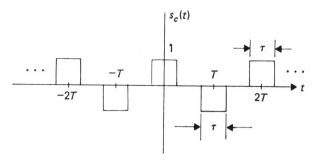

6.4. An information signal is of the form

$$s(t) = \frac{\sin \pi t}{\pi t}$$

Find the Fourier Transform of the waveform resulting if each of the two carrier waveforms shown is pulse amplitude modulated with this signal.

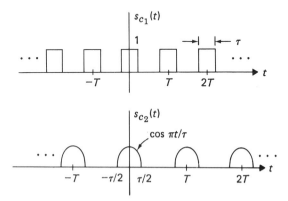

6.5. Sketch a block diagram of a typical pulse amplitude modulator and demodulator.

6.6. Consider the system shown below. We wish to compare $y(t)$ to $x(t)$ in order to evaluate the sample and hold circuit as a PAM demodulator. The comparison between $y(t)$ and $x(t)$ is performed by defining an error, e, as follows:

$$e = \frac{1}{T} \int_0^T [y(t) - x(t)]^2 \, dt$$

Find the value of this error term.

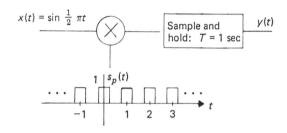

6.7. Show that a sample and hold circuit is an approximation to a lowpass filter provided that the sampling is done at the Nyquist rate or higher. (Hint: You may wish to find the step response of the sample and hold circuit and compare this to the step response of a lowpass filter.)

6.8. Consider a two channel TDM PAM system where both channels are used to transmit the same signal, $s(t)$, with Fourier Transform $S(f)$ as shown. Sample $s(t)$ at the minimum rate (1 sample per second). Sketch the Fourier Transform of the TDM waveform and compare it to the Fourier Transform of a single channel PAM system used to transmit $s(t)$. (Hint: Consider $s_p(t)$ as shown in Problem 6.6. The single channel system transmits $s(t)s_p(t)$, while the TDM system can be thought of as transmitting $s(t)\hat{s}_p(t)$ where $\hat{s}_p(t) = s_p(t) \times s_p(t - \frac{1}{2})$.)

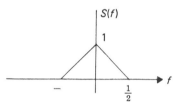

6.9. Given a bandpass signal $s(t)$, where $S(f) = 0$ for f outside of the range $f_1 < |f| < f_2$, is it possible to recover $s(t)$ from samples spaced at

$$T_s < \frac{1/2}{f_2 - f_1}$$

rather than $T_s < \frac{1}{2}f_2$? Discuss.

6.10. Three information signals are to be sent using PAM and TDM. Suppose that the maximum frequency of each of the first two signals is 5 kHz and that the maximum frequency of the third signal is 10 kHz. Can these signals be time division multiplexed? Is it possible to time division multiplex these signals while sampling each at its Nyquist rate? Sketch the block diagram of a system that might accomplish this.

6.11. Explain why a PPM system would require some form of time synchronization while PAM and PWM do not require any additional information. (We are talking about a single channel without TDM.)

6.12. Show that the system function of Fig. 6.25 is the Fourier Transform of the $h(t)$ of Eq. (6.8).

6.13. Ten signals are to be time division multiplexed and transmitted using PAM. Four of the signals have a maximum frequency of 5 kHz, two have a maximum frequency of 10 kHz, two have a maximum frequency of 15 kHz, and two have a maximum frequency of 20 kHz. Each of the signals is sampled at its Nyquist rate.

(a) Design a multiplex system.

(b) How may pulses per second must be transmitted?

Chapter 7
Digital Communication

The earlier portions of this text dealt with transmission of a continuous time signal. The vehicle for transmission was either a continuous time function or a discrete (pulsed) time function.

The present chapter adds another dimension to the techniques for transmission. The actual signal values transmitted will become members of a discrete set. Thus, instead of dealing with the transmission of a continuum of voltage values, we concentrate upon a finite set of discrete values. The advantages of this approach will shortly become evident.

As was the case with pulse modulation, we first consider transmission of a list of numbers. This could result from a sampling operation on a time function, or the original information could be in the form of a list. The essential difference between digital communication and pulse modulation is that the numbers in the list can only take on discrete values. Such forms of transmission are termed *digital* as distinct from *analog*.

Many signals are already in the form of a list of numbers drawn from a finite set. The output from a digital computer is in this format. Other common examples include the time of day (if we round to the nearest second or minute), quantities of a particular item manufactured each hour (e.g., number of cars off an assembly line), and Teletype messages (where only the alphanumeric characters are permitted). In contrast with these digital signals, the audio signal representing someone speaking is an important example of an analog signal. If digital transmission techniques are to be used for the latter signal, we must somehow transform to a digital format. The transformation from an analog to a digital format is performed, quite logically, by an *analog to digital converter* (A/D or ADC).

7.1   A/D AND D/A CONVERTERS

The first step in changing an analog continuous time signal into a digital form is to convert the signal into a list of numbers. This is accomplished by sampling the time function. The resulting list of numbers will represent a continuum of values. That is, although a particular sample may be stated as a rounded-off number (e.g., 5.758 volts), in actuality it should be continued as an infinite decimal. The list of numbers must therefore be *coded* into discrete *code words*. The first suggestion one might have as a way to accomplish this is to round off each number in the list. Thus, for example, if the samples ranged from 0 to 10 V, each sample could be rounded to the nearest integer. This would result in code words drawn from the 11 integers between 0 and 10.

In a majority of digital communication systems, the actual form chosen for code words is a binary number composed of 1's and 0's. The reasons for this choice will become clear when we discuss specific transmission techniques. With this binary restriction, we can envision one simple form of analog to digital converter for our "0- to 10-V" example (not necessarily the best). The converter would operate on the 0- to 10-V samples by first rounding each sample value to the nearest volt. It would then convert the resulting integer into a 4-bit binary number.

While practical A/D converters perform an operation similar to that outlined above, an explanation of this operation is best introduced by starting from scratch.

Analog to digital conversion is also known as *quantizing*. The goal is to change a continuous variable into one that has discrete values. In *uniform quantization*, the continuum of functional values is divided into uniform regions, and an integer code is assigned to each region. Thus, all functional values within a particular region are coded into the same number.

Figure 7.1 illustrates the concept of 3-bit quantization in two different ways. Figure 7.1(a) shows the range of functional values divided into 8 regions. Each of

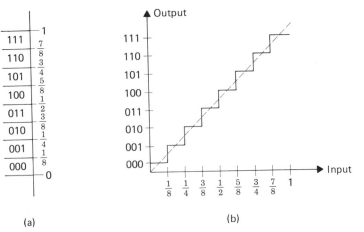

(a)   (b)

FIGURE 7.1   The concept of quantization.

these regions is assigned a 3-bit binary number. We have chosen 8 regions because this is a power of 2. All 3-bit binary combinations are used, leading to greater efficiency. Also note that the range of values is given as that between zero and unity. This may seem restrictive, but any function can be normalized to fall in this region through addition of a constant and scaling.

Before leaving this figure, we shall use it to make a general observation about binary counting. As you examine the binary numbers along the ordinate, you should note that the first bit is equal to one for the top half of the range, and zero for the bottom half. It oscillates with a period equal to the total range. The next bit alternates with a period equal to half of the range and is equal to one for the top half *of each half*, and zero for the bottom half *of each half*. This pattern continues with each successive bit subdividing the region by two, and indicating which half of the new subregion the sample value is in.

Figure 7.1(b) illustrates quantization by use of an input-output relationship. While the input is continuous, the output can take on only discrete values. The width of each step is constant, since the quantization is uniform. We shall explore non-uniform quantization in Section 7.4.

Figure 7.2 shows a representative $s(t)$ and the resulting digital form of the signal for both 2-bit and 3-bit ADC.

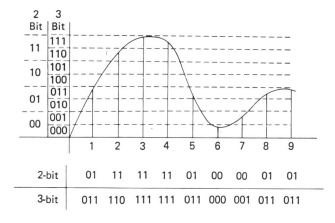

FIGURE 7.2 Result of ADC for sample $s(t)$.

Quantization (Coding)

There are several generic forms for the quantizer. These forms can be categorized into three groups:

1. *Counting quantizers*, which serially count through each quantizing level.

2. *Serial quantizers*, which generate a code word, bit by bit. That is, they start with the most significant bit and work their way to the least significant bit.

3. *Parallel quantizers*, which generate all bits of a complete code word simultaneously.

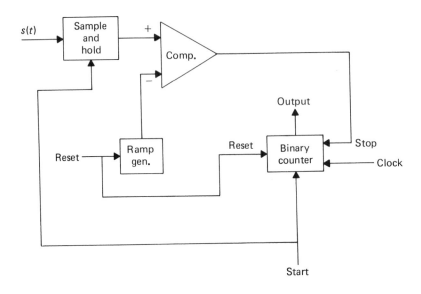

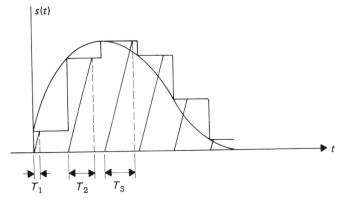

FIGURE 7.3   The counting quantizer.

Counting Quantizers

Figure 7.3 shows a counting quantizer. The ramp generator starts at each sampling point, and a binary counter is simultaneously started. The output of the sample and hold system is a staircase approximation to the original function, with stair steps that stay at the sample values throughout each sampling interval. A typical waveform is shown in the figure. The time duration of the ramp, and therefore the duration of the count, T_s, is proportional to the sample value. This is true because the ramp slope is kept constant. If the clock frequency is such that the counter has enough time to count to its highest count (all 1's) for a ramp duration corresponding to the maximum possible sample, then the ending counts on the counter will correspond to the quantizing levels.

Illustrative Example 7.1

Given that the ramp slope is 10^6 V/sec, signal amplitudes range from 0 to 10 V, and a 4-bit counter is used, what should the clock frequency be for a voice signal? Assume a maximum frequency of 3 kHz.

Solution

The only reason for worrying about the maximum frequency of the signal is to see if the ramp slope is sufficient to reach the maximum possible values within one sampling period. With a maximum signal frequency of 3 kHz, the minimum sampling rate for recovery (i.e., to avoid aliasing) is 6 kHz, so the maximum sampling period is $\frac{1}{6}$ msec. Since the ramp can reach the 10-V maximum in 0.01 msec, it is sufficiently fast to avoid overload problems. Therefore, the counter must be capable of counting from 0000 to 1111 in 0.01 msec. The clock frequency must be 1.6 MHz, since up to 16 counts are required in this sampling period.

Illustrative Example 7.2

It is desired to build a counting ADC to convert $s(t) = \sin 2\pi t$ into a 4-bit digital signal. Choose appropriate parameters for the ADC.

Solution

To comply with the sampling theorem, the sampling rate must be greater than 2 samples/sec. We can therefore assume a sampling rate of 2.5 samples/sec (there are many factors which go into choosing the rate, and a rate several times the theoretical minimum is often chosen). It is now necessary to set the slope of the ramp generator. The maximum slope of the signal is 2π. The slope of the ramp must be considerably greater than this value to assure that the intersection points will be close to the sample points. For this example, let us choose a ramp slope of 100 V/sec. Since the maximum value of the signal is unity, it will take the ramp function 0.01 sec to reach this maximum. The counter should therefore count from 0000 to 1111 in 0.01 sec. This requires a counting rate (clock) of 1 count/0.000625 sec, or 1,600 counts/sec.

Serial Quantizers

We now turn our attention to serial quantizers. The serial quantizer successively divides the ordinate into two regions. It first divides the axis in half and observes whether the sample is in the upper or lower half. The result of this observation generates the most significant bit in the code word. The half-region in which the sample lies is then subdivided into two regions, and a comparison is again performed. This generates the next bit. The process continues a number of times equal to the number of bits of encoding.

Figure 7.4 shows a block diagram form of this encoder for 3 bits of encoding and for inputs in the range 0 to 1. The diamond-shaped boxes are comparators. They

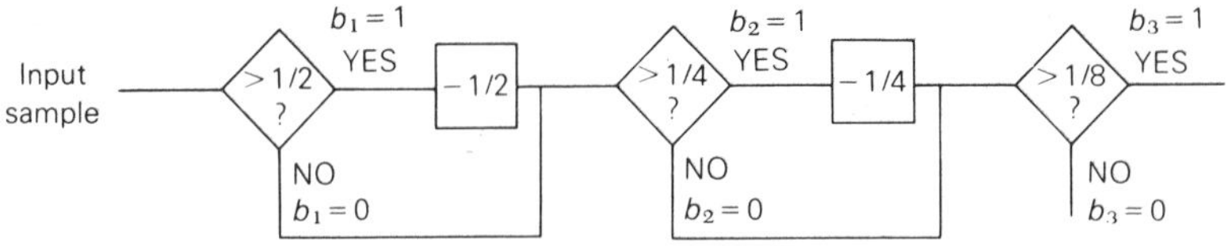

FIGURE 7.4   A/D converter.

compare the input to some fixed value and give one output if the input exceeds that fixed value and another output if the reverse is true. The block diagram indicates these two possibilities as two possible output paths labelled YES and NO. The figure is shown for 3-bit code words and for a range of input values between 0 and 1 V. If the input range of signal sample values were not 0 to 1, the signal could be scaled (amplified or attenuated) to achieve values within this range. If more (or fewer) bits are required, the appropriate comparison blocks can be added (or removed).

Note that b_1 is the first bit of the coded sample value, sometimes known as the *most significant bit* or msb. b_3 is the third and final bit of the coded sample value, known as the *least significant bit* or lsb. The reason for this terminology is that the weight associated with b_1 is 2^2 or 4, while the weight associated with b_3 is 2^0 or 1. The ADC of Fig. 7.1 is performing the exact operation described previously. It performs a successive approximation to the sample value. First we ask whether the sample value is in the upper or lower half of the range. Then we investigate the half-interval to see if the sample is in the upper or lower half *of that half*. Thus, each successive bit (moving from left to right in the binary number) divides the previous region in half. This is true since each bit in a binary number is weighted by a power of 2, and as we move to the right in a binary number, the weight of each bit is one half of the previous weight.

Illustrative Example 7.3

Illustrate the operation of the system of Fig. 7.4 for the following two input sample values: 0.2 and 0.8 V.

Solution

For 0.2 V, the first comparison with $\frac{1}{2}$ would yield a NO answer. Therefore $b_1 = 0$. The second comparison with $\frac{1}{4}$ would yield a NO answer. Therefore, $b_2 = 0$. The third comparison with $\frac{1}{8}$ would yield a YES answer. Therefore, $b_3 = 1$, and the binary coding of 0.2 into 3 bits would be 001.

For the 0.8-V input, the first comparison with $\frac{1}{2}$ would yield a "YES" answer. Therefore $b_1 = 1$, and we subtract $\frac{1}{2}$, leaving 0.3. The second comparison with $\frac{1}{4}$ would yield a YES answer. Therefore, $b_2 = 1$, and $\frac{1}{4}$ is subtracted, leaving 0.05. The third comparison with $\frac{1}{8}$ would yield a NO answer. Therefore, $b_3 = 0$, and the binary coding of 0.8 into 3 bits would be 110.

A simplified system can be realized if, at the output of the block marked "$-\frac{1}{2}$" in Fig. 7.4, a multiplication by 2 is performed and the result is fed back into the

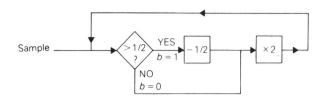

FIGURE 7.5   Simplified ADC of Fig. 7.4.

comparison with $\frac{1}{2}$. All blocks to the right can then be eliminated, as in Fig. 7.5. The signal sample can be cycled through as many times as desired to achieve any number of bits of code word length.

Parallel Quantizers

The parallel quantizer is the fastest in operation, since it develops all bits of the code word simultaneously. It is also the most complex, requiring a number of comparators that is only one less than the number of levels of quantization. We illustrate this with an example using 3 bits of encoding. Figure 7.6 shows a block diagram of the parallel 3-bit encoder. The block labeled "coder" observes the output of the seven comparators. It is a relatively simple combinational logic circuit. If all seven outputs are 1 (YES), the coder output is 111, since the sample value had to be greater than $\frac{7}{8}$. If comparator outputs 1 through 6 are 1, the coder output is 110, since the sample had to be between $\frac{6}{8}$ and $\frac{7}{8}$. We continue through all levels and finally, if all comparator outputs are low, the sample had to be less than $\frac{1}{8}$, and the coder output is 000. (Readers familiar with binary logic design will note that of the 128 input possibilities, only 8 are legal. The design can thus be simplified by using 120 don't care conditions.)

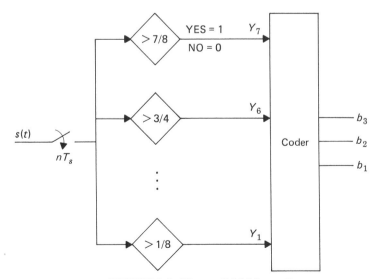

FIGURE 7.6   The parallel 3-bit quantizer.

Decoding (DAC)

We now shift our attention to the conversion of a digital signal to an analog signal. This is performed by a *digital to analog converter* (D/A converter or DAC). To perform this conversion, we need simply associate a value with each binary code word. If the code word represents a region of sample values, the actual value chosen for the conversion is usually the center point of the region. If the A/D conversion is performed as previously described, the reverse operation is equivalent to assigning a weight to each bit position. In Illustrative Example 7.2, the interval was divided into 16 regions. Therefore, for example, the lower limit of the region corresponding to the code word "0011" is $\frac{3}{16}$ V. The lower limit of the region corresponding to the code word "1111" is $\frac{15}{16}$. To reach the midpoint of the region, we add $\frac{1}{32}$ to each of these lower limits.

We see that conversion to the analog sample value is accomplished by converting the binary number to decimal, dividing by 16, and adding $\frac{1}{32}$. The conversion of the binary number to the decimal equivalent is accomplished by recognizing that each column in a binary number has associated with it a power of 2. Thus, a "1" in the leftmost position indicates 2^3, or 8. A "1" in the next position indicates 2^2, or 4. An analog voltage can be reconstructed from a digital code word by using the 1's and 0's to switch appropriate sources and adding the results together. If a 1 appears in the first position, a $\frac{1}{2}$-V battery is switched in. The second position controls a $\frac{1}{4}$-V battery, and the third position, a $\frac{1}{8}$-V battery. An idealized circuit that accomplishes this conversion is shown in Fig. 7.7. The addition of $\frac{1}{32}$ V to arrive at the midpoint of the region is included in the figure.

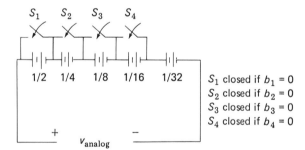

S_1 closed if $b_1 = 0$
S_2 closed if $b_2 = 0$
S_3 closed if $b_3 = 0$
S_4 closed if $b_4 = 0$

FIGURE 7.7   Idealized D/A converter—4-bit.

The ideal decoder of Fig. 7.7 is analogous to the serial quantizer, since each bit is being associated with a particular component of the sample value. Figure 7.8 shows a practical realization of this idealized system.

The gain of the operational amplifier shown in Fig. 7.8 is $R/5R_{in}$, where R_{in} is the parallel combination of the resistors included in the input circuit (i.e., those with associated switches closed). In terms of input conductance, the gain is $RG_{in}/5$, where G_{in} is the sum of the conductances associated with closed switches. Therefore, closing S_1 contributes $\frac{8}{5}$ to the gain. Closing S_2 contributes $\frac{4}{5}$ to the gain; S_3 contributes $\frac{2}{5}$; S_4 contributes $\frac{1}{5}$. In order to calculate the gain due to more than one switch being closed,

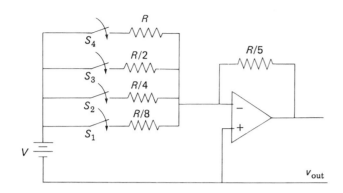

FIGURE 7.8   DAC using operational amplifier.

we simply add the associated gains. Therefore if the input voltage source, V, is made equal to 5 V, we see that S_1 contributes 8 V to the output; S_2, 4 V; S_3, 2 V; and S_4, 1 V. The output voltage is an integer between 0 and 15 corresponding to the decimal equivalent of the binary number controlling the switches. To convert this system into a DAC analogous to that of Fig. 7.7 we need simply divide the output by 16 and add $\frac{1}{32}$ V. The division can be performed by scaling V or by scaling the resistor values. The addition of $\frac{1}{32}$ V can be accomplished by adding a fifth resistor (unswitched) to the input. The value of this resistor would be $2R$.

A more complex decoder results when an analogy to the counting operation is attempted. Figure 7.9 shows the counter decoder.

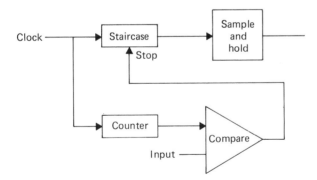

FIGURE 7.9   The counting decoder (digital to analog converter).

A clock feeds a staircase generator and, simultaneously, a binary counter. The output of the binary counter is compared to the binary digitized input. When a match occurs, the staircase generator is stopped. The output of the generator is sampled and held until the next sample value is achieved. The final staircase approximation result is smoothed by a lowpass filter to recover the original signal.

7.2   CODED COMMUNICATION

The previous section described techniques for converting a continuous analog signal into a digital signal. The digital signal was composed of a list of numbers, where the numbers could take on only a finite number of values. We saw that the list of numbers did not exactly equal the original sample values but represented rounded-off versions of these values. Therefore in addition to the necessity of doing extra work in converting an analog signal into a digital signal, the digital signal cannot be used to perfectly reconstruct the original analog signal. Why, then, might somebody wish to change an analog signal into a ditigal signal? The present section seeks to answer this question.

The earliest forms of communication were by telegraph. A series of dots and dashes were transmitted to spell out each word. This was done since the complex audio signals could not be distinguished from the background static along a line, while a long and short duration of a tone could be distinguished.

In early telegraph, the coding and decoding of the signals were done by a human operator. The operator read (or listened to) the message and changed each letter into its Morse coded version. In receiving a message, the operator reversed this procedure. The speed of transmission had to be very carefully controlled so that it did not exceed the speed of the keyer (operator). Although the telegraph transmission line, by virtue of its bandwidth limitations, did place a maximum limit on the speed of transmission, this was never a serious consideration. The human operator never approached this speed limit even though the early telegraph lines were highly inefficient by today's standards.

After many years of favoring non-coded communication, we are rapidly returning to the coded forms for two major reasons. The first is related to the original (telegraph) reason. The air is being polluted with so many electrical signals that noise-free communication is virtually impossible. Signals are constantly corrupted by static and other forms of interfering signal.

Unless a great deal is known about the kind of signal expected at the receiver, no effective techniques of correcting the noise corruption are known. That is, when the signal is changed by noise and the altered signal is received, there must exist some way of separating the noise from the signal. This requires that the signal have some distinguishing form. Such a form is provided through the use of coding.

The second reason for the re-emphasis upon digital coded communication is the changing format of information signals. For many years the only type of information signal encountered was the audio signal limited to frequencies below the cutoff for human hearing. Today we find machines communicating data to other machines using information signals that are considerably different from audio waveforms. The demands placed upon a communication system are much more complicated than those resulting from voice communication.

In a coded communication system, the signal can take on only one of several discrete forms. In most cases when the set of discrete forms is small, an acceptable signal will not change into another acceptable signal when corrupted by noise.

Using more acceptable terminology, we transmit a word from a *dictionary* of acceptable message words. The received word will usually not exactly match any entry in our dictionary.

If the noise corruption is not too great, we can make an educated guess as to which dictionary word was sent. This is the essence of coded communication.

7.3   PULSE CODE MODULATION (PCM)

Pulse code modulation (PCM) is a direct application of the analog to digital converter discussed in Section 7.1. Rather than simply present it as such, we will rederive the results using a slightly different approach. We shall begin this derivation with a pulse modulation system in order to illustrate some physical concepts and to develop some of the relevant terminology.

Suppose that in a PAM system the amplitude of each pulse is rounded off to one of several possible levels. Alternatively, suppose the original continuous time function is first rounded off to several possible levels (yielding a staircase-type function) and then this rounded-off function is sampled for PAM transmission. An example of this rounding-off operation is shown in Fig. 7.10.

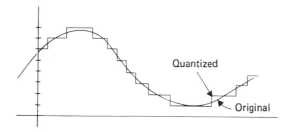

FIGURE 7.10   Example of rounding off to discrete levels.

The rounding-off process is known as *quantization*, and its performance introduces an error known as *quantization noise*. That is, the staircase approximation is not identical to the original function, and the difference between the two is an error signal. This will be discussed in detail in Section 7.6.

The dictionary of possible signals (pulse heights) has been reduced to include only those quantization levels used. A received pulse is compared to the possible transmitted pulses, and it is decoded into the dictionary entry which most closely resembles the received signal. In this way, small errors (up to one half of the quantization step size) can be corrected.

This principle of error correction is the major reason for the popularity of PCM. Suppose, for example, it were desired to transmit a signal from California to New York over coaxial cable. If the signal were transmitted via PAM, the resulting noise would make the received signal unintelligible. That is, noise would be added along the entire transmission path, with additional noise being added in each amplifier. Many amplifiers would be required to cancel attenuation along the way.

If this same signal were now coded using PCM, under certain conditions most errors could be corrected. If the amplifiers were spaced in such a way that the noise introduced between any two amplifiers is, on the average, less than one half of the quantization step size, each amplifier could restore the function to its original form before amplifying and sending it on its way. That is, each amplifier would round the received pulses to the nearest acceptable level and then retransmit, thereby essentially eliminating all but the strongest noise. In such applications, the amplifiers are known as *repeaters* since they are regenerating the original PCM signal.

The more quantization round-off levels used, the closer the staircase function resembles the desired signal. The number of levels then determines the signal *resolution*, that is, how small a change in signal level can be detected by looking at the quantized version of the signal.

If high resolution is required, the number of quantizing levels must increase. At the same time, the spacing between levels decreases. As the dictionary words get closer and closer together, the noise advantages of this noncoded type of transmission decrease.

If resolution could be improved without increasing the dictionary size (i.e., without putting dictionary words closer together), the error correcting advantages could be maintained. PCM is a method of accomplishing this.

In a PCM system, the dictionary of possible transmitted signals contains only two entries, a "0" and a "1." The quantization levels are coded into binary numbers. Thus, if there were eight quantization levels, the values could be coded into 3-bit binary numbers. Three pulses would be required to send each quantization value. Each of the three pulses would represent either a 0 or a 1. Note that we do not necessarily send a 0-V pulse to indicate a coded "0" and 1-V pulse for a coded "1." The actual techniques used for sending the binary digits are discussed in Chapter 8.

When the above description of PCM is compared with the operation of the ADC as described in Section 7.1, we see that if the digital output of the ADC is transmitted, we have PCM. Therefore, a binary ADC is actually a PCM modulator.

Figure 7.11 repeats the representative $s(t)$ of Fig. 7.2 and shows the 2-bit and 3-bit PCM forms of the encoded waveform. For illustrative purposes, a 1-V pulse is

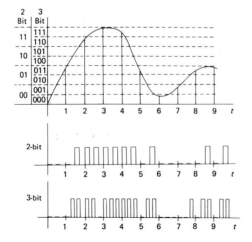

FIGURE 7.11   Example of 2-bit and 3-bit PCM.

used to represent a "1" and a 0-V pulse represents a "0." Again, we emphasize that the PCM signal is not necessarily transmitted in this form.

In a similar manner, a PCM demodulator is simply a binary digital to analog converter, as discussed in Section 7.1. The modulator and demodulator are available as LSI chips and are given the name *CODEC* (for coder-decoder).

Time Division Multiplexing (TDM)

The concept of time division multiplexing was introduced in the previous chapter for pulse modulation systems. We need only modify that discussion since each sample, instead of needing one pulse for transmission, now requires a number of pulses equal to the number of bits of quantization. Thus, for example, for 6-bit PCM, 6 pulses must be transmitted during each sampling period.

If the channel bandwidth is such that the 6 pulses require less than the sampling period to transmit, other pulses can be interspersed with these. For example, if transmission of the 6 pulses requires $\frac{1}{20}$ of the sampling period, 20 channels can be multiplexed. The 120 binary pulses then transmitted during each sampling period can be interleaved in a number of different ways. One simple technique is to send the 6 bits of channel 1's sample followed by 6 bits of channel 2's sample, and so on until the 6 bits representing the sample of channel 20 are sent. We then return to channel 1 in time to get the 6 bits representing the second sample.

We shall illustrate this concept with a practical example, the T-1 carrier system.

T-1 Carrier System

In the early 1960's, the need developed to provide capability for commercial and public digital communications. This need arose first with voice channels. The Bell System already had an elaborate carrier system in operation, with cables that could handle bandwidths up to several megahertz. The major advantages of digital communication were well known. That is, greater noise immunity is possible. As a bonus, advances in electronics have made digital equipment less expensive than analog. In addition, the signalling information required to control telephone switching operations could be economically transmitted in digital form.

The *Bell System T-1 digital carrier communication system* was formulated to be compatible with existing carrier communication systems. The existing equipment was designed primarily for interoffice trunks within an exchange, and therefore was suited for relatively short distances of about 15 to 65 kilometers.

The system develops a 1.544 megabit/second pulsed digital signal for transmission down the cable. The signal is developed by multiplexing 24 voice channels. Each channel is first sampled at a rate of 8000 samples/second. The samples are then quantized into 127 discrete levels, 63 positive, 63 negative, and zero. Thus, seven bits are required to send each sample value. The coding for each

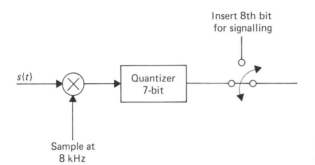

FIGURE 7.12   Coding for each voice channel in T-1 carrier.

channel is illustrated in Fig. 7.12. An 8th bit is added to each encoded sample for *signalling* information. The 24 channels are then time division multiplexed as shown in Fig. 7.13. The interleaved order of channels is due to the method of combining which was designed to use existing multiplex equipment. The quantized samples from the 24 channels form what is known as a frame, and since sampling is done at a rate of 8 kHz, each frame occupies the reciprocal of this, or 125 microseconds.

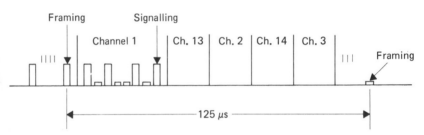

FIGURE 7.13   Time division multiplexing in T-1 carrier system.

There are 192 information and signalling bits in each frame (24 × 8), and a 193rd bit is added for *frame synchronization*. This framing signal is a fixed pattern of alternating 1's and 0's in every 193rd pulse position. Such an alternating pattern will rarely be found in any other position for more than 2 or 3 consecutive frames, so synchronization is fairly easy to obtain. Lack of frame synchronization would lead to unacceptable crosstalk between adjacent channels.

With 193 bits per frame and 8,000 frames per second, the product yields a transmission rate of 1,544,000 bits per second.

We see that the signalling information for each channel can contain 8,000 bits/second of information. If this is not sufficient, as is the case in some foreign dialing, the least significant bit of the information can be used to double the number of bits of signalling information. This, of course, lowers the quantization resolution from 127 to 63 levels.

The T1 carrier system can be used to send data signals as well as PCM, since the basic system is simply a data communication system transmitting 1.544 Mbps. In reality, it is not quite that simple. It is necessary to maintain compatibility with the voiced oriented T1 system. This means developing the various timing signals to control the receiver clocks. The framing pulses must be sent as with the voice system. The data are therefore usually transmitted in blocks of 8-bit characters.

7.4   NON-UNIFORM QUANTIZATION

Quantization need not be uniform. That is, we could replace the function of Fig. 7.1(b) with that shown in Fig. 7.14. The function of Fig. 7.14 has the property that the spacing between quantization levels is no longer uniform, and the output levels are

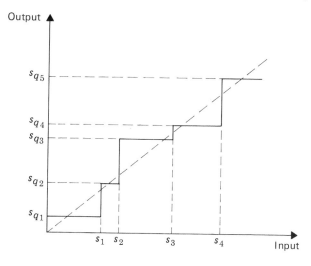

FIGURE 7.14   Input-output relationship for nonuniform quantization.

no longer in the center of each interval. Let us first see intuitively why non-uniform quantization may be preferable to uniform. Consider a musical piece where the voltage waveform ranges from -2 to $+2$ volts. Suppose further that 3-bit uniform quantization is used. Therefore all voltages between 0 and $\frac{1}{2}$ volt will be coded into the same code word, 100, corresponding to an output function of $\frac{1}{4}$ V. Likewise all samples between 1.5 and 2 volts will be coded into a single code word, 111, corresponding to an output function of $\frac{7}{4}$ V. During soft music passages, where the signal may not exceed $\frac{1}{4}$ volt for long periods, a great deal of music definition would be lost. The quantization provides the *same resolution* at high levels as at soft, even though the human ear is less sensitive to changes at higher levels. The response of the human ear is non-linear. It would appear to be desirable to use small quantization steps at the lower levels and larger steps at the higher levels.

One way to evaluate the effectiveness of a quantization scheme is to examine the quantization error, or quantization noise. This will be done in Section 7.6. For now, we observe that the *probability distribution* of analog input signal values affects

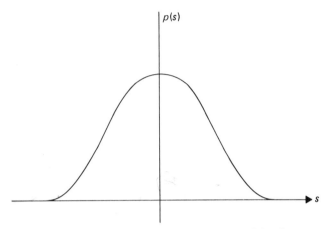

FIGURE 7.15   Probability density function of signal.

the amount of quantization noise. As a trivial example, suppose that you had a very unusual voice, where the waveform ranged from -2 V to $+2$ V, but amplitudes were always concentrated around odd multiples of $\frac{1}{4}$ V. Then the uniform quantization outlined at the beginning of this section would yield hardly any quantization noise.

It can be shown that the uniform quantization is optimum (yields minimum quantization error) in the case of uniform input signal distributions. Suppose now that the input, $s(t)$, is distributed according to a probability density, $p(s)$, as shown in Fig. 7.15. The statistics of the quantization error can be found by transforming this input density according to the function illustrated in Fig. 7.14.

We will show later that once quantization regions have been chosen (s_i's in Fig. 7.14), the s_{qi}'s are chosen to be the center of gravity of the corresponding portion of the probability density. Figure 7.16 shows the approximate quantization levels, given

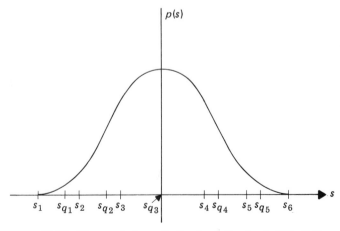

FIGURE 7.16   Placement of quantization levels for signal of Fig. 7.15.

the density of Fig. 7.15 and the division of the *s*-axis as shown. We note that the quantization level, rather than being in the center of each interval as in the case of uniform quantization, is skewed toward the higher probability end of the interval.

If in addition to the specific quantization levels we also make the actual intervals variable, a complicated optimization problem results and a complex system would be required to implement the result. Indeed, the coders (quantizers) presented in Section 7.1 would become extremely complicated with each individual quantization level requiring separate comparators. For this reason, and also because systems should be applicable to a variety of input signals, sub-optimum systems are almost always used. The usual approach is to first distort the analog signal according to some pre-determined nonlinear function. The distorted signal is then quantized using uniform quantization. At the receiver, the inverse of the nonlinear operation is performed after the decoding operation.

The most common form of non-uniform quantization is known as *companding*, which is derived from the words *compressing-expanding*. Prior to quantization, the signal is distorted by a function similar to that shown in Fig. 7.17. This operation tends to compress the extreme values of the waveform while enhancing the small values, much as the logarithm is used to permit viewing of very large and very small values on the same set of axes. If the analog signal forms the input to this compressor and the output is uniformly quantized, the result is equivalent to quantizing with steps that start out small and get larger for higher signal levels. This is not an optimum form of non-uniform quantization, but its shortcomings are often balanced by the ease of implementation.

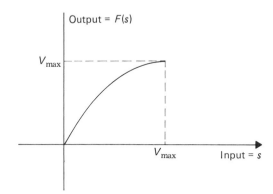

FIGURE 7.17 Typical companding function.

7.5   ALTERNATE MODULATION TECHNIQUES

Delta Modulation

PCM is an excellent method to send data or analog signals when correction for noise errors is critical. However, PCM requires somewhat complex coders and decoders, and, as a resolution is increased, the number of bits per sample increases. This may severely limit the number of channels that can be time division multiplexed. A continuing search is therefore taking place to devise coded transmission systems which possess noise immunity without having to send as many bits per second as is required by PCM. *Delta modulation* is one such discovery.

In developing PCM, each sample value was taken and coded into a series of binary pulses. Each set of binary pulses gave sufficient information to enable evaluation of the corresponding quantized sample value.

In delta modulation, the knowledge of past information is used to simplify the coding technique and the resulting signal format. The signal is first quantized into discrete levels, but the size of each step in the staircase approximation to the original function is kept constant. That is, the quantized signal is constrained to move by only one quantization level at each transition instant. The location of the steps is controlled to correspond to the sampling instants. Thus, at each sampling instant the quantized waveform must either increase by the standard step size or decrease by this amount. The quantized signal *must* change at each sampling point. It cannot remain constant since this would result in three possible actions, thus negating the possibility of using binary communication techniques. An example of this quantizing process is shown in Fig. 7.18.

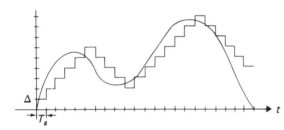

FIGURE 7.18 Example of quantized signal for delta modulation.

Once the quantization is performed, transmission of the quantized signal becomes trivial. We simply transmit a string of 1's and 0's. A 1 can indicate a positive transition, and a 0 indicates a negative transition. The transmitted bit train for the example of Fig. 7.18 would then be

$$1\ 1\ 1\ 1\ 1\ 0\ 0\ 0\ 0\ 1\ 1\ 1\ 1\ 1\ 1\ 1\ 0\ 0\ 0\ 0\ 0$$

A delta modulator (encoder) can be constructed using a staircase generator. At each sampling instant, the generator output is compared to the input waveform. If the input is greater than the staircase function, a positive step is initiated. If the input is smaller than the staircase, a negative step results.

Figure 7.19 shows a block diagram of the modulator.

The demodulator for a delta modulated waveform is simply a staircase generator. If a 1 is received, the staircase increments positively. If a 0 is received, the staircase increments negatively.

The key to effective use of delta modulation is the intelligent choice of the two parameters, *step size* and *sampling period*. These must be chosen such that the

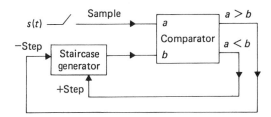

FIGURE 7.19 Delta modulator.

staircase signal is a close approximation to the actual analog waveform. Since the signal has a definable upper frequency cutoff, we know the fastest rate at which it can change. However, to account for the fastest possible change in the signal, the sampling frequency and/or the step size must be increased. Increasing the sampling frequency results in the delta modulated waveform requiring a larger bandwidth. Increasing the step size increases the quantization error. That is, the step approximation to the function becomes poorer as the step size increases. This is most obvious during periods when the function is almost constant. Figure 7.20 shows the consequences of a poor choice of step size.

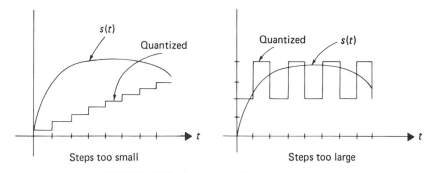

FIGURE 7.20   Importance of proper step size.

Some improvement is possible over the basic delta modulation system without increasing bandwidth or increasing quantization noise. Such improvement is realized if the step size *adapts* to the situation rather than remaining constant. Thus, during periods when the signal is changing slowly, the step size decreases, thus reducing the quantization error. During periods when the signal is changing rapidly, the step size increases. Such a system is known as *adaptive delta modulation* (ADM). This is also a form of *variable-slope delta modulation* (VSDM). It is, of course, necessary for the receiver to be able to reconstruct the staircase function being transmitted, including the knowledge of the size of each step. Therefore the factor controlling the step size must be derivable both at the transmitter and at the receiver.

Expanding upon this last point, consider one possible adaptive system. The step size is made exactly equal to the difference between the step function and the actual signal. A representative example is shown in Fig. 7.21. This is seen to be the sample

FIGURE 7.21   Adaptive quantization example.

and hold system and can be implemented with the switch and capacitor as shown. Transmission of the staircase approximation is accomplished by sending a 1 if a positive step occurs and a 0 if a negative step occurs.

For the example shown, the transmitted bit sequence is

$$1\ 1\ 1\ 0\ 0\ 0$$

This system is not workable for the following reason. The receiver sees the train of 0's and 1's and has no way of extracting the information regarding the step size. There is no way of reconstructing the original staircase function at the receiver.

We require a system whose step sizes depend only upon the sequence of binary numbers being transmitted. We observe that during periods of slow change, the sequence of transmitted bits tends to oscillate between 0 and 1 as the quantized waveform continually compensates for overshoot. During periods of rapid change, the staircase function is trying to catch up. Therefore the transmitted bit train contains strings of either 0's or 1's depending upon whether the original function is decreasing or increasing.

The above observation is the key to a workable adaptive system. If a given length string of bits contains an almost equal number of 1's and 0's, the step size is decreased. If the string contains many more 1's than 0's (or vice versa), the step size increases. Since the step size is dependent only upon the sequence of bits, the exact same adaptive operation can be reproduced at the receiver.

The actual step size control is performed by a digital integrator. The integrator sums the bits (0's and 1's) over some fixed period. If the sum deviates from one half of the number of bits in the period, the step size is increased. As the sum approaches one half of the number of bits, the step size approaches zero. In practice, the bit sum is translated into a voltage which is then fed into a variable gain amplifier. The amplification is a minimum when the input voltage corresponds to an equal number of 1's and 0's in the period. The amplifier controls the step size.

With the use of adaptive delta modulation, acceptable telephone transmission has been achieved using 32-kilobit/sec transmission. This compares to the 64 kilobits/sec (8 bits/sample $\times$ 8,000 samples/sec) required for one channel of telephone transmission using PCM. This nominal two-to-one reduction in bit rate is the reason for considerable attention being given to delta modulation.

Delta PCM

In addition to PCM, DM, and adaptive DM, there are numerous other methods for coding analog information into a digital format. The goal of each system is to send the information with maximum reliability and minimum bandwidth.

Delta modulation can be viewed in a way different from what we used previously. This new approach will point the way to several variations. In delta modulation, a continuous waveform was approximated with a staircase waveform. At each sampling point, we defined an error term as the difference between the signal and the staircase function. This error was then quantized to develop a correction term, which was then added to the staircase function. In the case of basic DM, the quantization of the error was performed as 1-bit. This can be viewed as taking *sgn* (error), where *sgn* is the sign function which is $+1$ for positive argument and -1 for negative argument. In the case of *delta PCM*, the error is coded into more than 1-bit of quantization, and this quantized error term is added to the previous staircase value. This process is illustrated in Fig. 7.22.

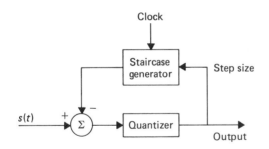

FIGURE 7.22   The delta PCM modulator.

Therefore, instead of the stair steps being of only one possible magnitude, they can now be one of 2, 4, 8, or any power of 2 sizes. At each sampling point, we must send more than 1 bit of information, the various bits representing the PCM code of the error term. The advantage of delta PCM over ordinary PCM is that, with proper choice of sampling interval, the error being quantized has a much smaller dynamic range than that of the original signal. Better resolution can therefore be achieved for the same number of bits of quantization. The price of this increased resolution is complexity of the modulator. It can be shown that if the signal were always at or near its maximum frequency (the one that determines sampling rate in PCM), the performance of delta PCM would be essentially the same as that of PCM. However, since the signal is usually distributed over a range of frequencies, it is possible to get better performance from this system than from a PCM system with the same bit transmission rate. This is true since, during periods when the instantaneous frequency is relatively small, the function is changing slowly. The amplitude changes are therefore small numbers, so the quantization error decreases.

Differential PCM (DPCM)

Differential PCM is another technique for sending information about *changes* in the samples rather than the sample values themselves. The differential approach includes an additional step which is not part of delta PCM. The modulator does not send the difference between adjacent samples, but instead uses the difference between a sample and its *predicted* value. The prediction is made on the basis of previous samples. This is illustrated in Fig. 7.23. The symbol $\hat{s}(nT_s)$ is used to denote the

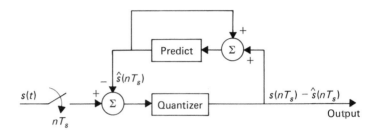

FIGURE 7.23   The differential PCM modulator.

predicted value of $s(nT_s)$, where T_s is the sampling period. The simplest form of prediction is *linear prediction*, where the estimate is a linear function of the measured samples. Thus, if only one sample is used,

$$\hat{s}(nT_s) = As([n-1]T_s) \tag{7.1}$$

where A is a constant. The *Predict* block in Fig. 7.23 is therefore simply a multiplier of value A.

The problem is to choose A to make this prediction as *good* as possible. A measure of goodness results from defining a prediction error as the difference between the sample and its estimate. Thus,

$$e(nT_s) = s(nT_s) - \hat{s}(nT_s)$$
$$= s(nT_s) - As([n-1]T_s) \tag{7.2}$$

The mean square value of this error is

$$\begin{aligned} mse &= E\{e^2(nT_s)\} \\ &= E\{s^2(nT_s)\} + A^2 E\{s^2([n-1]T_s)\} \\ &\quad - 2AE\{s(nT_s)s([n-1]T_s)\} \\ &= R(0)[1 + A^2] - 2AR(T_s) \end{aligned}$$

where $R(t)$ is the autocorrelation of $s(t)$. The error derived above can be minimized with respect to A by setting the derivative equal to zero.

$$\frac{d(mse)}{dA} = 2AR(0) - 2R(T_s) = 0$$

or from the earlier expression,

$$E\{s([n - 1]T_s)[s(nT_s) - As([n - 1]T_s)]\} = 0 \qquad (7.3)$$

thus yielding

$$A = \frac{R(T_s)}{R(0)}$$

Equation (7.3) indicates that the expected value of the product of the error with the measured sample is zero. That is, the error has no component in the direction of the observation. The error and the observation are *orthogonal*. This makes sense since, if the error did have a component in the direction of the observation, that component could be reduced to zero by readjusting the constant, A.

The predictor of Fig. 7.23 would take the most recent sample value (which it forms by adding the prediction to the difference term) and weight it by $R(T_s)/R(0)$. Note that the input process must have been observed sufficiently long to enable estimation of its autocorrelation.

Illustrative Example 7.4

Find the weights associated with a predictor operating on the *two* most recent samples of a waveform. Also evaluate the performance of this predictor.

Solution

The prediction is given by

$$\hat{s}(nT_s) = As([n - 1]T_s) + Bs([n - 2]T_s)$$

where the object is to make the best choice of A and B. For this best choice, we saw in Eq. (7.3) that the error is orthogonal to the measured quantities. Thus,

$$E\{[s(nT_s) - As([n - 1]T_s) - Bs([n - 2]T_s)]s([n - 1]T_s)\} = 0$$
$$E\{[s(nT_s) - As([n - 1]T_s) - Bs([n - 2]T_s)]s([n - 2]T_s)\} = 0$$

Expanding these yields

$$R(T_s) - AR(0) - BR(T_s) = 0$$
$$R(2T_s) - AR(T_s) - BR(0) = 0$$

yielding values of A and B.

$$A = \frac{R(T_s)[R(0) - R(2T_s)]}{R^2(0) - R^2(T_s)}$$

$$B = \frac{R(0)R(2T_s) - R^2(T_s)}{R^2(0) - R^2(T_s)}$$

The mean square error using these values is

$$mse = E\{[s(nT_s) - \hat{s}(nT_s)]^2\}$$

$$= E\{s^2(nT_s)\} - E\{\hat{s}(nT_s)s(nT_s)\}$$

$$= R(0) - \frac{R(0)[R^2(T_s) + R^2(2T_s)] - 2R(2T_s)R^2(T_s)}{R^2(0) - R^2(T_s)}$$

$$(7.4)$$

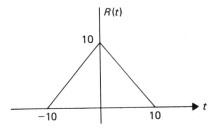

FIGURE 7.24 Autocorrelation function for Illustrative Example 7.4.

The first term in Eq. (7.4) is the error that would arise if we guessed at zero for the prediction. The remaining term represents a correction to that error.

As a specific example, if the autocorrelation function of $s(t)$ is as shown in Fig. 7.24, and the sampling period is $T_s = 1$ second, then the mean square error is given by Eq. (7.4) and is equal to

$$mse = 1.895$$

For comparison, if $s([n - 1]T_s)$ and $s([n - 2]T_s)$ were not measured and zero were guessed at for the estimate of the value, the mean square error would have been 10. This is simply the mean square value of the function.

It can be further shown that since

$$\frac{R(2T_s)}{R(T_s)} \approx \frac{R(T_s)}{R(0)}$$

the performance of the single sample predictor would have been almost the same as the performance we found for this double sampler predictor. The added information doesn't appreciably improve performance.

For a speech signal, it has been shown experimentally that a differential PCM system using the most recent sample in the predictor can save one bit per sample over PCM. That is, the DPCM system can achieve the same error performance as the PCM system with one less bit. This frees the channel for other uses during the time that bit would be sent. The extra effort of DPCM may therefore prove cost effective.

7.6   QUANTIZATION NOISE

The process of digital communication begins with a quantization, or rounding off, of the sample values to discrete levels or transitions. It is these discrete values which are transmitted, and the receiver (decoder) recovers the quantized signal. Once the quantization process is performed at the transmitter, the specific values of the samples are *lost forever*, and they cannot be reconstructed at the receiver. The quantizing operation can be viewed as a function operating upon each sample value. Since the inverse of this function is not single valued, the original signal cannot be uniquely recovered. The error is known as *quantization noise*.

PCM

Uniform Quantization

The first type of quantization we discussed in this chapter had uniform spacing between levels, and is therefore called uniform quantizing. It is characterized by the transfer function illustrated in Fig. 7.1 and repeated as Fig. 7.25. If an analog signal forms the input to this quantizer, the output will be a staircase approximation to that signal. If the staircase function is sampled in time, the result is the PCM samples ready for binary encoding.

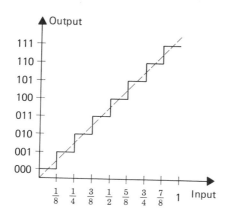

FIGURE 7.25   Transfer function of uniform quantizer.

Quantization noise, or error, is defined as the time function which is the difference between $s_q(t)$, the quantized waveform, and $s(t)$:

$$e(nT_s) = s(nT_s) - s_q(nT_s) \qquad (7.5)$$

An example of this error term is illustrated in Fig. 7.26. The magnitude of this error never exceeds one half of the spacing between quantization levels. Using the relationship in Fig. 7.25, $e(nT_s)$ can be plotted as a function of $s(nT_s)$. This plot will

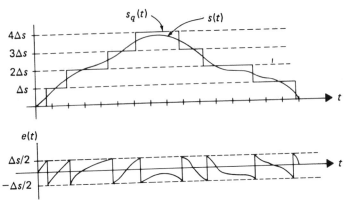

FIGURE 7.26   Quantization noise in PCM.

be a sawtooth waveform as shown in Fig. 7.27. The advantage of viewing the error in this manner is that if the distribution of $s(nT_s)$ is known, the distribution of $e(nT_s)$ and all of the statistical averages of the error can be found. This becomes a straight-forward application of functions of a random variable.

In order to provide a feel for the statistics of the quantization error, a simplifying assumption is often made. This assumption is that the original time function is *uniformly distributed* along its entire range of values. That is, the signal spends an equal fraction of time in any constant width range of values. If we divide the range into N equal width strips, the function spends $1/N$ of the total time within each strip. The error in each sample is given by $e_n = s(nT_s) - s_n$, where s_n is the nearest quantization level.

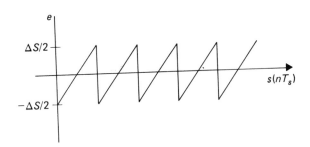

FIGURE 7.27   Error as a function of signal sample for uniform quantizer.

Figure 7.26 is included to help illustrate the concept. It is important, however, to note that it is the *sample values* which are quantized and not the *continuous function*. This is critical, since the quantized version of the function, $s_q(t)$, is not bandlimited and therefore cannot be sampled without aliasing error. To avoid confusion, the following derivation will use nT_s in place of t to indicate sample values of time.

With these assumptions, the quantizing error would be uniformly distributed between $-\Delta S/2$ and $+\Delta S/2$ as shown in Fig. 7.28.

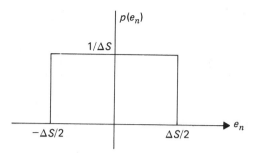

FIGURE 7.28 Probability density of quantizing error for uniform quantization.

Let us briefly examine the uniform density approximation. The exact formula for transforming a random variable by a multi-valued function is

$$p(e) = \sum_i \frac{p(s_i)}{|de_i/df_i|} \tag{7.6}$$

where the s_i are the various values of s corresponding to e. With N bits of quantization, there are 2^N ramp segments within the range -1 to $+1$. The magnitude of the slope is then equal to 2^N.

If Eq. (7.6) is applied to either a uniform or a triangular density, the result is that $p(e)$ is uniformly distributed between -1 and $+1$, with amplitude of 0.5. Some thought will show that the result is *exactly* uniform provided the input density is piecewise linear over the step increment. This restriction will almost always approximately obtain in a real life situation, so the uniform error density is a reasonable approximation in almost all cases.

The average value of this error is zero due to the symmetry of the probability density. The mean square value of the error is given by Eq. (7.7).

$$mse = \int_{-\infty}^{\infty} p(e_n) e_n^2 de_n$$

$$= \frac{1}{\Delta S} \int_{-\Delta S/2}^{\Delta S/2} e_n^2 de_n = \frac{\Delta S^2}{12} \tag{7.7}$$

This number, by itself, is not the desired final result. It merely yields the mean square error of each time sample. The desired result is the effect this error has on the reconstructed analog signal waveform at the receiver. At best, the receiver decoding operation returns the quantized sample values. A lowpass filter is then used to change this staircase approximation into a smooth time function. The input to the lowpass filter, assuming ideal impulse sampling, can be written as in Eq. (7.8).

$$x(t) = \sum_{n=-\infty}^{\infty} s(nT_s)\delta(t - nT_s) + \sum_{n=-\infty}^{\infty} e_n \delta(t - nT_s) \tag{7.8}$$

where the $s(nT_s)$ are the original unquantized sample values and e_n is the quantization noise associated with the sample at nT_s.

What was found in Eq. (7.7) is that the e_n have a mean square value of $\Delta S^2/12$. This must somehow be transformed into spectral information about the summation in Eq. (7.8). In that manner, the portion of the error signal that gets through the lowpass filter can be determined.

The first summation in Eq. (7.8) represents an ideally sampled version of $s(t)$ and therefore, when put through the lowpass filter, generates $s(t)/T_s$ at the output. The second summation constitutes a form of *shot noise*. Since the e_n are identically distributed and the transform of a single impulse is a constant, the power spectral density of the sum of impulses is a constant (white noise). Since the mean square value of each impulse coefficient is $\Delta S^2/12$, and these impulses are spaced by T_s, the average power per unit of frequency is $\Delta S^2/12T_s$. Since this white noise forms the input to the lowpass filter with cutoff frequency f_m, the output power is given by Eq. (7.9).

$$P_{N_q} = 2f_m \frac{\Delta S^2}{12T_s} \tag{7.9}$$

If the Nyquist (minimum) sampling frequency is used, $T_s = 1/2f_m$, and the quantization noise at the output of the filter has power equal to $\Delta S^2/12T_s^2$. This can now be compared to the signal power to yield a ratio. Note that because the sampling and recovery system introduces a multiplying factor of $1/T_s$, the units of the power expression appear to be watts/sec^2.

Illustrative Example 7.5

Consider an audio signal comprising three sinusoidal terms.

$$s(t) = 3 \cos 500t + 3 \cos 700t + 2 \cos 1{,}000t$$

This is quantized using 10-bit PCM and the Nyquist sampling rate is used. Find the signal to quantization noise ratio.

Solution

The Nyquist rate is $f_s = 1{,}000/\pi$ and $T_s = \pi/1{,}000$. The extreme amplitudes of the signal are ± 8 volts for a total swing of 16 volts. If we use 10-bit PCM, there are 1,024 quantization levels and $\Delta S = \frac{1}{64}$ volts. The quantization noise power is then given by Eq. (7.9).

$$P_{N_q} = \frac{2f_m \Delta S^2}{12T_s} = \frac{\Delta S^2}{12T_s^2} = 23.48 \text{ W}$$

The output signal would be $s(t)/T_s$, so the signal power is given by

$$P_s = \frac{3^2 + 3^2 + 2^2}{2T_s^2} = \frac{11}{\pi^2} \text{ MW}$$

$$\frac{P_s}{P_{N_q}} = 4.75 \times 10^4 = 46.8 \text{ dB}$$

Non-Uniform Quantization

Suppose the input, $s(t)$, is now assumed to be distributed according to a probability density, $p(s)$, as shown in Fig. 7.15 and repeated as Fig. 7.29 with values

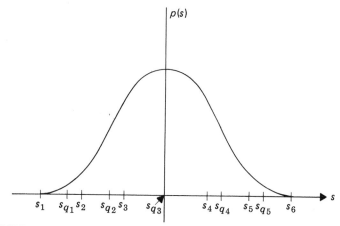

FIGURE 7.29   Probability density of $s(t)$. Placement of quantization levels is shown.

added. The figure shows a *Gaussian* density with standard deviation of $\frac{1}{3}$. The statistics of the quantization error can be found by transforming the input density according to the quantizing function illustrated in Fig. 7.14 (repeated as Fig. 7.30). The output mean square error is then given by

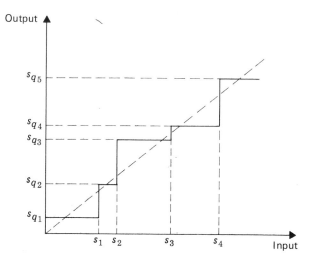

FIGURE 7.30   Nonuniform quantizing function.

$$mse = E\{[s(nT_s) - s_q(nT_s)]^2\}$$

$$= \int_s (s - s_q)^2 p(s) \, ds$$

where s_q is the quantized signal. Referring to Fig. 7.30, we see that this can be rewritten as

$$mse = \sum_i \int_{s_i}^{s_{i+1}} (s - s_{q_i})^2 p(s) \, ds \tag{7.10}$$

where s_{q_i} are the various quantization levels indicated as the ordinate of Fig. 7.30. To minimize the error, the integral in Eq. (7.10) is differentiated and the derivative is set equal to zero. This yields

$$\int_{s_i}^{s_{i+1}} (s - s_{q_i}) p(s) \, ds = 0 \qquad \text{for all } i \tag{7.11}$$

Equation (7.11) indicates that once quantization regions have been chosen (s_i in Fig. 7.29), the s_{q_i} are chosen to be the center of gravity of the corresponding portion of the probability density. Given the density of Fig. 7.29 and the division of the s-axis as shown, the approximate quantization levels are indicated on the figure.

The quantization levels, rather than being in the center of each interval as in the case of uniform quantization, are skewed toward the higher probability end of each interval.

Illustrative Example 7.6

Given the Gaussian density of Fig. 7.29, and three bits of quantization, evaluate the quantization error using uniform quantization, and using the following proposed non-uniform scheme. Suppose that the quantization regions are chosen to yield *equal area* under the probability density function over each region. That is, the probability of the function being within any particular interval is the same as that of any other interval.

Solution

Equation (7.7) can be used to approximate the mean square error in the uniform quantization case (recall that this equation is an approximation to Eq. (7.10), and use of the exact equation should yield a similar result). The resulting mean square error is

$$mse = 0.00625 \qquad \text{for uniform quantization}$$

For non-uniform quantization, we must resort to Eq. (7.10) and Eq. (7.11). We must first find the quantization intervals using the assumption stated in the problem. Assuming first that the Gaussian density is approximately equal to zero beyond the 3 sigma points, we divide the region between -1 and $+1$ into 8 equal area regions. Reference to a table of error functions gives the following dividing points:

$$s_i = -1, \quad -0.38, \quad -0.22, \quad -0.1, \quad 0$$

with a symmetrical repeat to the right of the axis.

Equation (7.11) is now used to find the optimum quantization point within each interval. A numerical approximation was used to find the following quantization points.

$$s_{q_i} = -0.54, \quad -0.3, \quad -0.16, \quad -0.05$$

Again, this repeats symmetrically.

Finally, the mean square error is found from Eq. (7.10). A numerical approximation yields the following result:

$$mse = 0.00534$$

Thus, the non-uniform quantization has yielded some improvement in quantization error.

Illustrative Example 7.6 suggested one possible algorithm for selection of quantization regions. In fact, this is not optimum and will even cause degradation of performance when compared to uniform quantization in some cases. The equation for mean square error weights the probability by the square of the deflection from the quantized value before integration. In general, the problem is to minimize the error of Eq. (7.10) as a function of two variables, s_i and s_{q_i}. The s_{q_i} are constrained to satisfy Eq. (7.11). Unless the probability density can be expressed in closed form, this problem is computationally difficult.

The *mean value theorem* can be used in Eq. (7.10) to get an approximation which improves as the number of bits of quantization increases. Using this approximation, the following rule for choosing the quantization regions results: Choose the quantization regions to satisfy the uniform moment constraint,

$$(s_{i+1} - s_i)^2 p(\text{midpoint}) = \text{constant} \tag{7.12}$$

We will explore this more fully in the problems at the end of this chapter.

Companded Systems

The comparison of the performance of companded systems to that of uniform quantization systems will be illustrated for several examples.

Illustrative Example 7.7

A signal, $s(t)$, has probability density function as shown in Fig. 7.31. This signal is quantized using 5 bits of resolution.

(a) Evaluate the mean square error using uniform quantization.

(b) Evaluate the mean square error if the signal is first compressed according to the following equation,

$$F(s) = \sqrt{|s|} \; sgn(s)$$

and is then uniformly quantized.

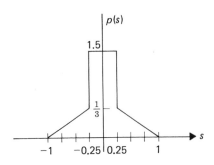

FIGURE 7.31 Probability density function of signal for Illustrative Example 7.7.

Solution

(a) If uniform quantization is used, each interval is $\frac{1}{16}$ wide with the quantization point in the middle of the interval. The quantization error can be found directly from Eq. (7.10), or from its approximation, Eq. (7.7).

Due to symmetry, the summation can be evaluated for either the positive or negative function values. The result is approximately

$$mse = 3.3 \times 10^{-4}$$

If the signal is now compressed according to the equation $F(s) = \sqrt{|s|}\ sgn(s)$ and then uniformly quantized, equation (7.10) can again be used, but the s_i and s_{q_i} change from their previous values. For example, the s_{17} of $(\frac{1}{16})$ is replaced by an s_{17} of $(\frac{1}{16})^2$, and all values between 0 and $(\frac{1}{16})^2$ are quantized to the level $(\frac{1}{32})^2$. Making the appropriate changes and reevaluating the summed integrals yields

$$mse = 2.90 \times 10^{-4}$$

This represents an improvement over that attained using uniform sampling. The improvement results since the compression makes intervals close to the axis smaller while spreading those nearer the edges. Since the probability of the values closer to zero is higher, the weighting tends to reduce the error integral.

Illustrative Example 7.8

Repeat Illustrative Example 7.7 for an $s(t)$ with probability density function as shown in Fig. 7.32.

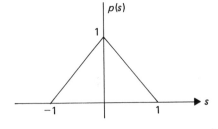

FIGURE 7.32 Probability density function of signal for Illustrative Example 7.8.

Solution

Using the same approach as in the previous problem, we find

$$\text{mean square error for companded system} = 4.41 \times 10^{-4}$$

$$\text{mean square error for uniform quantizing} = 3.3 \times 10^{-4}$$

Therefore, uniform quantizing yields a smaller error than that of the compressed quantizing system. It should therefore be clear that optimum compression is a function of the probability density. Both the compression function and the placement of the quantizing levels within each interval must be derived based upon this probability density. The square root curve used in these examples is by no means optimum for any given density and its associated mean square error may either improve or degrade performance.

Illustrative Example 7.9

Repeat the previous examples for a function which is uniformly distributed in the range between -1 and $+1$.

Solution

5-bit uniform quantizing results in the same mean square error as in the previous problems.

$$mse = 3.3 \times 10^{-4}$$

Using the square root compression curve results in a mean square error of

$$mse = 6.64 \times 10^{-4}$$

Thus, the error is about twice as large for the compressed system. In fact, for the uniform density function, uniform quantization is optimum. It should be noted that the mean square error for uniform quantization is about the same for each of the three probability density functions. This reinforces the earlier discussion which showed that the error was approximately uniformly distributed as long as the quantization intervals were small compared to changes in the density function. Since 5 bits of quantization is being used, this approximation is very good, so the same error is expected. Note that this does not imply that uniform quantization is optimum—only that the resulting mean square error is almost independent of functional probability density.

As in the earlier discussion of non-uniform quantization, it would be desirable to find the *optimum compression function* for any given probability density. This is, as before, a complex numerical optimization problem. Some approximations are possible as the following analysis will show.

The results can be generalized starting with the equation for mean square error in terms of quantization levels, Eq. (7.10). This expression for mean square error can be simplified if it is assumed that the quantization levels fall in the middle of each interval. This is, of course, an approximation and will not be the optimum choice yielding minimum mean square error unless the probability density of the function is constant within each interval. As the number of bits of quantization increases, the approximation will become better and better (i.e., the quantization steps become smaller and smaller). Further assuming that the intervals are sufficiently small that $p(s)$ can be approximated throughout the interval by its value at the mid point, Eq. (7.10) can be reduced to Eq. (7.13).

$$\overline{e^2} = \sum_i p(s_{q_i}) \left. \frac{(s - s_{q_i})^3}{3} \right|_{s_i}^{s_{i+1}}$$

$$= \sum_i p(s_{q_i}) \frac{(s_{i+1} - s_i)^3}{12} \tag{7.13}$$

As a check, <u>if</u> we let the quantization step size be uniform and equal to ΔS, Eq. (7.13) reduces to $\overline{e^2} = \Delta S^2/12$ as derived earlier for the uniform quantization case. If $s(t)$ is distorted according to a non-linear function $F(s)$, and the distorted samples are then quantized using uniform quantization, the result is equivalent to quantizing the original samples with non-uniform quantization levels.

If ΔS is the uniform step size in the quantizing operation, then

$$(s_{i+1} - s_i)F'(s_i) \approx \Delta S$$

where $F'(s) = dF/ds$. Equation (7.13) then becomes

$$\overline{e^2} = \frac{1}{12} \sum_i p(s_{q_i})(s_{i+1} - s_i)(s_{i+1} - s_i)^2 \tag{7.14}$$

and in the limit as the step size approaches zero,

$$\overline{e^2} = \frac{\Delta S^2}{12} \int_s \frac{p(s)}{(F'(s))^2} ds \tag{7.15}$$

Once again, we can check this result by referring back to uniform quantization. If we set $F(s) = s$ as is the case for uniform quantization, the expression reduces to, $\Delta S^2/12$, as expected. In fact, since this term appears explicitly in Eq. (7.15), the reciprocal of the integral can be defined as an *improvement factor*,

$$I = \left(\int_s \frac{p(s)}{(F'(s))^2} ds \right)^{-1} \tag{7.16}$$

and

$$\overline{e^2} = \frac{e_u^2}{I} \tag{7.17}$$

where e_u^2 is the mean square error assuming uniform quantizing. Of course, $F(s)$ should be chosen to make the improvement factor greater than unity. Otherwise uniform quantization would be preferable.

The optimization problem can now be stated in a way different from that used previously. First assume that the original function is normalized to lie in the range between -1 and $+1$. The maximum improvement factor corresponds to choosing $F'(s)$ to minimize

$$\int_s \frac{p(s)}{(F'(s))^2}\,ds \qquad (7.18)$$

The normalization yields the constraint that

$$\int_0^1 F'(s)\,ds = 1 \qquad (7.19)$$

It is clear from Eq. (7.18) that $F'(s)$ should increase over regions where $p(s)$ increases and decrease over regions where $p(s)$ decreases. Thus, the slope of the compression function should be greater for those regions of s which are most highly probable. Viewing Eq. (7.18) indicates that a reasonable first choice of $F'(s)$ is

$$F'(s) = K\sqrt{p(s)} \qquad (7.20)$$

where K is chosen to satisfy the constraint of Eq. (7.19). That is,

$$K = \left(\int_0^1 \sqrt{p(s)}\,ds \right)^{-1}$$

The improvement factor then becomes

$$I = \frac{1}{K^2} = \left(\int_0^1 \sqrt{p(s)}\,ds \right)^2 \qquad (7.21)$$

Since performance only depends upon the value of the compression function at discrete points (the boundaries of the quantization regions and the center of each region), any compression curve which goes through the same values at these points will result in the same performance.

A common algorithm for companding is *logarithmic companding* with $F(s)$ defined as in Eq. (7.22).

$$F(s) = V_{\text{MAX}} \frac{\log(1 + |s|\mu/V_{\text{MAX}})}{\log(1 + \mu)}\, sgn(s) \qquad (7.22)$$

This is plotted in Fig. 7.33 for various values of μ.

Illustrative Example 7.10

The input to a logarithmic compressor of Eq. (7.22) is a sinusoid. Find the improvement factor.

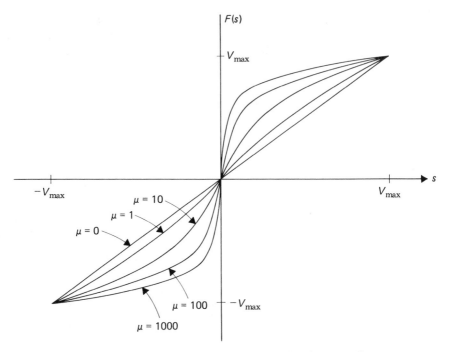

FIGURE 7.33   Functional relationship for logarithmic companding.

Solution

The density of the signal lies between V_{MAX} and $-V_{MAX}$ as the reciprocal of the derivative of sine (i.e., the fraction of time the function spends in any infinitesimal range of the ordinate is inversely proportional to the derivative). Therefore, the density of the signal is given by

$$p(s) = \frac{K}{1 + (K/V_{MAX})^2}$$

where K is chosen to make the density integrate to unity. Thus,

$$K = \frac{1}{\pi V_{MAX}}$$

The improvement factor is found by plugging $p(s)$ and $F(s)$ into Eq. (7.16) to yield

$$I^{-1} = 2 \int_0^{V_{MAX}} \frac{K}{1 - (s/V_{MAX})^2} \frac{\ln(1 + \mu)(1 + \mu s/V_{MAX})^2}{\mu^2} \, ds$$

$$= \frac{2 \ln^2(1 + \mu)}{\mu^2 \pi} \int_0^1 \frac{(1 + \mu x)^2}{\sqrt{1 + x^2}} \, dx$$

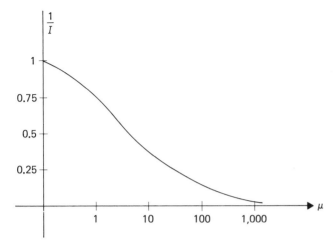

FIGURE 7.34 Improvement factor for logarithmic compressor.

The improvement factor can now be evaluated as a function of μ. This is plotted in Fig. 7.34.

Illustrative Example 7.11

Find the improvement factor for the square root compression function introduced earlier in this section. The function was characterized by

$$F(s) = \sqrt{|s|}\ sgn(s)$$

Solution

The derivative of this function is

$$\frac{dF}{ds} = \frac{\frac{1}{2}}{\sqrt{|s|}}$$

and the improvement factor is given by

$$I = \frac{1}{4 \displaystyle\int_0^1 sp(s)\, ds} \tag{7.23}$$

As a check of this result, the uniform density function will be substituted for $p(s)$ in Eq. (7.23). This yields $I = \frac{1}{2}$. Earlier in this section (*see* Illustrative Example 7.9) the mean square error was calculated as 3.3×10^{-4} and 6.64×10^{-4} for uniform quantization and for square root compression, respectively. Note that these two performance levels vary by approximately a factor of 2, thus verifying the approach using improvement factor for this example. This verification is significant since the earlier result was obtained numerically without any approximations relating to placement of quantization levels within each region.

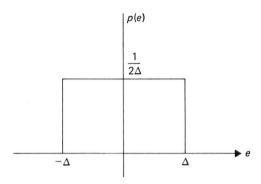

FIGURE 7.35 Probability density of error for delta modulation.

Delta Modulation

As in the case of PCM, the difference between the quantized (staircase approxima-tion) function and the original $s(t)$ is again defined as the quantization noise, $e(t)$.

$$e(t) = s(t) - s_q(t)$$

We will assume that the sampling rate and step size are chosen so as to avoid overloading. Under this assumption, the magnitude of the quantization noise never exceeds the step size, Δ. If we make the simplifying assumption that all signal amplitudes are equally likely, we conclude that the error is uniformly distributed over the range between $-\Delta$ and $+\Delta$ as shown in Fig. 7.35. The mean square value of quantization noise is therefore given by

$$E\{e^2\} = \int_{-\Delta}^{\Delta} \frac{1}{2\Delta} e^2 de$$

$$= \frac{\Delta^2}{3} \qquad\qquad (7.24)$$

As in the case of quantization noise in PCM, this is not yet in a meaningful form for us. The demodulator includes a lowpass filter to smooth the staircase into a continuous curve. We must therefore somehow translate the quantization noise statistics into a *power spectral density*. This is not a simple analytical task, and it requires that a specific form be assumed for $s(t)$. We will attempt a rather intuitive approach to get an approximate result.

For simplicity, assume the original $s(t)$ was a sawtooth wave. The wave, its quantized version, and the quantization noise are shown in Fig. 7.36. This noise function is almost periodic with period T_s (the sampling period). It would be exactly periodic with period T (the period of $s(t)$) only if T_s was an integer multiple of T. That is, if a sampling point coincides with the start of each period of $s(t)$, the entire sequence will repeat. We are assuming that the step size and sampling period are chosen to avoid overloading, and in this case, to give perfect symmetry. The power

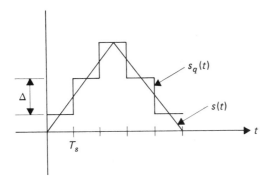

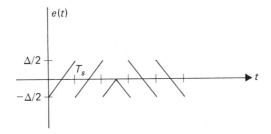

FIGURE 7.36 Quantization noise for sawtooth signal using delta modulation.

spectrum of $s_q(t)$ can be found precisely. Since we are not necessarily interested in the particular $s(t)$ illustrated, we shall simply sketch the result.

Each ramp segment can be considered to be the integral of a corresponding square pulse. Thus the power spectrum of the sawtooth will be $\frac{1}{4}\pi^2 f^2$ multiplied by the spectrum of a pulse train. The spectrum of a pulse of duration T reaches its first zero at $f = 1/T_s$, and decreases rapidly beyond that point. The multiplication by $\frac{1}{4}\pi^2 f^2$ reduces this even more rapidly. We can therefore assume that the power spectrum of the ramp is approximately zero beyond $1/T_s$.

We now turn our attention to a general $s(t)$. If the quantization step sizes are carefully chosen, $e(t)$ will almost always go from a negative peak to a positive peak (or vice versa) in each sampling interval, T_s. That is, the continuous $s(t)$ will intersect the staircase approximation in almost every interval. (Think about this and try some examples if necessary.) Therefore, although $e(t)$ will not be the sawtooth of Fig. 7.36, it can be expected to possess similar almost periodic properties, and its spectrum will be approximately zero outside the range $1/T_s$.

We shall not examine proper choice of sampling period and step size in detail. Since this is dependent upon the specific time function, general results are difficult to obtain. We simply state that a good choice, for a wide class of well behaved functions, is to set the step slope approximately equal to the maximum derivative of the function.

If $s(t)$ is a general time function, human speech for example, it is reasonable to assume that over a period of time, every conceivable waveform segment will appear for the T_s length intervals. It is therefore often assumed that the power spectral density of this quantization noise is that of bandlimited white noise. That is, it is constant up to a frequency of $1/T_s$, and zero beyond that point. We have previously found that the mean square value of the quantization noise (*see* Eq. (7.24)) is $\Delta^2/3$, so this must be equal to the integral of the power spectral density. Finally, if we put this noise through a lowpass filter with cutoff frequency f_m, the output noise power is

$$P_{N_q} = \frac{\Delta^2}{3} \cdot \frac{f_m}{1/T_s} = \frac{\Delta^2}{3} \cdot \frac{f_m}{f_s}$$

where f_s is the number of samples per second.

Illustrative Example 7.12

Given an audio waveform which is the sum of three sinusoids,

$$s(t) = 3 \cos 500t + 4 \cos 1{,}000t + 4 \cos 1{,}500t$$

Find the signal to quantization noise level if this is coded using delta modulation.

Solution

We must first choose a step size and a sampling frequency for this waveform. The Nyquist rate is $T_s = \pi/1{,}500$. Suppose for purposes of illustration we choose 32 times this rate, that is, $T_s = \pi/48{,}000$. At the fastest frequency the function can change ± 4 volts in $\pi/3{,}000$ seconds. In this amount of time, we have 16 sampling periods, so if a step size of $\frac{1}{4}$ volt is chosen, the staircase will only overload in extreme circumstances. The quantization noise power is given by,

$$P_{N_q} = \frac{\Delta^2}{3} \cdot \frac{f_m}{f_s} = \frac{1}{48} \cdot \frac{1}{64} \approx 0.3 \text{ mW}$$

The signal power is

$$\frac{3^2 + 4^2 + 4^2}{2} = 20.5 \text{ watts}$$

for a signal to noise ratio of

$$S/N = 6.29 \times 10^4 \approx 48 \text{ dB}$$

Comparison of PCM and DM

Now that general results have been derived for quantization noise in delta modulation and in PCM, it is possible to compare the performance of a delta modulator to that of PCM.

The signal to quantization noise ratio for PCM with uniform quantization is given by Eq. (7.25).

$$\mathrm{SNR} = \frac{P_s}{T_s^2} \cdot \frac{12 T_s^2}{\Delta S^2} \qquad (7.25)$$

In this equation, P_s is the average signal power. If sampling is done at the Nyquist rate, the number of bits per second is

$$BPS = 2^{N+1} f_m$$

where N is the number of bits of quantization. The step size can then be written as

$$\Delta S = \frac{2 V_{\mathrm{MAX}}}{2^N}$$

$$= \frac{4 V_{\mathrm{MAX}} f_m}{BPS}$$

where V_{MAX} is the maximum amplitude of the signal. Combining these expressions yields a signal to noise ratio of

$$\mathrm{SNR} = \frac{P_s \times 3 BPS^2}{V_{\mathrm{MAX}}^2 \times 4 f_m^2}$$

The equivalent result for delta modulation is found by starting with the expression for quantization noise power,

$$P_{N_q} = VAR \times \frac{f_m}{f_s}$$

where VAR is the variance of individual sample errors, f_s is the sampling frequency, and f_m is the maximum frequency of the signal. We shall assume that the slope of the staircase is greater than or equal to the maximum slope of $s(t)$ in order to avoid overloading. Thus, the step size and sampling rate are related by

$$\frac{\Delta}{T_s} > f'_{\mathrm{MAX}}$$

The maximum slope can be approximated by

$$f'_{\mathrm{MAX}} = V_{\mathrm{MAX}} \, 2 \pi f_m$$

so

$$f_s > \frac{2 \pi V_{\mathrm{MAX}} f_m}{\Delta}$$

and the noise power is bounded by

$$P_{N_q} < \frac{VAR \times \Delta}{2 \pi V_{\mathrm{MAX}}}$$

Assuming a uniform error probability distribution, the variance is given by

$$VAR = \frac{\Delta^2}{3}$$

Finally,

$$P_{N_q} < \frac{V_{\text{MAX}}^2 \; 8\pi^2 f_m^3}{6BPS^3}$$

and the signal to noise ratio is bounded by

$$SNR > \frac{6P_s BPS^3}{8\pi^2 V_{\text{MAX}}^2 \; f_m^3}$$

Comparing this result to the result found for PCM in Eq. (7.25) yields

$$\frac{SNR_{\text{DM}}}{SNR_{\text{PCM}}} > \frac{BPS}{\pi^2 f_m}$$

Finally, with the bit rate given by

$$BPS = 2^{N+1} f_m$$

this becomes

$$\frac{SNR_{\text{DM}}}{SNR_{\text{PCM}}} > \frac{2^N}{\pi^2}$$

Thus, under the assumptions made, as the number of bits increases, the performance of delta modulation is superior to that of PCM for the same bit transmission rate.

It should be noted that the measure of performance, SNR, only considers *quantization noise*. In the next chapter, we will add the dimension of *transmission errors*. Therefore, one must exercise caution and not get too optimistic about the virtues of delta modulation versus PCM.

PROBLEMS

7.1. Design a counting quantizer for 6-bit PCM and a voice signal as input. Assume the voice signal has a maximum frequency of 5 kHz. Choose all parameter values and justify your choices.

7.2. Find the output of the 3-bit ADC of Fig. 7.4 for the following voltage inputs:
(a) 0.327 V
(b) 0.631 V
(c) 0.751 V
(d) 1.000 V

7.3. Design a 5-bit serial quantizer which would operate upon a voice signal. The voice signal has maximum frequency of 5 kHz and an amplitude swing of ± 10 volts.

Illustrate the operation of the quantizer for sample values of -2 and $+6.7$ volts.

7.4. Design an OP AMP DAC for 8-bit digital signals and analog input amplitudes from 0 to 6 V.

7.5. You are given a time signal $s(t) = \sin \pi t$. This signal is transmitted via 4-bit PCM. Find the PCM wave for the first 5 sec, and sketch the result.

7.6. A PCM wave is shown below where voltages of $+1$ and -1 are used to send a 1 and a 0, respectively. Two-bit quantization has been used. Sketch a possible analog information signal, $s(t)$ that could have resulted in this PCM wave. Now assume that a pulse of $+1$ V represents a binary 0 and -1 V represents a 1. Sketch a possible (t).

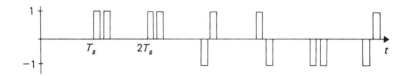

7.7. Two voice signals (3-kHz upper frequency) are transmitted using PCM with eight levels of quantization. How many pulses per second are required to be transmitted, and what is the minimum bandwidth of the channel?

7.8. The input to a delta modulator is $s(t) = 5t + 1$. The sampler operates at 10 samples/sec, and the step size is 1 V. Sketch the output of the delta modulator.

7.9. Delta modulation is used to transmit a voice signal with maximum frequency of 3 kHz. The sampling rate is set at 20 kHz. The maximum amplitude of the analog voice signal can be assumed to be 1 V. Discuss the choice of appropriate step size.

7.10. An adaptive delta modulator is to be used to transmit $s(t) = 5t + 1.1$. The sampler operates at 10 samples/sec. The step size depends upon the number of 1's in the 4 most recent samples. The step size is either 0.5 V, 1 V, or 1.5 V depending upon the number of 1's. That is, if the number of 1's is 0 or 4, the largest step size is used. If it is 1 or 3, the intermediate size is used. An equal number of 1's and 0's results in the smallest step size.

Sketch the staircase approximation to $s(t)$ for the first 3 sec and also sketch the transmitted digital waveform. Assume that all binary transmissions prior to $t = 0$ were 0.

7.11. Suppose in a T-1 carrier system, the synchronizer views 10 successive frame synchronization bits looking for a perfect match. Recall that a perfect match means alternating 1's and 0's.

Find the probability of locking onto the incorrect frame synchronization. What is the probability of continuing this incorrect lock for 3 successive frames? You may assume that the information part of the signal is random, with equal probabilities of a 1 or 0 for any particular bit.

7.12. White noise of power spectral density, $N_0/2$, forms the input to a predictor. The predictor is set to predict T seconds into the future based upon a single sample. Thus,

$$\hat{n}(t + T) = An(t)$$

(a) Find the optimum value for A.
(b) Now assume the noise is lowpass filtered prior to prediction. Find A and the mean square error as a function of the filter cutoff, f_m.

7.13. A signal is triangularly distributed as shown below. Show that Eq. (7.6) will yield a uniform density for the quantization error.

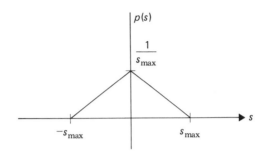

7.14. Find the mean-square quantization noise for the PCM modulation described in Problem 7.5.

7.15. A signal is Gaussian distributed with zero mean and variance, σ^2. The signal is to be sent with 4-bit PCM, where the quantization steps are uniformly distributed over the range $\pm 3\sigma$. Find the mean square quantization error.

7.16. A signal is derived by passing a 1-kHz square wave through an ideal lowpass filter with cutoff at 5.1 kHz. The resulting signal is quantized using 6-bit PCM. Find the signal to quantization noise ratio.

7.17. A signal is given by

$$s(t) = 10 \cos 100t + 17 \cos 500t$$

How many bits of quantization are required so that the signal to quantization noise ratio is greater than 40 dB?

7.18. Repeat Illustrative Example 7.7 for 2 bits of quantization.

7.19. Repeat Illustrative Example 7.8 for 2 bits of quantization.

7.20. Repeat Illustrative Example 7.9 for 2 bits of quantization.

7.21. Non-uniform quantization is to be used for 2-bit PCM where the signal is triangularly distributed as shown below.

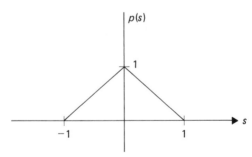

Equation (7.12) defines the quantization. Find the mean square error and compare this to the mean square error that would occur using uniform quantization.

7.22. Find the mean square quantization error when a signal, $s(t)$, with probability density as shown below, is
(a) uniformly quantized.

(b) compressed according to the formula

$$F(s) = \sqrt{|s|}\ sgn(s)$$

and then uniformly quantized. Assume 4-bit quantization is used.

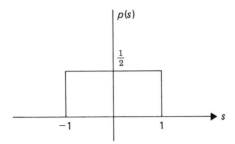

7.23. Repeat Problem 7.22 for 2-bit quantization. Then see if you can find a non-uniform quantizer that achieves lower mean square error.

7.24. A compressor operates along a sinusoidal curve,

$$F(s) = V_{MAX} \sin \frac{\pi s}{2 V_{MAX}}$$

as shown below.

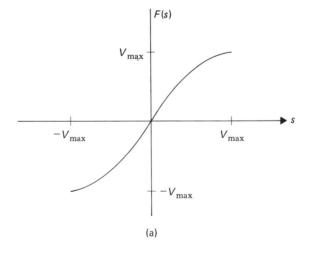

(a)

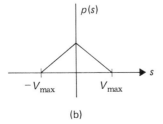

(b)

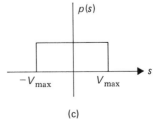

(c)

Find the improvement factor if
(a) the signal is a sinusoid.
(b) the signal is triangularly distributed as shown in the figure.
(c) the signal is uniformly distributed as shown in the figure.

7.25. A signal is derived by passing a 1-kHz square wave through a lowpass filter with cutoff at 5.1 kHz. Choose appropriate parameters for quantization by Delta modulation and find the signal to quantization, noise ratio.

7.26. A signal to be transmitted is of the form

$$s(t) = 10 \cos 1{,}000t + 5 \cos 1{,}500t$$

This signal is to be quantized using Delta modulation.
(a) Choose an appropriate sampling rate and step size.
(b) Using the values found in (a), find the signal to quantization noise ratio.

7.27. A signal is given by

$$s(t) = 5t + 1$$

It is sampled at 10 times per second, and quantized using 3-bit delta PCM. Sketch the output of the modulator.

7.28. Use linear prediction to approximate $s(t_0 + T)$ as a linear combination of $s(t_0)$ and its derivative, $s'(t_0)$. (Your answer will be in terms of the autocorrelation of the process.) Also find the mean square error in this approximation.

7.29. Use linear prediction to estimate the value of $s(NT + \Delta)$ as a linear combination of $s(nT)$ for all n. Find the mean square error in this approximation. Show that if the conditions of the sampling theorem are met, the mean square error goes to zero.

PROJECT: Write a computer program to approximate Eqs. (7.10) and (7.11). Use this program to verify the results of Illustrative Example 7.6. Then vary the parameters and observe the effect upon the error.

Chapter 8

Digital Transmission and Reception

Chapter 7 illustrated techniques for changing an analog signal into a sequence of binary digits. The next job is to transmit these bits from the source to the receiver.

The bits can be transmitted by reading them into a telephone or other audio device or by writing them on a piece of paper which is then transferred from source to receiver (e.g., through the mail). Of course, we prefer more direct electrical techniques. Ironically, we will find analog signals being used to transmit these digital signals.

This chapter will first present four techniques for assigning analog signals to binary digits. We will then examine two classes of receivers. Finally, we will assess performance of these digital receivers.

8.1 BASEBAND SYSTEMS

A *baseband digital system* is a system which transmits relatively low frequency signals in order to communicate digital information. The baseband design we shall present is a digital version of analog PAM (pulse amplitude modulation) as discussed in Chapter 6.

The simplest proposal for transmitting a sequence of 1's and 0's might be to transmit a 1-volt signal for a digital one and a 0-volt signal for a digital zero. This is known as *unipolar* transmission, since the signal deflects from zero in only one direction. A slight modification of this natural choice is usually made. In fact, we consider sending a constant of $+V$ volts for a digital 1 and $-V$ volts for a digital zero. This is called *bipolar* transmission. V is chosen on the basis of available power and other physical considerations. We use $+V$ and $-V$ rather than $+V$ and zero since, as we shall see in Section 8.7, it is advantageous to make the two signals as different as

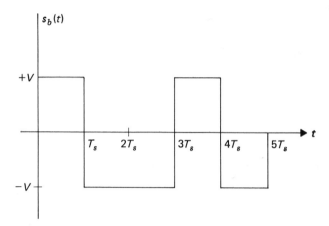

FIGURE 8.1   Example of bipolar baseband signal used to transmit 1 0 0 1 0.

possible. Figure 8.1 shows a signal waveform that might be used to send the digital sequence 1 0 0 1 0.

The specific coding scheme illustrated is the *NRZ*, or *non-return to zero*. Figure 8.2 illustrates the three most common pulse coding schemes. In NRZ, the pulse stays at its transmitted value until the next bit is ready for transmission, at which time the new value immediately takes over.

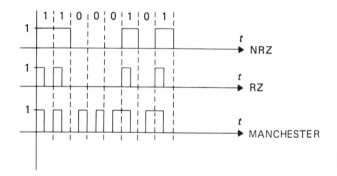

FIGURE 8.2   Three pulse transmission schemes.

In *RZ*, or *return to zero*, the pulse returns to the "0" state before reaching the end of the sampling interval. Thus, the pulses are not as wide as in the NRZ case, and this change increases the bandwidth. The advantage is the presence of a transition in each interval containing a "1" which can aid synchronization. This transition occurs even if there is a long string of 1's. In the case of NRZ, a long string of identical bits would result in a constant being transmitted. Such an occurrence could have serious consequences upon timing.

For the *Manchester* code, a "0" is transmitted by sending one value during the first half of the interval and the other value during the second half. A "1" is transmitted with the reverse sequence. The advantage is that the Manchester code produces a transition in the center of *each* sampling period. An additional advantage is that the average value of the transmitted waveform is independent of the actual sequence of bits.

If we return now to the signal of Fig. 8.1, the receiver would have to determine whether the unperturbed signal was $+V$ or $-V$ within each sampling interval. Methods of making this decision are described in Section 8.5.

Baseband transmission is used primarily where the communication channel is a coaxial cable. Other types of channels do not pass low frequencies, so information would be lost with these channels.

In order to assess the distortion effects of channels upon the baseband signal, we need some idea of the frequency spectrum of the waveform. We can use the same approach as presented in Chapter 6 by considering the unperturbed waveform as the output of the idealized system illustrated in Fig. 8.3. This system starts with a train of

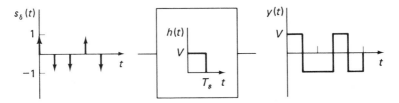

FIGURE 8.3 A filter can be used to generate the baseband signal from an impulse sampled waveform.

ideal impulses with strengths of $+1$ or -1 depending upon the data bit to be transmitted. This train of impulses then forms the input to a filter whose impulse response is a pulse of height V and width T_s. The output of the filter is the required baseband signal.

The Fourier Transform of the output is found by multiplying the Fourier Transform of the impulse train with the transfer function of the filter.

$$Y(f) = S_\delta(f)H(f)$$

$$= S_\delta(f)\, Ve^{-j\pi fT_s}\, \frac{\sin \pi fT_s}{\pi f}$$

The impulse train can be written as

$$s_\delta(t) = \sum_n a_n \delta(t - nT_s)$$

where the a_n are either $+1$ or -1 depending upon the corresponding information bit. The Fourier Transform of a single term in this sum is given by

$$a_m e^{-j2\pi mfT_s}$$

Therefore, the Fourier Transform of the impulse train is

$$S_\delta(f) = \sum_n a_n e^{-j2\pi nfT_s} \qquad (8.1)$$

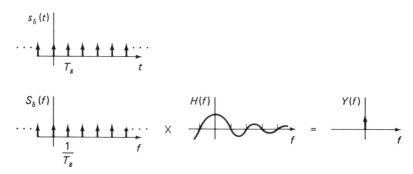

FIGURE 8.4   Generation of output Fourier transform for bipolar baseband with all samples equal.

We start by saying as much about Eq. (8.1) in general as possible. If the a_n are random and zero mean, then the expected value of $S_\delta(f)$ is 0 except for $f = k/T_s$. At these points in frequency, we are summing an infinity of zero mean variables. More importantly, we note that the transform is periodic with basic period $f_0 = 1/T_s$.

Evaluation of $S_\delta(f)$ is simple for some periodic $s_\delta(t)$ examples. Suppose first that $a_n = 1$ for all n. We see that when $f = k/T_s$, all of the terms in the summation of Eq. (8.1) are unity and the sum is infinity. This is not surprising, since the Fourier Transform of any periodic time function is a train of impulses at harmonics of the fundamental frequency. The time function, its transform, and the transform of the output, Y(f), are illustrated in Fig. 8.4. The result may appear surprising until we observe that the baseband signal corresponding to this choice of a_n's is a constant. Thus, the energy is concentrated at a frequency of $f = 0$.

If we now change the sequence to an alternating train of 1's and -1's, that is,

$$a_n = (-1)^n$$

we have the result of Fig. 8.5. We note that the impulses follow a $(\sin f)/f$ type

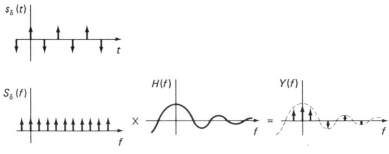

FIGURE 8.5   Generation of output Fourier transform for bipolar baseband with alternating samples.

envelope, and we can therefore define a bandwidth. The exact definition of bandwidth depends upon the application. We shall choose the distance out to the first zero of the envelope to specify bandwidth. The envelope observation is fairly general since the transform of the impulse train is periodic. That is, although a different sequence would lead to a different transform, the transform will always repeat at intervals of $1/T_s$, and therefore periods beyond the $1/T_s$ basic period will be reduced according to the $(\sin f)/f$ multiplication.

The channel distorts the waveform and causes *time dispersion*. This, in turn, causes *intersymbol interference*.

Intersymbol Interference

In Chapter 6, we discussed intersymbol interference for pulse modulation systems where the pulse heights were analog. We shall now apply this analysis to binary pulses. Intersymbol interference is caused not only by channel distortion (which spreads or disperses the pulses) but also by *multipath* effects (echo).

At present, we illustrate a simple case in order to develop the concept of eye patterns. Suppose binary information is being transmitted using a unipolar pulse waveform. That is, a 1-volt pulse is used to send a binary 1 and a 0-volt pulse for a binary 0. When this waveform goes through the system it is distorted and, among other effects, any sharp corners of the wave are rounded. Therefore, the values in previous sampling intervals affect the value within the present interval. If, for example, we send a long string of 1's, one would expect the channel output to eventually settle to being a constant, 1. Likewise if we send a long string of 0's, the output should eventually settle toward 0. If we alternate ones and zeros, the output might resemble a sine wave.

If we examine a single interval in which a binary 1 is being transmitted, the output waveform within that interval will depend upon the particular sequence which preceded the interval in question. In an ideal non-causal system, the sequence following the interval also affects the current values. If we now plot all possible waveforms within the interval, including those for a 1 and those for a 0 in the interval, we get a pattern which resembles a picture of an eye. Figure 8.6 shows some representative transmitted waveforms and the resulting received waveform. The partial eye pattern is sketched. The eye pattern is the superposition of many waveforms within one sampling interval, the components of this composite waveform being the signals due to all possible preceding data strings.

If a random binary pulsed waveform forms the input to a system and the output is observed on an oscilloscope synchronized to the sampling interval, the eye pattern results. The eye opening (blank space in the center) is most significant. Suppose you were at the receiver and wished to decide whether a 1 or a 0 was being transmitted. A reasonable approach would be to sample the received waveform in the middle of each interval and compare the sample value to the midpoint between 0 and 1. A sample greater than the midpoint would lead to a decision that a 1 was being transmitted.

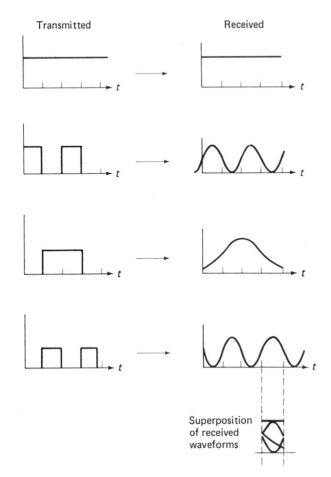

FIGURE 8.6 Intersymbol interference and generation of eye pattern.

Similarly, a sample less than the midpoint would cause one to decide that a 0 was being sent. As long as all components of the eye due to a 1 in the interval lie above the middle at the sampling point, and vice versa for the 0's, this receiving scheme will reproduce the transmitted signal with no errors. However, if the eye closes, errors will be made in receiving certain sequences. Note that for the distortionless channel, the eye consists of two horizontal lines, one at the top and one at the bottom of the interval.

8.2   AMPLITUDE SHIFT KEYING (ASK)

Chapter 6 introduced the concept of using pulses of varying amplitude in order to send discrete information. This technique was called pulse amplitude modulation, or PAM. In finding the Fourier Transform of the PAM waveform, we noted that significant signal information resided in the band around DC, or zero frequency. The same was true for the baseband signals of Section 8.1. Since most real communication channels do not pass very low frequency signals, frequency translation techniques are commonly used to shift frequencies into a desirable range. Amplitude modulation, as used extensively in analog communication, finds application in digital communication as one simple way of translating frequencies. The reasons for using this technique for digital communication are essentially the same as those applying to analog communication.

An AM waveform can be generated using either of two distinct approaches. One technique starts with the baseband signal and uses this signal to amplitude modulate a sinusoidal carrier. Since the baseband signal consists of discrete waveform segments, the AM wave will also consist of discrete modulated segments. This observation suggests a second approach to modulation. It is possible to generate the AM wave directly without first forming the baseband signal. In the binary case, the generator would only have to be capable of formulating one of two distinct AM wave segments.

We begin by examining the baseband approach, since the results derived in the previous section can be borrowed to gain insight into the properties of the AM wave.

The baseband signal is multiplied by a sinusoid at the carrier frequency f_c. The system is illustrated in Fig. 8.7. We have started with impulse samples so that the filter, $H_1(f)$, can be used to develop any baseband signal shape.

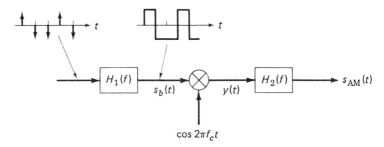

FIGURE 8.7   Generation of ASK waveform.

The filter, $H_2(f)$, following the modulator is used to shape the signal prior to transmission through the channel. For example, it can be used to shape the square pulses shown in the figure in order to limit the bandwidth. The filter can also be used to eliminate portions of the signal thus producing single sideband or vestigial sideband waveforms. It can also be used for shaping the signal in order to reduce intersymbol interference. This filter function can be moved in front of the modulator. That is, the baseband signal can be shaped prior to modulation. The required filter would be a frequency shifted version of $H_2(f)$, and could be combined with $H_1(f)$. For that reason, we will assume without loss of generality that $H_2(f) = 1$ in the following analysis.

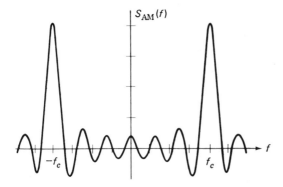

FIGURE 8.8 Fourier Transform of ASK waveform.

The Fourier Transform of the output of the modulator is

$$S_{AM}(f) = \frac{H_1(f)}{2}[S_b(f - f_c) + S_b(f + f_c)] \tag{8.2}$$

where $S_b(f)$ is the Fourier Transform of the baseband signal. This transform is sketched in Fig. 8.8.

Before we can talk extensively about transmission techniques, it is necessary to know the band of frequencies occupied by this waveform. Equation (8.1) provides the key for a simple extension of the baseband analysis. The Fourier Transform of the AM wave is simply a shifted version of the transform of the baseband waveform.

Illustrative Example 8.1

Find the bandwidth of a unipolar ASK signal where the sampling period is 1 msec and the carrier frequency is 1 MHz. Assume that an alternating series of 1's and 0's is sent.

Solution

The ASK waveform consists of 1-msec bursts of the carrier alternating with 1-msec segments of zero signal as shown in Fig. 8.9. The Fourier Transform of this waveform is most

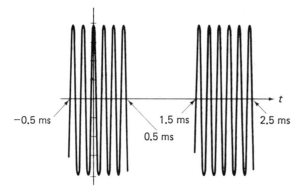

FIGURE 8.9 Waveform for Illustrative Example 8.1.

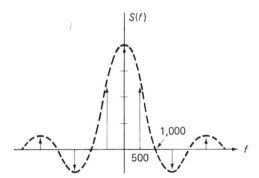

FIGURE 8.10   Fourier Transform of waveform of Fig. 8.9.

easily found by first evaluating the transform of the baseband signal. This baseband signal is a periodic pulse train. One way to do this is to first find the Fourier Series. The resulting transform is evaluated as shown in Fig. 8.10. The Fourier Transform of the ASK waveform is found by translating the transform of Fig. 8.10 to a center frequency of 1 MHz. Since the transform never goes identically to zero, the exact bandwidth depends upon the definition used.

Modulators

As observed earlier, generation of the amplitude shift keyed waveform can be accomplished in two ways. One technique is to first generate the baseband signal and then use any of the amplitude modulation techniques which are common to analog forms of AM communication. The second way is to actually do a keying operation. Thus, for example, in binary communication the transmitted waveform is one of two possible signal waveforms. In the baseband case, these two possible signals are low frequency, while in the ASK case, they are modulated sinusoidal bursts. A particularly simple realization occurs if the baseband signal before modulation is unipolar. That is, the baseband signal (before shaping) is a positive constant to transmit a binary 1 and zero to transmit a binary 0. The ASK waveform then becomes a sinusoidal burst to transmit a 1 and zero to transmit a 0. This can be implemented as an oscillator which is turned on for a specified length of time in order to send a 1 and it is left off to send a 0. This technique is known as *ON-OFF keying* and is illustrated in Fig. 8.11. Although this is instructive and a good starting point, ON-OFF keying is not widely used, since it does not provide very effective error performance.

If a level other than zero is used to transmit a binary 0, we can resort to an oscillator feeding a step-variable attenuator. A conceptualized version of this using resistive voltage dividers is shown in Fig. 8.12.

FIGURE 8.11   ON-OFF keying modulator for ASK.

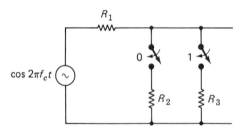

FIGURE 8.12 Simplified modulator for non-unipolar ASK.

8.3   FREQUENCY SHIFT KEYING (FSK)

As we did in the previous section in discussing ASK, we again look at a system which starts with the baseband signal. Let us assume that the baseband signal is composed of piecewise constant segments. That is, a binary 1 is transmitted with a square pulse of voltage V_1 and a binary 0 is transmitted with a pulse of V_0 volts. In some of our previous analyses, V_1 was $+V$ volts and V_0 was $-V$ volts (we called this bipolar), but that special case is not needed for the present analysis.

We observed in Section 8.1 that the baseband waveform contains relatively low frequencies, and can therefore not be transmitted through a bandpass channel. In Section 8.2, we resorted to amplitude modulation techniques in order to shift the frequencies of the waveform for efficient transmission. Frequency modulation is another way of superimposing the baseband information upon a carrier, thereby shifting the frequency content from a band around DC to a band around the carrier frequency. In frequency modulation it is the frequency of the carrier rather than its amplitude which varies in accordance with the information baseband signal. Recall that if we denote the modulated carrier as

$$s_m(t) = A \cos 2\pi\theta(t) \qquad (8.3)$$

then the frequency of this is given by the derivative of the phase

$$f(t) = \frac{d\theta}{dt}$$

The modulation is now performed by varying $f(t)$ in accordance with the baseband signal as follows:

$$f(t) = f_c + Ks_b(t) \qquad (8.4)$$

$s_b(t)$ is the baseband signal and K is a constant which, together with the amplitude of the baseband signal, determines the size of the swings in frequency.

Since we are currently assuming that the baseband signal only takes on one of two values, the frequency of the modulated waveform will also take on one of two values, and the modulation process can be thought of as a keying operation. Figure 8.13 illustrates a simple modulator consisting of two oscillators and a switch (key).

FIGURE 8.13   Simplified FSK modulator.

This form of FM for digital information transmission is referred to as *frequency shift keying (FSK)*.

In order to assess the transmission characteristics of this signal, we will first find the band of frequencies occupied. One would expect this band to span the space between the two frequencies of the oscillators. Indeed, due to the switching operation and the transients it causes, the bandwidth is greater than the difference, $f_1 - f_0$, as the following analysis shows. Recall that in the analog case, the bandwidth of the FM waveform was not simply the total swing of the instantaneous frequency curve. It also depended upon the rate at which the instantaneous frequency was changing.

The output of the modulator of Fig. 8.13 can be considered as the superposition of two Amplitude Shift Keyed signals. One of these is the ASK signal resulting from

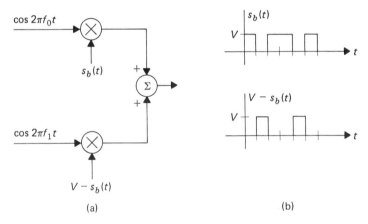

(a)   (b)

FIGURE 8.14   (a) FSK as superposition of two ASK waveforms; (b) sample waveforms.

modulating a carrier of frequency f_1 with an ON-OFF keying from the baseband signal. The other results from amplitude shift keying a carrier of f_0 frequency with the complement of the baseband signal. Figure 8.14 shows the system that would accomplish this. We can therefore borrow the ASK results of Section 8.2 to find the bandwidth of the FSK signal. Each of the two components has a frequency spectrum which is of the general shape $(\sin f)/f$ shifted to the carrier frequency. The Fourier Transform of the FSK waveform is a sum of two of these as sketched in Fig. 8.15. We have exaggerated the frequency difference, $f_1 - f_0$, for purposes of the diagram. Thus

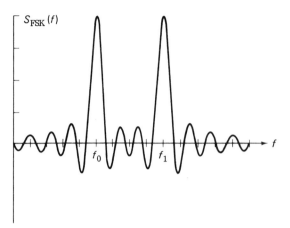

FIGURE 8.15 Fourier Transform of FSK waveform.

the bandwidth out to the first zero of the FSK waveform is

$$BW = f_1 - f_0 + \frac{2}{T_s}$$

Illustrative Example 8.2

An FSK signal consists of bursts of frequency 999 kHz and 1 MHz, the higher frequency being used to send a binary 1. The sampling period is 1 msec. Find the bandwidth of the FSK signal.

Solution

Looking first at the bursts representing the 1's in the binary signal, the frequency spectrum is centered at 1 MHz and has a $(\sin f)/f$ shape. Since the bursts are of 1 msec duration, the first zero of the spectrum is at 1 MHz $\pm$ 1 kHz. The spectrum of the binary zeroes is centered at 999 kHz, with the first zeroes of the spectrum at 999 kHz $\pm$ 1 kHz. The bandwidth out to the first zero is therefore 1,001 kHz $-$ 998 kHz or 3 kHz.

If the baseband signal is shaped prior to frequency modulation in order to decrease intersymbol interference, the keying concept is no longer directly applicable. It can still be used if the sources are changed from sinusoidal to shaped sinusoids, where the shaping function is the pulse shape.

Illustrative Example 8.3

Find the bandwidth of the FM signal where the pulses are shaped to be $\frac{1}{2}$ cycle of a sine wave. Assume that an alternating train of 0's and 1's is transmitted and that bipolar transmission is used.

Solution

We shall assume that the shaping is done on the amplitude and not on the frequency of the modulated waveform. Thus, the signal is a burst during each period, where the envelope is

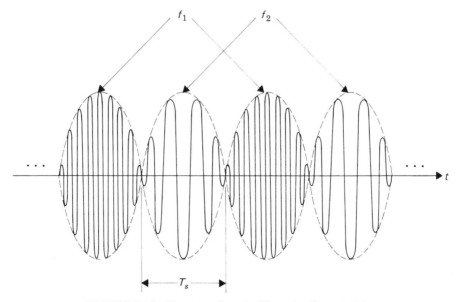

FIGURE 8.16   Signal waveform for Illustrative Example 8.3.

sinusoidally shaped. The signal can therefore be thought of as a superposition of 2 waveforms differing only in the carrier frequency, as shown in Fig. 8.16. The frequency spectrum of this waveform is found by first evaluating the Fourier Transform of the unmodulated waveform. This wave is periodic with period $2T_s$. The Fourier Series can be evaluated as

$$s_b(t) = a_0 + \sum_{n=1}^{\infty} a_n \cos \frac{n\pi t}{T_s}$$

where

$$a_n = \frac{2}{T_s} \int_0^{T_s/2} \cos \frac{\pi t}{T_s} \cos \frac{n\pi t}{T_s} \, dt$$

$$a_0 = \frac{1}{T_s} \int_0^{T_s/2} \cos \frac{\pi t}{T_s} \, dt = \frac{1}{\pi}$$

Evaluating the integral for a_n, we find

$$a_n = \frac{(-1)^{n/2}}{\pi} \left[\frac{1}{n+1} - \frac{1}{n-1} \right]$$

The spectrum of the wave shown in Fig. 8.16 consists of lines at multiples of $1/T_s$ away from f_c, where the height of each line is the corresponding value of a_n. Looking at the magnitudes of the a_n for n even, we have

$$a_n = 1, 0.880, 0.133, 0.06, 0.03, 0.014, 0.010, 0.008$$

If we define bandwidth by the point at which the amplitude has decreased to 1% (a very rigid definition), we see that it is necessary to go out 6 harmonics of the double frequency. The bandwidth of the FSK waveform is therefore

$$BW = f_1 - f_0 + \frac{12}{T_s}$$

In dealing with FM, there is a second kind of shaping which can arise. Rather than having abrupt frequency changes and shaped amplitude as in Illustrative Example 8.3, we can instead have constant amplitude and shaped frequency. This is more analogous to the analog FM case. Thus, we might replace the abrupt frequency transitions by sinusoidal variations as shown in Fig. 8.17. We will borrow analog FM

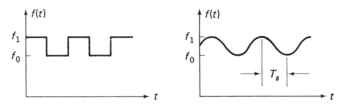

FIGURE 8.17   Example of FSK with sinusoidally shaped frequency transitions.

analysis techniques to find the bandwidth of the resulting waveform. Let us designate the middle frequency as f_c. We then have

$$f(t) = f_c + \Delta f \cos 2\pi f_0 t$$

where

$$\Delta f = \frac{f_1 - f_0}{2}$$

and

$$f_0 = \frac{1}{2T_s}$$

Using techniques from Chapter 5, we find that the bandwidth is approximately

$$BW = 2(\Delta f + f_0)$$

Modulators

Frequency modulation can be performed using the keying system of Fig. 8.13 provided that the baseband is composed of square pulses. For continuous baseband signals, we can borrow the results from analog FM. That is, we use the baseband signal as the information signal in any of the modulators described in Chapter 5. One representative system is presented as Fig. 8.18.

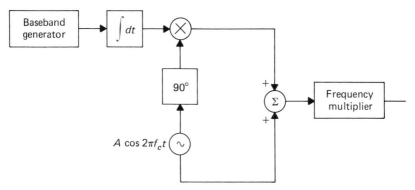

FIGURE 8.18 FSK modulator using baseband generator and analog FM modulator.

8.4   PHASE SHIFT KEYING (PSK)

In the case of analog signals, phase modulation is not significantly different from frequency modulation and would probably not warrant an extensive examination. This will prove not to be the case in digital communication.

Phase modulation starts with a sinusoidal carrier of the form $A \cos(2 f_c t + \theta(t))$ just as in AM and FM. In frequency modulation the derivative of $\theta(t)$, which we called instantaneous frequency, followed the baseband or information signal. In phase modulation, it is the phase itself, $\theta(t)$, which follows this baseband signal. Thus,

$$\theta(t) = 2\pi f_c t + K s_b(t)$$

Note that there is no difference between phase modulating a carrier with $s_b(t)$ and frequency modulating the same carrier with the derivative, $2\pi ds_b/dt$. For this reason, the form of *analog* phase modulation does not look significantly different from that of analog frequency modulation.

In the case of digital transmission where the baseband signal is piecewise constant, the resulting simplification makes phase modulation look considerably different from frequency modulation. In binary frequency shift keying (FSK) we switched back and forth between sinusoids of two different frequencies depending upon whether a 1 or a 0 was being transmitted. In binary phase shift keying (PSK), the frequency of the waveform stays constant while the *phase shift* has one of two constant values.

Thus, for example, the two signals used to transmit a 0 and 1 can be expressed as

$$s_0(t) = A \cos (2\pi f_c t + \theta_0)$$
$$s_1(t) = A \cos (2\pi f_c t + \theta_1) \qquad 0 < t \le T_s \qquad (8.5)$$

where θ_0 and θ_1 are constant phase shifts. We will see when performance is analyzed later in this chapter that the best choice of phase angles is such that $\theta_1 - \theta_0 = 180°$.

For this choice of phase angles, the signals take on a particularly simple form since a 180° phase shift represents a polarity reversal. That is,

$$s_1(t) = -s_0(t)$$

As in the case of FSK, the PSK signal can be considered as a superposition of two ASK waveforms, and the bandwidth of the resulting waveform can be found by examining the component parts. If 180° is used for the phase separation, phase shift keying is equivalent to bipolar amplitude shift keying. That is, the two signals are

$$A \cos(2\pi f_c t + \theta_0) \quad \text{and} \quad -A \cos(2\pi f_c t + \theta_0)$$

We found in Section 8.3 that the bandwidth of the FSK waveform was given by the difference between the two frequencies used plus a factor representing the spreading caused by the switching operation. In the case of PSK, only one frequency is used, so the bandwidth is entirely due to the switching harmonics. Thus, the bandwidth of the PSK waveform, out to the first zero, is $2/T_s$. The argument for deriving this result is identical to that used in Section 8.3 and reflected by Fig. 8.15.

If a continuous baseband signal is used instead of the pulse waveform to reduce distortion (intersymbol interference), the bandwidth is then given by

$$BW = 2\left(f_m + 2\pi K \max \frac{ds_b}{dt}\right) \qquad (8.6)$$

where f_m is the maximum frequency component of the baseband signal. It should be noted that this is the same as the expression for frequency modulation, except that the derivative of the baseband signal has been substituted for the baseband signal. The maximum frequency of $s_b(t)$ is the same as the maximum frequency of the derivative, ds_b/dt, since differentiation is a linear process.

Illustrative Example 8.4

Find the bandwidth of a PSK waveform used to transmit an alternating train of zeros and ones. The frequency is 1 MHz and the phase varies sinusoidally from 0° to 180°. The bit rate is 1 kbps.

Solution

We need simply evaluate the expression in Eq. (8.6). f_m is the maximum frequency of the baseband. In this case, the baseband is a sinusoid of frequency 500 Hz (one half of the bit rate). The second term in Eq. (8.6) requires that we find an expression for the baseband signal. If we carefully match the given information to Eq. (8.6), we have

$$\theta(t) = 2\pi f_c t + \tfrac{1}{2}\pi \sin 2\pi \times 10^3 t$$

Thus, $Ks_b(t)$ is given by

$$Ks_b(t) = \tfrac{1}{2}\pi \sin 2\pi \times 10^3 t$$

and

$$\max K \frac{ds_b}{dt} = \pi^2 \times 10^3$$

Finally, the bandwidth is given by Eq. (8.6):

$$BW = 2(500 + 500\pi) \text{ Hz}$$

$$= 4.14 \text{ kHz}$$

Reception of PSK requires exact knowledge of absolute phase (timing) to know which of the two signals in Eq. (8.5) is being received. This exact knowledge often does not exist. However it will be possible to identify phase *changes* even if the absolute phase is not known. A variation of PSK is sometimes used which takes advantage of this last observation.

DPSK

Differential phase shift keying (DPSK) is a system which uses *changes* in phase to indicate values of the data. If a zero is to be transmitted, the transmitter changes the phase from its present value to the other value. Thus, if the phase were θ_0 at the time a 0 is input to the transmitter, the phase changes to θ_1.

If, on the other hand, the original phase is θ_1, a binary 0 causes the transmitter to change phase to θ_0. To transmit a 1, the phase is *not* changed during the interval. Observation of the absolute phase during any given interval is not sufficient to detect which of the two possible data bits is being sent. One must compare the phase in the interval to the phase in the previous interval. This leads to a particularly simple detector (receiver) implementation. Figure 8.19 shows the detector.

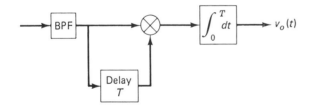

FIGURE 8.19 Detector for DPSK waveform.

This receiver compares the received signal from one interval to the next. If the signals being compared are identical, the input to the integrator is the square of the signal segment. If no change has occurred, the output is

$$v_0(T) = \int_0^T A^2 \cos^2 (2\pi f_c t + \theta_i) \, dt$$

$$= \frac{A^2 T}{2} - \frac{A^2}{4\pi f_c} \sin (4\pi f_c T + \theta_i)$$

If $f_c T = n/2$ or $T \gg 1/f_c$,

$$v_0(T) = A^2 T/2$$

If a change has occurred, the output is

$$A^2 \int_0^T \cos(2\pi f_c t + \theta_0) \cos(2\pi f_c T + \theta_1) \, dt$$

$$= \frac{A^2}{2} \left[\int_0^T \cos(4\pi f_c t + \theta_1 + \theta_0) \, dt + \int_0^T \cos(\theta_0 - \theta_1) \, dt \right]$$

$$= \frac{A^2 T}{2} \cos(\theta_0 - \theta_1)$$

Now if $\theta_0 - \theta_1 = 180°$, the last term becomes $-\frac{1}{2} A^2 T$, which is the negative of the output when no change occurs. The output is therefore a bipolar baseband signal.

Modulators

The modulators for PSK exactly parallel those for FSK. We start with the keyed modulator which is applicable if the baseband signal is piecewise constant. This is illustrated in Fig. 8.20, and consists of two oscillators (or a single oscillator with one delayed path), and a switch which cycles back and forth depending upon which bit is to be transmitted.

FIGURE 8.20 Simplified PSK modulator.

If the baseband signal is continuous, we use the modulator that was used for FM, except that the baseband signal is differentiated prior to modulating. This is illustrated in Fig. 8.21.

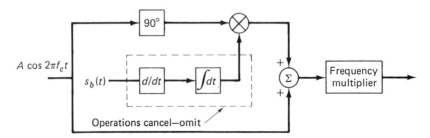

FIGURE 8.21 PSK modulator using baseband generator and analog modulator.

8.5   MATCHED FILTER DETECTORS

The requirements placed upon a digital receiver are radically different from those of an analog receiver. The analog receiver must reconstruct a waveform which is as close as possible to the information signal present at the transmitter. *The digital receiver is not the least bit concerned with reconstructing the information waveform that existed at the transmitter.* Instead, it is content to decide at each sampling point which of the possible symbols in the transmission coding alphabet is being received. This opens up processing possibilities which could not be considered in the analog system. Stated another way, the receiver can distort the signal as much as it wishes if such distortion helps to make a decision as to which symbol is being communicated.

The actual mechanism of making a decision at the receiver is usually one of processing the incoming signal in some prescribed manner, and then viewing a sample value of the output of this processor. This sample value is then manipulated in some manner to yield a decision. The processing prior to sampling is such that the sample value is most likely to yield the correct decision. This involves decreasing the effects of additive noise.

Use of a matched filter represents one technique for processing the received signal. The criterion for designing the filter is that the output at a sample time T is to have maximum signal to noise ratio. We designate the input to the filter as $s(t) + n(t)$, where $s(t)$ is the signal and $n(t)$ is the noise. The filter output is $s_o(t) + n_o(t)$, where $s_o(t)$ is the signal portion and $n_o(t)$ is the noise portion of the output. The filter is chosen to maximize $s_o^2(T)/n_o^2(T)$. Since the denominator of this expression is usually random, we must use the average of the random variable. This average of the square is the *average power* of the noise.

The output signal to noise ratio is given by

$$\rho = \frac{s_o^2(T)}{n_o{}^2(T)} = \frac{\left| \int_{-\infty}^{\infty} S(f)H(f)e^{j2\pi fT}\,df \right|^2}{\int_{-\infty}^{\infty} |H(f)|^2 G_n(f)\,df} \tag{8.7}$$

The numerator in Eq. (8.7) is the square of the inverse Fourier Transform of the product of the input transform with the system function. Thus it is the square of the deterministic time sample, $s_o(T)$. The $G_n(f)$ in the denominator is the power spectral density of the input noise. Thus the denominator integrand is the power spectral density of output noise. (See Section 2.9 for the derivations of this power expression.) We wish to choose $H(f)$ to maximize the expression of Eq. (8.7). The choice is simplified if we apply *Schwartz's inequality* to the numerator. Schwartz's inequality states that for all functions, $f(x)$ and $g(x)$,

$$\left| \int f(x)\,g(x)\,dx \right|^2 \le \int |f^2(x)|\,dx \int |g^2(x)|\,dx$$

Textbook writers enjoy giving an elegant proof of this theorem, and although it breaks

the continuity, I shall not break with tradition. We start by noting that for real
functions,

$$\int [f(x) - Tg(x)]^2 \, dx \geq 0$$

for all $f(x), g(x)$, and T. This is true because the integrand, being a square, is non-
negative. Expanding this, we find

$$T^2 \int g^2(x) \, dx - 2T \int f(x) g(x) \, dx + \int f^2(x) \, dx \geq 0 \qquad (8.8)$$

The left side of Eq. (8.8) is quadratic in T. Since the value is constrained never to go
negative, the quadratic cannot have distinct real roots. Therefore, its discriminant
must be non-positive. Thus,

$$4 \left[\int f(x) g(x) \, dx \right]^2 - 4 \int f^2(x) \, dx \int g^2(x) \, dx \leq 0$$

and the inequality is established for real functions. The extension to complex
functions is left as an exercise at the end of this chapter.

We now wish to apply the inequality to Eq. (8.7). We modify the numerator of
the equation in a seemingly unmotivated way that will factor out a term identical to
the denominator.

$$\left| \int_{-\infty}^{\infty} S(f) H(f) e^{j2\pi f T} df \right|^2 = \left| \int_{-\infty}^{\infty} \frac{S(f)}{\sqrt{G_n(f)}} H(f) \sqrt{G_n(f)} \, e^{j2\pi f T} \, df \right|^2$$

The square root operation is unambiguous due to the fact that $G_n(f)$ can never be
negative since it is a power spectrum. Now applying Schwartz's inequality, we find

$$\rho \leq \frac{\displaystyle\int_{-\infty}^{\infty} |H(f)|^2 G_n(f) \, df \int_{-\infty}^{\infty} |S(f)|^2 / G_n(f) \, df}{\displaystyle\int_{-\infty}^{\infty} |H(f)|^2 G_n(f) \, df}$$

$$\rho \leq \int_{-\infty}^{\infty} \frac{|S(f)|^2}{G_n(f)} \, df$$

An upper bound has therefore been established on the signal to noise ratio at the
output of the filter. If by some hocus pocus we can guess at an $H(f)$ which yields this
maximum, we need look no further. Starting at Eq. (8.7) makes the guess almost
obvious. By observation, if

$$H(f) = C e^{-j2\pi f T} \frac{S^*(f)}{G_n(f)} \qquad (8.9)$$

is plugged into Eq. (8.7), the maximum signal to noise ratio is achieved. C is an arbitrary constant and factors out, since it appears in both the numerator and denominator of Eq. (8.7).

Equation (8.9) gives the system function of the optimum filter. In order to find the impulse response, the inverse Fourier Transform of $H(f)$ is evaluated. Let us assume that $n(t)$ is white noise, so $G_n(f)$ is a constant, $N_o/2$. The filter impulse response then becomes

$$h(t) = \tfrac{1}{2} CN_o s(T - t)$$

This inverse transform is found by recognizing that the transform of $s(-t)$ is the complex conjugate, $S^*(f)$. The term $e^{-j2\pi fT}$ represents a time shift. A filter with this impulse response is called the *matched filter*. Note that this equation might represent a nonrealizable filter.

Illustrative Example 8.5

Find the impulse response of the matched filter for the two time functions shown in Fig. 8.22.

FIGURE 8.22   Time functions for Illustrative Example 8.5.

Solution

The $h(t)$ is derived directly from Eq. (8.9). The result is sketched in Fig. 8.23. Note that the first $h(t)$ is noncausal, and would have to be implemented with a time shift included.

FIGURE 8.23   Resulting matched filter impulse responses for Illustrative Example 8.5.

The actual time function at the output of the matched filter can be found by convolving the input time function with the impulse response. Therefore,

$$s_o(t) + n_o(t) = [s(t) + n(t)] * h(t)$$

$$= C \int_{-\infty}^{\infty} [s(\tau) + n(\tau)] s(T - t + \tau) d\tau$$

and at time T,

$$s_o(T) + n_o(T) = C \int_0^T [s(\tau) + n(\tau)] s(\tau) d\tau$$

Thus the matched filter is equivalent to the system of Fig. 8.24. The operation being performed by the system in Fig. 8.24 is called *correlation* (i.e., multiply two time

FIGURE 8.24   Equivalent form of matched filter as correlator.

functions together and integrate the product). For that reason, the matched filter is sometimes referred to as a *correlator*. In a generalized sense, the filter is finding the projection of the input signal in the direction of the information, $s(t)$. Since the system is aligned in the $s(t)$ direction, the output signal to noise ratio is thereby maximized.

Illustrative Example 8.6

Find the output signal to noise ratio of a matched filter where the signal is

$$s(t) = A \quad \text{for } 0 < t \leq T$$

in white noise with power spectral density $N_o/2$.

Solution

The appropriate matched filter simply multiplies the received waveform by a constant, A, and then integrates this product between 0 and T. We note that the multiplying constant could be any value without affecting the signal to noise ratio, since it multiplies both the signal and the noise waveform.

The output signal is

$$s_o(T) = \int_0^T s^2(t) dt = AT$$

The output noise is

$$n_o(T) = A \int_0^T n(t) dt$$

and the noise power is the mean square value of this expression. That is, we take the expected value of the square of the integral (note that the square of an integral is *not* the integral of the square—look at Schwartz's inequality). We will evaluate the noise power in the time domain. The particular approach we use will be needed later.

$$E\{n_o^2(T)\} = E\left\{A^2 \int_0^T \int_0^T n(t)n(\tau)\, dtd\tau\right\}$$

We now move the expected value operation inside the integral since the expected value of a sum is the sum of the expected values. The expected value of $n(t)n(\tau)$ is the autocorrelation of the noise whi h, for white noise, is an impulse. Thus,

$$E\{n_o^2(T)\} = A^2 \int_0^T \int_0^T E\{n(t)n(\tau)\}\, dt\, d\tau$$

$$= \frac{A^2 N_o}{2} \int_0^T \int_0^T \delta(t-\tau)\, dtd\tau$$

$$= \frac{A^2 N_o}{2} \int_0^T 1\, dt$$

$$= \frac{A^2 N_o T}{2}$$

The signal power is simply the square of the signal value, or $A^4 T^2$. The signal to noise ratio is therefore given by

$$\rho = \frac{A^4 T^2}{A^2 T N_o/2} = \frac{2A^2 T}{N_o}$$

The Matched Filter Detector for Baseband

In a unipolar baseband transmission system, we are attempting to detect a signal similar to that of Illustrative Example 8.6. The detector is then of the form shown in Fig. 8.25. The output of the integrator consists of a signal part and a noise part. The input signal has an amplitude of $+V$ volts, so the signal part of the output is $V^2 T_s$ when a 1 is being sent, and 0 when a 0 is being sent.

The noise perturbs these values so that when a 1 is sent, we will not necessarily measure exactly $V^2 T_s$ at the output. Likewise, when a 0 is sent, the output is not exactly equal to 0. Lacking any further information, it would seem reasonable to call the received bit a "1" if the measured value at the output is greater than $V^2 T_s/2$, and

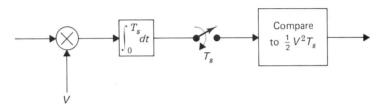

FIGURE 8.25   Matched filter detector for unipolar baseband.

call it a "0" if less. That is, we are comparing the received value to the midpoint of the interval.

In order to assess the error probabilities of this detector, we will need the noise statistics. The noise part of the output is given by

$$n = \int_0^{T_s} V n(t) \, dt$$

Since $n(t)$ is assumed to be *white Gaussian noise* with zero mean, n will be a Gaussian random variable. This is true since integration is a linear operation. The mean value of n is zero. This is true since we assume $n(t)$ has zero mean, and the mean of a sum (integral) is equal to the sum of the means. The variance is given by

$$\sigma^2 = E \left\{ \left[\int_0^{T_s} V n(t) \, dt \right]^2 \right\}$$

$$= E \left\{ V^2 \int_0^{T_s} \int n(t) n(\tau) \, dt \, d\tau \right\}$$

We use the same approach as earlier to find

$$\sigma^2 = V^2 \int_0^{T_s} \int R_n(t - \tau) \, dt \, d\tau$$

$$= V^2 \int_0^{T_s} \int \frac{N_o}{2} \delta(t - \tau) \, dt \, d\tau$$

$$= V^2 \frac{N_o}{2} \int_0^{T_s} 1 \, dt = V^2 \frac{N_o T_s}{2}$$

The signal to noise ratio at the output is

$$\text{SNR} = \frac{V^4 T_s^2}{V^2 N_o T_s / 2} = \frac{2 V^2 T_s}{N_o}$$

V in the detector of Fig. 8.25 multiplies both the signal and the noise components. Therefore, if the receiver had been modified by eliminating the pre-integration multiplier, the signal to noise ratio would not change.

The Binary Matched Filter Detector

We now consider the general case of the matched filter receiver for binary signals in noise. We shall not restrict the two signal waveforms to any particular format.

Therefore, the results of this section will be applicable to any of the modulation schemes which were discussed in this chapter.

Figure 8.26 shows the binary matched filter detector. $s_0(t)$ and $s_1(t)$ are the two signals which are assumed to be known completely and of equal energy. We illustrate the receiver using correlators, but an exact equivalent receiver is that using matched filters to replace the multiplication and integration.

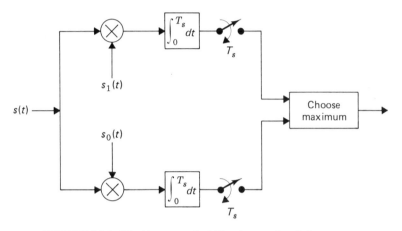

FIGURE 8.26   The binary matched filter (or correlator) detector.

We need to justify choosing of the maximum. The detector must somehow try to pull one of two separate signals out of noise. If a single matched filter is matched to the difference, $s_1(t) - s_0(t)$, we get the system of Fig. 8.27. The reason we use the

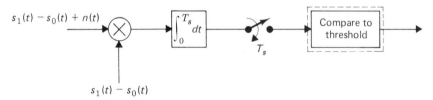

FIGURE 8.27   Correlator detector matched to difference between two signals.

difference is that it conveys all of the useful information. This point needs expansion. We can add or subtract any constant signal to each of the waveforms in a binary communication system without changing the performance of that system. Thus, in the binary matched filter detector, we can subtract $s_0(t)$ from both input signals. This changes the system to a unipolar system where one of the signals is zero and the other is $s_1(t) - s_0(t)$.

We return now to the system of Fig. 8.27. If a 1 is sent, the output of the system is

$$\int_0^{T_s} (s_1^2 - s_1 s_0)\, dt$$

If a 0 is sent, the output becomes

$$\int_0^{T_s} (s_1 s_0 - s_0^2)\, dt$$

Since the signals are assumed to have equal energy, these two quantities are the negative of each other. In the baseband case, we put the threshold in the middle of the interval. If we do the same in this case, the middle becomes zero, and the system in the dashed box of Fig. 8.27 implements this. But this is the same as the operation shown in Fig. 8.26.

An equivalent form to that of Fig. 8.27 is shown in Fig. 8.28.

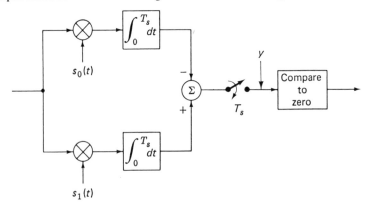

FIGURE 8.28   Equivalent form to detector of Fig. 8.27.

8.6   QUADRATURE DETECTION

Suppose we decided to build a binary matched filter for FSK. The detector would be as shown in Fig. 8.29. Now suppose that the locally generated sinusoids are not matched in phase to the received signals. This is analogous to the synchronous demodulator situation in analog AM, where the phase and frequency of the local oscillator had to perfectly match those of the received carrier. Such a mismatch could cause serious consequences since the phase mismatch translates to an attenuation factor. This could cause the signal to get lost in the noise.

There is a class of detectors that do not require reproduction of the phase of the sinusoid at the receiver. As in the analog case, these detectors are known as *incoherent*, while the perfectly synchronized matched filter detectors are *coherent*.

One type of incoherent detector is the *quadrature detector* which forms an incoherent sum of two basically coherent signals. The receiver is illustrated in Fig. 8.30. It is used when the frequency of the carrier is known, but the phase is not known. The detector shown in the figure is used for detection of ASK. The signals at points A_c and A_s are found in the same manner as for the synchronous demodulator.

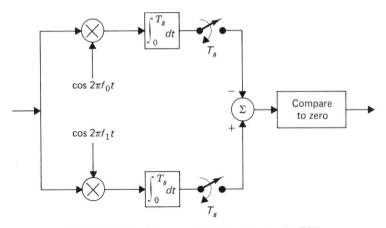

FIGURE 8.29 Binary matched filter detector for FSK.

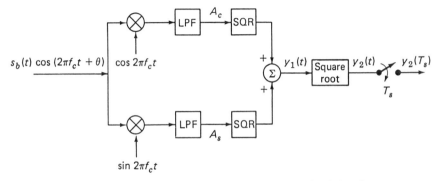

FIGURE 8.30 Quadrature detector illustrated for ASK detection.

$$A_c(t) = \frac{1}{2} s_b(t) \cos \theta$$

$$A_s(t) = \frac{1}{2} s_b(t) \sin \theta$$

$A_c(t)$ is called the *in-phase* (or I) part of the baseband signal and $A_s(t)$ is called the *quadrature* (or Q) part. The word quadrature is used since cosine and sine are in phase quadrature (90° difference). The receiver of Fig. 8.30 is known as a *quadrature receiver*.

The signal at $y_1(t)$ is

$$y_1(t) = \frac{1}{4} s_b^2(t) [\cos^2 \theta + \sin^2 \theta]$$

$$= \frac{1}{4} s_b^2(t)$$

$s_b(t)$ must be non-negative in order for the final square root operation to yield an unambiguous result. This is equivalent to the incoherent detector requirement for transmitted carrier analog AM where the signal had to be non-negative. A signal which is bipolar can be transformed by adding a constant, which can later be subtracted if necessary.

8.7 PERFORMANCE

So far in this chapter, we have examined a number of ways of transmitting digital information. We have also briefly examined distortion and intersymbol interference and seen how these effects can cause errors in digital communication.

We are now ready to consider the various transmission and reception techniques in the presence of additive noise. A thorough analysis has as a prerequisite a study of *decision theory*. This forms the framework for making a decision based upon the observed detector output. We shall not study decision theory in this text. Instead, we shall make assumptions which permit rigorous analysis without the additional theory.

Single Sample Detector

Let us first look at an idealized baseband system in the absence of any distortion. A $+V$ or $-V$ volt bipolar pulse is used to transmit 1's and 0's respectively. We assume that the received signal is sampled at some intermediate point within each sampling interval. If the sample value is larger than zero, we decide that a 1 was being sent, and if less than zero, we decide a 0. Since the noise-free pulse has constant amplitude over the interval, that performance is independent of the exact point at which the sample is taken. In most practical situations, the sampling point will be either at the midpoint or at the endpoint of the interval in order to achieve the best performance.

Each sample is composed of a signal ($\pm V$) plus a sample of the additive noise, $n(t_o)$ is the sampling point. We assume that the noise is a sample of a zero-mean Gaussian noise process. The result of adding a signal, $\pm V$ to a zero-mean Gaussian random variable is a Gaussian random variable with mean $\pm V$ and variance equal to the variance of the noise sample. Therefore, under each possible signal condition, the sample value is Gaussian distributed, and the two densities are as shown in Fig. 8.31. $p_1(y)$ is the density of the sample assuming that a binary 1 is transmitted, and $p_0(y)$ is the density assuming a 0 is sent.

There are four possible combinations of transmission and reception bits. If we denote S_i as the bit being sent and D_i as the bit detected (i.e., the decision), the four combinations are

$$(S_0, D_0); \quad (S_1, D_1); \quad (S_0, D_1); \quad (S_1, D_0)$$

The first two combinations represent correct decisions, while the latter two are errors. (S_0, D_1) is called an *error of the first kind*, or *false alarm*. (S_1, D_0) is an *error of the*

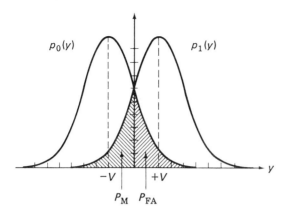

FIGURE 8.31 Conditional densities of received sample assuming a 1 or 0 is transmitted.

second kind, or *miss*. The terminology of miss and false alarm is borrowed from the radar situation where the two possible binary signals are a reflected return from a target (a "1") and no return (a "0").

It will prove easier to evaluate *conditional* probabilities of the form

$$P_{FA} = Pr(D_1/S_0)$$

$$P_m = Pr(D_0/S_1)$$

If we know these conditional probabilities, one additional piece of information is needed to find the actual probability of that combination. This information is the probability of sending a 0 or a 1. These probabilities, $P(S_0)$ and $P(S_1)$, are known as *a-priori probabilities*. Since we must either send a zero or a one, the two probabilities add to unity. Additionally, since one of the four possible pairs of transmitted/decision bits must occur, the sum of the four probabilities is unity.

$$Pr(S_0, D_0) + Pr(S_1, D_1) + Pr(S_0, D_1) + Pr(S_1, D_0) = 1$$

The conditional probability of the two types of error is given by the integral of the appropriate tail of the Gaussian density. Thus, the probability of an error of the first kind or false alarm, P_{FA}, is the area indicated as P_{FA} in Fig. 8.31. The probability of an error of the second kind, or miss, is also labelled on the diagram. These are conditional error probabilities, and the total probability of error is given by the conditional error probabilities weighted by the a-priori probabilities of transmission of the two binary digits.

$$P_e = Pr(D_1, S_0) + Pr(D_0, S_1)$$

$$= P_{FA} P(S_0) + P_M P(S_1)$$

Because of the symmetry of the densities, if the threshold is at zero, the two conditional error probabilities are equal, and therefore the probability of error is the same as either of the conditional error probabilities. This is true since the probability of a 1 plus the probability of a 0 is unity. That is,

$$P_e = P_{FA}P(S_0) + P_M P(S_1)$$
$$= P_{FA}[P(S_0) + P(S_1)]$$
$$= P_{FA} = P_M$$

Illustrative Example 8.7

Assume that in a digital communication system pulses of either $+1$ V or -1 V are transmitted to represent a binary 1 and binary 0, respectively. Gaussian noise with zero mean and unit variance is added to the signal during transmission. Find the probability of making a bit error, known as the *bit error rate, BER*.

Solution

We assume that the receiver samples the received signal and compares the sample value to 0. If the sample is greater than 0, the receiver decides that a 1 was sent. If the sample is less than 0, a 0 is decoded.

If a 1 is transmitted, the sum of the signal plus noise will be Gaussian with mean 1 and variance 1. The probability of this being erroneously detected as a 0 is simply the probability that this Gaussian variable is negative. This is given by

$$\int_{-\infty}^{0} p(s)\, ds = \frac{1}{\sqrt{2\pi}} \int_{-\infty}^{0} e^{-(s-1)^2/2}\, ds$$

$$= 0.159$$

The last result was found by using a table of error functions. This quantity represents the probability of miss. Due to symmetry, it is also the probability of false alarm. When these two conditional probabilities are weighted by the a-priori probabilities, we find

$$0.159P(S_0) + 0.159P(S_1) = 0.159[P(S_0) + P(S_1)]$$
$$= 0.159$$

Now that the simple case has been analyzed, we observe that it is not very realistic. In order for the received signal to consist of perfectly square pulses, the channel would have to admit frequencies up to infinity. If the channel did so, the received noise, assuming additive white channel noise, would have *infinite variance*, and the probability of error would be 0.5. If we instead consider a channel with minimum Nyquist bandwidth, $f_m = 1/2T_s$, we can assume that the signal is of the form

$$s(t) = \pm \frac{T_s V \sin \pi t/T_s}{\pi t}$$

This is the result of starting with an impulse and passing this through an ideal lowpass filter. We assume that sampling is done at the midpoint of each interval, yielding the peak signal value of $\pm$V. The noise power, or variance, would be $N_o f_m$, where N_o is the noise power in the channel per hertz of bandwidth. Therefore, the probability of error is given by integrating the tail of the Gaussian density.

$$P_e = \frac{1}{\sqrt{2\pi}} \frac{1}{\sqrt{N_o f_m}} \int_0^\infty \exp\left(\frac{-(y+V)^2}{2N_o f_m}\right) dy \tag{8.10}$$

We can make a change of variables to simplify this expression. Letting

$$x = \frac{y+V}{\sqrt{2}\sqrt{N_o f_m}}$$

the integral becomes

$$P_e = \frac{1}{\sqrt{\pi}} \int_{V/\sqrt{2N_o f_m}}^\infty e^{-x^2} dx$$

$$= \frac{1}{2} \text{erfc}\left(\frac{V}{\sqrt{2N_o f_m}}\right) \tag{8.11}$$

We now identify $V^2/f_m N_o$ as the signal to noise ratio, so finally

$$P_e = \frac{1}{2} \text{erfc}\sqrt{\frac{S}{2N}} \tag{8.12}$$

This error probability is plotted as a function of the signal to noise ratio in Fig. 8.32.

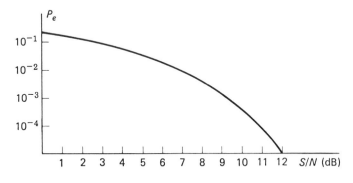

FIGURE 8.32 Error probability as function of signal to noise ratio for the baseband detector.

Illustrative Example 8.8

A 1-volt bipolar pulsed signal at 1 kbps goes through a channel with ideal lowpass filter characteristic, and cutoff frequency of 4 kHz. Noise of 10^{-5} watts/hertz is added in the channel. Find the probability of error using a receiver that samples at the midpoint of each interval.

Solution

The signal portion of the output due to a single pulse is as shown in Fig. 8.33. Since the ripple has frequency of about 4 kHz, we can assume that it has essentially died out at the next

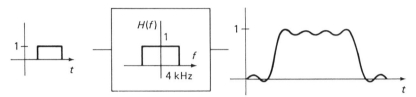

FIGURE 8.33   Signal portion of output due to input pulse for system of Illustrative Example 8.8.

sample point. That is, the intersymbol interference is approximately zero. The signal portion of each sample is therefore ± 1 volt, with average power of 1 watt. The noise power is $4{,}000 \times 10^{-5}$ or 0.04 watts, yielding a signal to noise ratio of 25 or 14 dB. The corresponding probability of error can be read directly from Fig. 8.32. (The curve must be extrapolated.)

$$P_e = 3 \times 10^{-7}$$

The Binary Matched Filter Detector

The binary matched filter detector was illustrated in Fig. 8.28. When the input to this system is $s(t) + n(t)$, the input to the comparison device (y on diagram) becomes a random variable. Since the random part is a weighted sum of samples of the Gaussian noise process, n(t), y will be Gaussian. If a 0 is sent, the mean of y is

$$E\{y/0\} = m_0 = \int_0^{T_s} s_0(t)[s_1(t) - s_0(t)]dt \tag{8.13}$$

Similarly, if a 1 is sent,

$$E\{y/1\} = m_1 = \int_0^{T_s} s_1(t)[s_1(t) - s_0(t)] \, dt \tag{8.14}$$

The variance of y is the same under both cases (prove it) and is equal to

$$\sigma_y^2 = E\left\{ \left[\int_0^{T_s} n(t)[s_1(t) - s_0(t)] \, dt \right]^2 \right\}$$

$$= E\left\{ \int_0^{T_s}\int_0^{T_s} n(t)n(\tau)[s_1(t) - s_0(t)][s_1(\tau) - s_0(\tau)] \, dt d\tau \right\}$$

$$= \frac{N_o}{2} \int_0^{T_s} [s_1(t) - s_0(t)]^2 \, dt \tag{8.15}$$

The last relationship is derived by moving the expected value inside of the integral and substituting the impulse autocorrelation of the noise, as we have done several times before.

The probability of error is found as in the previous analysis for the single sample detector. The only thing that changes is the mean and variance of the Gaussian distribution.

The probability of error is given either by the probability of false alarm or the probability of miss due to symmetry.

$$P_{FA} = \frac{1}{\sqrt{2\pi}\sigma_y} \int_0^\infty \exp\left(\frac{-(y - m_o)^2}{2\sigma_y^2}\right) dy$$

Letting

$$x = \frac{y - m_0}{\sqrt{2}\sigma_y}$$

this becomes

$$P_{FA} = \frac{1}{\sqrt{\pi}} \int_{-m_0/\sqrt{2}\sigma_y}^\infty e^{-x^2}\, dx = \frac{1}{2}\, \text{erfc}\left(\frac{-m_0}{\sqrt{2}\,\sigma_y}\right)$$

In order to draw some general conclusions from this error equation, we define the energy of the signals and the correlation between the two signals as follows.

$$E = \int_0^{T_s} s_1^2(t)\, dt = \int_0^{T_s} s_0^2(t)\, dt$$

$$\rho = \frac{\displaystyle\int_0^{T_s} s_1(t)s_0(t)\, dt}{E}$$

Before plugging these new definitions into the error expressions, we pause to show that the correlation, ρ, is bounded by 1. We begin with the inequality

$$\int [s_1(t) \pm s_0(t)]^2\, dt \geq 0$$

This is true since the integrand is non-negative. Expanding the square, we find that

$$\int s_1^2(t)\, dt + \int s_0^2(t)\, dt \pm 2 \int s_1(t)s_0(t)\, dt \geq 0$$

$$E + E \pm 2E\rho \geq 0$$

and solving for ρ, we have

$$|\rho| < 1$$

Now substituting ρ and E into the expressions for the mean and variance of y gives

$$\text{VAR}(y) = \frac{N_o}{2} \int_0^{T_s} [s_1(t) - s_0(t)]^2 \, dt = N_o E(1 - \rho)$$

$$m_0 = \int_0^{T_s} s_0(t)[s_1(t) - s_0(t)] \, dt = \rho E - E = E(\rho - 1)$$

Note that since ρ is bounded by 1, the variance is positive, and m_0 is negative. Finally, the probability of error is given as

$$P_e = \frac{1}{2} \, \text{erfc} \left(\frac{E(1 - \rho)}{\sqrt{2N_o E(1 - \rho)}} \right)$$

$$= \frac{1}{2} \, \text{erfc} \sqrt{\frac{E(1 - \rho)}{2N_o}} \qquad (8.16)$$

Equation (8.16) is plotted for several values of ρ in Fig. 8.34. The abscissa is the signal to noise ratio, E/N_o. Note that as the correlation increases, the probability of error also increases. When $\rho = 1$, the signals are identical, and measurement of the received variable gives no information about what was sent. We may as well flip a coin, which explains why the error probability is 0.5. The best performance occurs when $\rho = -1$. This occurs when the signals are the negative of each other.

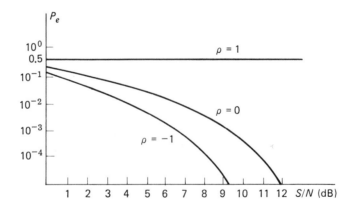

FIGURE 8.34 Error probability for binary matched filter detector as a function of input signal to noise ratio and correlation between two signals.

Matched Filter Detector for FSK

The analysis of the binary matched filter detector which we just completed was perfectly general. The only assumption was that the two signals have equal energy and that their a-priori probabilities were equal. We therefore need not reinvent the wheel, Eq. (8.16) can be used as a starting point.

$$P_e = \frac{1}{2} \text{erfc} \sqrt{\frac{E(1-\rho)}{2N_o}}$$

where E was the signal energy and ρ was the correlation coefficient. This result can be applied directly to the FSK system. The two signals are

$$s_0(t) = A \cos 2\pi f_0 t \qquad 0 < t < T_s$$
$$s_1(t) = A \cos 2\pi f_1 t \qquad 0 < t < T_s$$

Therefore, E and ρ are given by

$$E = \int_0^{T_s} A^2 \cos^2 2\pi f_0 t \, dt$$

$$= \int_0^{T_s} \left[\frac{A^2}{2} + \frac{A^2}{2} \cos 4\pi f_0 t \right] dt$$

$$= \frac{A^2 T_s}{2} \qquad \text{if } T_s \gg 1/f_0$$

$$\rho = \frac{A^2 \int_0^{T_s} \cos 2\pi f_0 t \cos 2\pi f_1 t \, dt}{A^2 T_s / 2}$$

$$= \frac{\sin 2\pi(f_0 + f_1)T_s}{2\pi(f_0 + f_1)T_s} + \frac{\sin 2\pi(f_1 - f_0)T_s}{2\pi(f_1 - f_0)T_s} \qquad (8.17)$$

It is desirable to make ρ as close to -1 as possible in order to realize the best possible performance. The first term in Eq. (8.17) is usually much smaller than the second since $f_0 + f_1 \gg f_1 - f_0$. We can therefore approximately minimize ρ by minimizing the second term in Eq. (8.17). Letting $f_1 - f_0 = \Delta f$ and setting the derivative equal to zero, we have

$$\frac{\partial \rho}{\partial \Delta f} = 2\pi \Delta f T_s^2 \cos (2\pi \Delta f) T_s - T_s \sin (2\pi \Delta f) T_s = 0$$

$$2\pi \Delta f T_s = \tan (2\pi \Delta f T_s)$$

and

$$2\pi \Delta f T_s \approx 0.715 \times 2\pi \text{ radians}$$

This yields a correlation coefficient,

$$\rho \approx -0.22$$

Note that ρ can be made equal to zero to setting

$$f_1 - f_0 = \frac{n}{T_s}$$

It would therefore appear that choosing $2\pi \Delta f T_s = 0.715 \times 2\pi$ yields better performance than $\Delta f T_s = n$. Caution is, however, advised in a practical situation. Suppose we are more realistic and modify s_0 and s_1 as follows:

$$s_0(t) = A \cos (2\pi f_0 t + \theta_0)$$
$$s_1(t) = A \cos (2\pi f_1 t + \theta_1)$$

That is, we admit that matching phases is virtually impossible. Equation (8.17) then becomes

$$\rho \approx \frac{1}{T_s} \int_0^{T_s} \cos (2\pi \Delta f t + \Delta \theta) \, dt$$

$$= \frac{1}{2\pi \Delta f T_s} [\sin (2\pi \Delta f T_s + \Delta \theta) - \sin (\Delta \theta)] \qquad (8.18)$$

If $\Delta f T_s$ is set equal to 0.715, there is no longer any assurance that ρ will be a minimum. It may even be positive.

However, if $\Delta f T_s = n$, the expression in Eq. (8.18) still yields $\rho = 0$. Therefore, with this choice of frequency separation, performance is independent of the exact phase.

If we use $\rho = 0$, the probability of error is given by

$$P_e = \frac{1}{2} \text{erfc} \sqrt{\frac{E}{2N_o}} = \frac{1}{2} \text{erfc} \left(\frac{A}{2} \sqrt{\frac{T_s}{N_o}} \right)$$

Again, $N_o/2$ is the height of the power spectral density of the noise. The error is plotted as a function of signal to noise ratio (E/N_o) in Fig. 8.35.

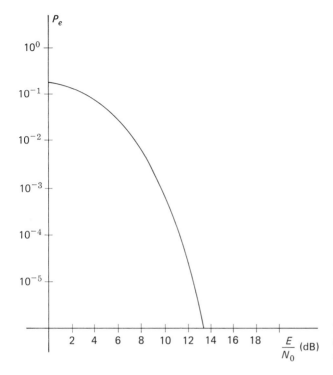

FIGURE 8.35   Performance of binary matched filter detector for FSK with zero signal correlation.

Illustrative Example 8.9

Find the probability of error for the FSK system where

$$s_0(t) = 1.414 \cos 1{,}000t$$

$$s_1(t) = 1.414 \cos 1{,}010t$$

Noise of power spectral density

$$\frac{N_o}{2} = 0.01$$

is added, and a matched filter detector is used. Assume that the phase can be exactly reproduced and that the sampling period is equal to 1 second.

Solution

The expression for error is

$$P_e = \frac{1}{2} \operatorname{erfc} \sqrt{\frac{E(1-\rho)}{2N_o}}$$

where

$$E = \frac{1}{2}(1.414^2) \times 1 \text{ second} = 1$$

Since $T_s \gg 1/f_0$, ρ is given by Eq. (8.17) as

$$\rho = \frac{\sin 2{,}010}{2{,}010} + \frac{\sin 10}{10}$$

$$= -0.054$$

The probability of error is then

$$P_e = \frac{1}{2} \text{ erfc} \sqrt{\frac{1.054}{2(0.02)}}$$

$$= \frac{1}{2} \text{ erfc}(5.1)$$

Most tables of the error function do not have arguments going above 5, since the erfc(5) is extremely small. For larger arguments, asymptotic expansions can be used. One such expansion is

$$\sqrt{\pi z} \; e^{z^2} \text{ erfc}(z) \approx 1 + \sum_{m=1}^{\infty} (-1)^m \frac{1 \times 3 \times \ldots \times (2m-1)}{(2z^2)^m} \tag{8.19}$$

For $z = 5.1$, this yields

$$\text{erfc}(5.1) = 5.5 \times 10^{-13}$$

and the probability of error becomes

$$P_e = 2.75 \times 10^{-13}$$

Matched Filter Detector for PSK

Once again, we begin the analysis with Eq. (8.16) for the performance of the general binary matched filter detector.

$$P_e = \frac{1}{2} \text{ erfc} \sqrt{\frac{E(1-\rho)}{2N_o}}$$

The two signals are

$$s_0(t) = A \cos(2\pi f_0 t + \theta_0)$$

$$s_1(t) = A \cos(2\pi f_0 t + \theta_1)$$

The signal energy is $A^2 T_s / 2$ and the correlation coefficient is

$$\rho = \frac{1}{A^2 T_s / 2} \int_0^{T_s} A \cos(2\pi f_0 t + \theta_0) A \cos(2\pi f_0 t + \theta_1) \, dt$$

$$\approx \cos(\theta_0 - \theta_1)$$

The probability of error is monotonic increasing in ρ, so we can minimize the error probability by making $\rho = -1$. This value of ρ will obtain if the phase difference between the two signals is $180°$. This result is not surprising since the signals are then as different as possible—one is the negative of the other. The resulting error probability is

$$P_e = \frac{1}{2} \operatorname{erfc} \sqrt{\frac{E}{N_o}}$$

This error probability is plotted in Fig. 8.36. Also shown in the figure is a repeat of the curve of Fig. 8.35 showing performance for FSK. Note that the two curves are separated by a horizontal distance of 3 dB.

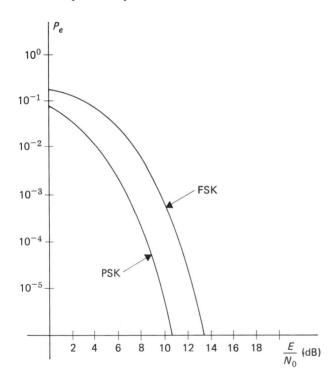

FIGURE 8.36   Performance of binary matched filter detector for PSK with signal correlation of -1. FSK curve of Figure 8.35 is repeated for comparison.

The Quadrature Detector

The quadrature detector was illustrated in Fig. 8.30. Recall that it is used in order to make the system insensitive to phase mismatch. We now examine performance assuming the input to the system is a unipolar ASK signal corrupted by additive bandlimited Gaussian noise. The reason the noise is bandlimited is that a bandpass filter precedes the system. The filter is necessary to prevent overloading of the correlators. The signals at points A_c and A_s on the diagram are given by

$$A_c(t) = \frac{1}{2}[s_b(t) + x(t)] \cos \theta$$

$$A_s(t) = \frac{1}{2}[s_b(t) + y(t)] \sin \theta$$

where $x(t)$ and $y(t)$ are quadrature components of the noise. The system output is given by

$$y_1(t) = \frac{1}{4}[s_b(t) + x(t)]^2 \cos^2 \theta + \frac{1}{4}[s_b(t) + y(t)]^2 \sin^2 \theta$$

$A_c(T_s)$ and $A_s(T_s)$ are Gaussian, and the final output, $y_2(T_s)$, is the square root of the sum of the squares of these two variables. If the variables had been zero-mean Gaussian, the resulting variable would be *Rayleigh* distributed. When a 1 is transmitted, $A_c(T_s)$ and $A_s(T_s)$ are not zero-mean, so the result is more complex. It is characterized by a *Ricean* density as follows:

$$p(y_2) = \frac{y_2}{\sigma^2} \exp\left(-\frac{1}{2\sigma^2}(y_2^2 + m^2)\right) I_o\left(\frac{my_2}{\sigma^2}\right) \qquad (8.20)$$

I_o is the *modified Bessel function of the first kind* (we first saw Bessel functions in our study of analog FM). The function cannot be expressed in closed form, and it is tabulated. The Ricean density is completely specified by its mean and variance. Note that if m is set equal to zero in Eq. (8.20), a Rayleigh density results.

Returning to the problem at hand, we need only find the means and variances of $A_c(T_s)$ and $A_s(T_s)$ in order to characterize the output. Assuming that a 0 is sent, both means are 0, and if a 1 is sent, the means are given by

$$E\{A_c\} = \frac{1}{2}s_b(T_s) \cos \theta$$

$$E\{A_s\} = \frac{1}{2}s_b(T_s) \sin \theta$$

Given any particular value of θ, the variance of either variable is given by

$$\mathrm{VAR}(A_c) = \mathrm{VAR}(A_s) = \frac{1}{4}E\{x^2(T_s) \cos^2 \theta\}$$

$$= \frac{N_o \pi f_m}{4} \cos^2 \theta$$

where f_m is the maximum signal frequency. Averaging this result over θ (that is, assume θ is uniformly distributed), we find

$$\text{Variance} = \frac{N_o \pi f_m}{8}$$

It can be shown that A_c and A_s are *independent* because they are Gaussian and in phase quadrature. Therefore, the joint density is the *product* of the individual densities. Assuming a 1 is sent, this is given by

$$p(A_c, A_s) = \frac{1}{2\pi\sigma_1\sigma_2} \exp\left(-\frac{(A_c - m_1)^2}{2\sigma_1^2} - \frac{(A_s - m_2)^2}{2\sigma_2^2}\right) \tag{8.21}$$

where

$$m_1 = \frac{1}{2} s_b(T_s) \cos\theta; \qquad m_2 = \frac{1}{2} s_b(T_s) \sin\theta \tag{8.22}$$

$$\sigma_1^2 = \sigma_2^2 = \frac{N_o \pi f_m}{8}$$

The square root of the sum of the squares is a distance function, and can be arrived at by converting Eq. (8.21) to polar coordinates. That is, we let

$$y_2 = \sqrt{A_c^2 + A_s^2}$$

$$\theta_0 = \tan^{-1}\left(\frac{A_s}{A_c}\right)$$

This yields

$$p(y_2, \theta_0) = \frac{y_2}{N_o \pi f_m / 4} \exp\left(\frac{-4}{N_o \pi f_m} [y_2^2 + s_b^2(T_s) - 2s_b(T_s) y_2 \cos(\theta - \theta_0)]\right)$$

We are only interested in y_2 at the output, so we integrate over θ_0. Also since θ is the phase of the incoming signal, we can again integrate over this variable assuming that it is uniformly distributed. The result is the Ricean density.

$$p_1(y_2) = \frac{y_2}{N_o \pi f_m / 8} \exp\left(\frac{y_2^2 + s_b^2(T_s)}{-N_o \pi f_m / 4}\right) I_o\left(\frac{y_2 s_b(T_s)}{N_o \pi f_m / 8}\right) \tag{8.23}$$

Now if a 0 is sent, $p_0(y_2)$ is found by setting $s_b(T_s) = 0$ in Eq. (8.23). This results in the Rayleigh density of Eq. (8.24).

$$p_0(y_2) = \frac{y_2}{N_o \pi f_m / 8} \exp\left(\frac{-y_2^2}{N_o \pi f_m / 4}\right) \tag{8.24}$$

Now that the two conditional probabilities are known, the probability of error becomes a simple matter of integrating the tails of these densities beyond the decision threshold.

Illustrative Example 8.10

A unipolar baseband signal transmits a half cycle of sin πt for a binary 1, and 0 for a binary 0. The sampling period is $T_s = 1$. ASK is used with a carrier frequency of 1 MHz. Noise with power spectral density of N_o watts/Hz is added during transmission, and a quadrature detector is used. Find the probability of bit error.

Solution

We will assume that the sampling is performed at the midpoint of each interval, so $s_b(T_s) = 1$ or 0 depending upon whether the transmitted bit is 1 or 0 respectively. Further assume $f_m = 0.5$, the frequency of the continuous sinusoid.

The probability density of y_2 is given by

$$p_1(y_2) = \frac{y_2}{N_o\pi/16} \exp\left(\frac{-(y_2^2 + 1)}{N_o\pi/8}\right) I_o\left(\frac{y_2}{N_o\pi/16}\right)$$

$$p_0(y_2) = \frac{y_2}{N_o\pi/16} \exp\left(\frac{-y_2^2}{N_o\pi/8}\right)$$

Unfortunately, these densities are not symmetric as in our previous analyses. This has two consequences. The decision threshold is no longer in the middle of the interval, and both conditional error probabilities are needed to find P_e.

Results drawn from decision theory are needed in order to set the threshold. Since we have not presented decision theory in this text, we shall assume that the decision threshold is set in the middle. Its value is then 0.5. The conditional error probabilities are then given by

$$P_{FA} = \int_{1/2}^{\infty} p_0(y_2) \, dy_2$$

$$= \int_{1/2}^{\infty} \frac{y_2}{N_o\pi/16} \exp\left(-\frac{y_2^2}{N_o\pi/8}\right) dy_2$$

$$= -\exp\left(\frac{-y_2^2}{N_o\pi/8}\right)\Bigg|_{1/2}^{\infty} = \exp\left(-\frac{2}{N_o\pi}\right)$$

$$P_M = \int_0^{1/2} p_1(y_2) \, dy_2$$

Since $p_1(y_2)$ contains the Bessel function, this last integral cannot be evaluated in closed form. It is tabulated under the name *Marcum-Q function*. Once this is found, the probability of bit error is given by

$$P_e = P_{FA} Pr(0) + P_M \, Pr(1)$$

If the probabilities of transmitting a 1 and 0 are equal, the bit error rate is the average of the two conditional error probabilities.

$$P_e = \frac{1}{2}(P_{FA} + P_M)$$

PROBLEMS

8.1.   A binary sequence,

$$\ldots 0\ 0\ 1\ 1\ 0\ 0\ 1\ 1\ 0\ 0\ 1\ 1 \ldots$$

is transmitted using a bipolar baseband waveform. The bit transmission rate is 1 Mbps. Find the frequency distribution and bandwidth of the transmitted waveform. Repeat the analysis for unipolar baseband transmission.

8.2.   The signal

$$s(t) = 2 \cos(6\pi t + 10°) + 2 \sin(12\pi t + 20°)$$

is sampled at 12 samples per second and encoded into 6-bit PCM. The resulting signal is transmitted via bipolar baseband.
(a) Find the bandwidth of the transmitted signal.
(b) Design a detector for this system.

8.3.   Redo Problem 8.2 for unipolar baseband transmission.

8.4.   Find the bandwidth of a bipolar ASK signal where the sampling period is 5 ms and a carrier of 20 MHz is used. You may assume a random data sequence where all possible combinations of bits occur.

8.5.   Repeat Problem 8.4 for unipolar ASK.

8.6.   A unit amplitude bipolar baseband signal is developed by sampling a signal and encoding the samples. The bit rate after encoding is 1,000 bits per second. The baseband signal is shaped by a lowpass filter that passes frequencies up to 1,000 Hz. The shaped signal then frequency modulates a carrier of frequency 2 MHz. The value of k_f is 500. Find the bandwidth of the resulting signal.

8.7.   Find the bandwidth of a PSK waveform used to transmit an alternating train of zeros and ones. The transmission frequency is 1 MHz, and the phase varies sinusoidally from 0° to 90°. The bit rate is 1 kbps.

8.8.   A bipolar baseband signal with a transmission rate of 1,000 pulses per second is shaped by passing it through a lowpass filter with cutoff at 1 kHz. The resulting shaped signal phase modulates a carrier with maximum phase deviation of $\Delta\theta = 90°$. Find the bandwidth of the resulting PSK signal.

8.9.   Prove Schwartz's inequality for complex functions. (Hint: Parallel the proof in the text, but replace square quantities by magnitude square quantities.)

8.10.   (a) Find the matched filter for the signal

$$s(t) = \begin{cases} 5 \cos 2\pi \times 2{,}000t & 0 < t < 0.005 \\ 0 & \text{otherwise} \end{cases}$$

in white noise with power spectral density $N_o/2 = 0.1$.
(b) What is the output signal to noise ratio?

8.11.   (a) Find the matched filter for the signal

$$s(t) = \begin{cases} 10 \cos 2\pi \times 1{,}000t & 0 < t < 50\ \mu\text{sec} \\ 0 & \text{otherwise} \end{cases}$$

in white Gaussian noise with a power of 10^{-4} watts/Hz.

(b) Find the output signal to noise ratio for this filter.

8.12. You are given two baseband waveshapes as shown below.

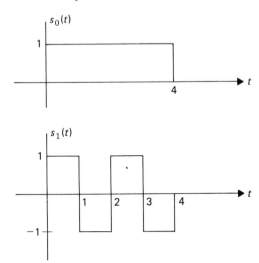

A matched filter detector is used to decide between the two signals. The noise is white with power spectral density $N_o/2 = 0.1$.

(a) Find the probability density of the detector output assuming that a 1 is being sent.

(b) Repeat part (a) for a 0 transmission.

(c) Find the output signal to noise ratio under each transmission assumption.

(d) Find the bit error rate.

8.13. Design (draw a block diagram of) a quadrature detector that could be used for reception of FSK in those cases where the incoming phase angle is not known.

8.14. Can you design a quadrature detector for PSK? If not, explain why. Repeat for DPSK.

8.15. A 1-volt bipolar pulsed signal at 2 kbps goes through a channel with ideal lowpass filter characteristic. The channel cuts off at 5 kHz. Noise of 10^{-6} watts/Hz is added in the channel. Find the probability of error using a receiver that samples at the midpoint of each interval.

8.16. Design a matched filter detector that could be used to choose between the two baseband signals shown below. (This is a Manchester code.)

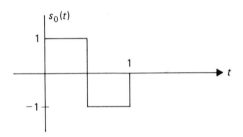

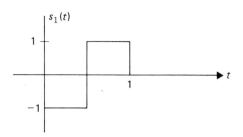

Assume that white Gaussian noise of power spectral density $N_o/2$ is added in the channel. Find the probability of error for $E/N_o = 4$.

8.17. (a) Design a detector for the two signals shown below. Assume that white Gaussian noise is added.

(b) Find the probability of bit error as a function of the noise power spectral density.

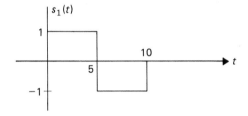

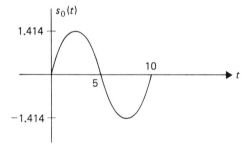

(c) Redesign the detector if $s_0(t) = 0$, and find the new probability of error.

8.18. Design a matched filter detector for the two signals shown below, and evaluate its performance as a function of Δ.

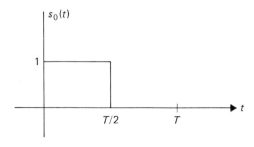

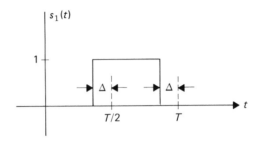

8.19.   (a) Design a binary matched filter detector for bipolar ASK where the baseband signal consists of 2-V flat pulses transmitted at 5,000 pulses per second, and the carrier frequency is 2 MHz.

      (b) Find the probability of error for this detector.

      (c) Assume that a "dead space" is defined in which no decision is made. That is, if the detector output is above some threshold, $+K$, we decide that a 1 was sent. If the output is less than $-K$, we decide in favor of a 0 transmission. Between $-K$ and $+K$, no decision is made, and a request is often made to repeat the transmission. Find the probability of error for this system and also find the probability that no decision is made (i.e., a request to retransmit occurs).

8.20.   Find the probability of bit error for a FSK system with

$$s_0(t) = 2 \sin (2\pi \times 1{,}000t + 45°)$$

$$s_1(t) = 2 \sin (2\pi \times 1{,}500t + 45°)$$

Assume $N_o/2 = 0.2$ and the sampling period is 5 seconds long. Assume that the matched filter detector is used.

8.21.   (a) Find the probability of error for a PSK system where

$$s_0(t) = \cos (2\pi \times 2{,}000t)$$

$$s_1(t) = \cos (2\pi \times 2{,}000t + \theta)$$

The power spectral density of the additive noise is 0.001, and the sampling period is 20 ms.

      (b) Plot the bit error probability as a function of θ for the matched filter detector.

Appendix I

Sources for Further Study

Note: Most of the following references are on a basic level and should give an alternate, or expanded, treatment of some of the topics in this text. Several (those marked with #) are not necessarily basic but are definitive works in a particular field. They are of particular interest in providing a perspective and framework. Finally, a few advanced sources (those marked with *) are included for those desiring to undertake an in-depth study.

PROBABILITY AND NOISE ANALYSIS

COOPER, G. R., and MCGILLEM, C. D., *Probabilistic Methods of Signal and System Analysis*, Holt, Rinehart and Winston, New York, 1971.

#DAVENPORT, W., JR., and ROOT, W., *Random Signals and Noise*, McGraw-Hill, New York, 1958.

*DOOB, J. L., *Stochastic Processes*, Wiley, New York, 1953.

FELLER, W., *Introduction to Probability Theory and Its Applications*, Vol. 1, 3rd Ed., Wiley, New York, 1968.

#PAPOULIS, A., *Probability, Random Variables, and Stochastic Processes*, McGraw-Hill, New York, 1965.

*VAN DER ZIEL, A., *Noise: Sources, Characterization, Measurement*, Prentice-Hall, Englewood Cliffs, N.J., 1970.

SIGNAL ANALYSIS

CHILDERS, D. G., and DURLENE, A. E., *Digital Filtering and Signal Processing*, West Publishing, New York, 1975.

GABEL, R. A., and ROBERTS, R. A., *Signals and Linear Systems*, 2nd Ed., Wiley, New York, 1980.

#HAMMING, R. W., *Digital Filters*, Prentice-Hall, Englewood Cliffs, N.J., 1977.

KUO, F., *Network Analysis and Synthesis*, 2nd Ed., Wiley, New York, 1966.

MILLMAN, J., and HALKIAS, C. C., *Integrated Electronics: Analog and Digital Circuits and Systems,* McGraw-Hill, New York, 1972.

#PAPOULIS, A., *The Fourier Integral and Its Applications*, McGraw-Hill, New York, 1962.

*RABINER, L.R., and RADER, C. M., eds., *Digital Signal Processing,* IEEE Press, New York, 1972.

#RICE, S. O., "Mathematical Analysis of Random Noise," *Bell System Technical Journal*, Vol. 23, pp. 282–333, July 1944; Vol. 24, pp. 96–157, January 1945.

SCHWARTZ, M., and SHAW, L., *Signal Processing, Discrete Spectral Analysis, Detection and Estimation*, McGraw-Hill, New York, 1975.

STANLEY, W. D., *Digital Signal Processing*, Reston Publishing, Reston, Va., 1975.

GENERAL COMMUNICATION SYSTEMS

CARLSON, A. B., *Communication Systems,* 2nd Ed., McGraw-Hill, New York, 1974.

COUCH, L. W., *Digital and Analog Communication Systems,* Macmillan, New York, 1983.

HAYKIN, S., *Communication Systems,* Wiley, New York, 1978.

*JAYANT, N. S., ed., *Waveform Quantization and Coding*, IEEE Press, New York, 1976.

LATHI, B. P., *Modern Digital and Analog Communication Systems*, Holt, Rinehart and Winston, New York, 1983.

*LINDSEY, W. C., and SIMON, M. K., *Telecommunication Systems Engineering,* Prentice-Hall, Englewood Cliffs, N.J., 1973.

RODEN, M. S., *Digital and Data Communication Systems,* Prentice-Hall, Englewood Cliffs, N.J., 1982.

SCHWARTZ, M., *Information Transmission, Modulation and Noise: A Unified Approach*, McGraw-Hill, New York, 1980.

SHANMUGAM, K. S., *Digital and Analog Communication Systems,* Wiley, New York, 1979.

*SPILKER, J. J., JR, *Digital Communication by Satellite,* Prentice-Hall, Englewood Cliffs, N.J., 1977.

STREMLER, F. G., *Introduction to Communication Systems,* 2nd Ed., Addison Wesley, New York, 1982.

TAUB, H., and SCHILLING, D. L., *Principles of Communication Systems*, McGraw-Hill, New York, 1971.

ZIEMER, R. E., and TRANTER W. H., *Principles of Communications*, Houghton Mifflin, Boston, 1976.

Appendix II
List of Symbols

In approximate order of appearance

SYMBOL	NAME
$\bar{a}_x, \bar{a}_y, \bar{a}_z$	Three rectangular unit vectors
$s^*(t)$	Complex conjugate of $s(t)$
$\|x\|$	Magnitude of the complex number x
$/x$	Phase angle of the complex number x
$\stackrel{\Delta}{=}$	Equal by definition
$\approx$	Approximately equal to
$\mathcal{F}[\ \ ]$	Fourier Transform of the quantity in brackets
$\mathcal{F}^{-1}[\ \ ]$	Inverse Fourier Transform of the quantity in brackets
$\leftrightarrow$	Quantities on left and right end of the arrow form a Fourier Transform pair
$r(t) * s(t)$	$r(t)$ convolved with $s(t)$
$U(t)$	The unit step function
$Re\{\ \}$	Real part of the term in braces
$Im\{\ \}$	Imaginary part of the term in braces
Hz	Hertz (cycles per second)
kHz	Kilohertz = 1,000 hertz
MHz	Megahertz = 1,000,000 hertz
$\delta(t)$	The unit impulse function
$sgn(t)$	Sign function. Equal to $+1$ for positive argument and to -1 for negative argument
f_m	Upper cutoff frequency above which the Fourier Transform is zero
f_s	Sampling frequency in samples/second
$r(t) \rightarrow s(t)$	$s(t)$ is the output when $r(t)$ is the input
DFT	Discrete Fourier Transform
FFT	Fast Fourier Transform
W	Complex exponential used with the DFT and FFT

SYMBOL	*NAME*
$S(z)$	z-Transform
$h(t)$	Impulse response of a linear system
$H(f)$	System function. The Fourier Transform of $h(t)$
$t_{\mathrm{gr}}(f)$	Group delay
$t_{\mathrm{ph}}(f)$	Phase delay
t_r	Rise time
E_s	Energy in a signal $s(t)$
P_s	Average power in a signal $s(t)$
$\psi_s(f)$	Energy spectral density of $s(t)$
$\phi_s(t)$	Energy autocorrelation of $s(t)$
$s_T(t)$	Truncated version of $s(t)$ limited to values of t between $-T/2$ and $+T/2$
$G(f)$	Power spectral density
$R(t)$	Power autocorrelation
$F(x_0)$	Probability distribution function
$p_X(x)$	Probability density function
m	Mean value of a random variable
σ^2	Variance of a random variable
$\mathrm{erf}(x)$	Error function of Gaussian density
$\mathrm{erfc}(x)$	Complementary error function
$m_x(t)$	Mean value of the process $x(t)$
$R_{xx}(t_1, t_2)$	Autocorrelation of the process $x(t)$
$P(A/B)$	Conditional probability
$x(t), y(t)$	Quadrature components of narrowband noise
S/N	Signal to noise (power) ratio
ΔSNR	Signal to noise improvement figure
dB	Decibel $= 10 \log_{10}(\;\;)$
$s_c(t)$	Unmodulated carrier signal
AM	Amplitude modulation
DSBSC	Double sideband suppressed carrier
$s_m(t)$	Either an AM or PAM modulated waveform
f_c	Carrier frequency
AMTC	AM transmitted carrier
η	Power efficiency
r.f.	Radio frequency
i.f.	Intermediate frequency
$z(t)$	Pre-envelope (analytic signal)
SSB	Single sideband
$s_{\mathrm{usb}}(t)$	Upper sideband SSB signal
$s_{\mathrm{lsb}}(t)$	Lower sideband SSB signal
VSB	Vestigial sideband
FM	Frequency modulation
$f_i(t)$	Instantaneous frequency
$\lambda_{fm}(t)$	FM signal
$\Lambda_{fm}(f)$	Transform of $\lambda_{fm}(t)$
k_f	Constant associated with FM
$J_n(\beta)$	Bessel function of the first kind
Δf	Maximum frequency deviation

SYMBOL	*NAME*
BW	Bandwidth
PLL	Phase Locked Loop
VCO	Voltage Controlled Oscillator
PM	Phase modulation
$\theta(t)$	Instantaneous phase
k_p	Constant associated with PM
$\lambda_{pm}(t)$	Phase modulated signal
$\Delta\theta$	Maximum phase deviation
PAM	Pulse amplitude modulation
PPM	Pulse position modulation
PWM	Pulse width modulation
TDM	Time division multiplexing
A/D	Analog to digital converter
D/A	Digital to analog converter
lsb	Least significant bit
msb	Most significant bit
PCM	Pulse code modulation
CODEC	Coder-decoder
T-1	Digital carrier communication system
s_i	Boundaries of quantization regions
s_{q_i}	Roundoff values in quantization regions
DM	Delta modulation
Δ	Step size in delta modulation
ADM	Adaptive delta modulation
VSDM	Variable-slope delta modulation
DPCM	Differential PCM
$\hat{s}(nT_s)$	Predicted value of sample
$s_q(t)$	Quantized version of waveform
$e(t)$	Quantization error
ΔS	Quantization step size
μ	Logarithmic companding parameter
P_{N_q}	Quantization noise power
RZ	Return to zero
NRZ	Non-return to zero
BER	Bit error rate
FSK	Frequency shift keying
PSK	Phase shift keying
DPSK	Differential phase shift keying
ASK	Amplitude shift keying
ρ	Signal to noise ratio
P_{FA}	Probability of false alarm
P_M	Probability of miss
P_e	Bit error probability
BER	Bit error rate
I_o	Modified Bessel function of first kind

Appendix III
Fourier Transform Pairs

GENERAL PROPERTIES

$$S(f) = \int_{-\infty}^{\infty} s(t)e^{-j2\pi ft}\, dt$$

$$s(t) = \int_{-\infty}^{\infty} S(f)e^{j2\pi ft}\, df$$

Function	*Fourier Transform*
$s(t - t_0)$	$e^{-j2\pi f t_0}\, S(f)$
$e^{j2\pi f_0 t}\, s(t)$	$S(f - f_0)$
$s(t)\cos 2\pi f_0 t$	$\frac{1}{2}\left[S(f - f_0) + S(f + f_0)\right]$
$\dfrac{ds}{dt}$	$j2\pi f S(f)$
$\displaystyle\int_{-\infty}^{t} s(\tau)\, d\tau$	$\dfrac{S(f)}{j2\pi f}$
$r(t) * s(t)$	$R(f)S(f)$
$r(t)s(t)$	$R(f) * S(f)$
$s(at)$	$\dfrac{1}{a}\, S\left(\dfrac{f}{a}\right)$
$\dfrac{1}{a}\, s\left(\dfrac{t}{a}\right)$	$S(af)$

SPECIFIC TRANSFORM PAIRS

Function	*Fourier Transform*	
$e^{-at}U(t)$	$\dfrac{1}{a + j2\pi f}$	$a > 0$
$te^{-at}U(t)$	$\dfrac{1}{(a + j2\pi f)^2}$	$a > 0$
e^{-at^2}	$\dfrac{\pi}{a}\exp\left(-\dfrac{\pi^2 f^2}{a}\right)$	$a > 0$
$\lvert t \rvert$	$\dfrac{1}{2\pi^2 f^2}$	
$\dfrac{\sin at}{\pi t}$	$\begin{cases} 1 & \lvert f \rvert < a \\ 0 & \text{otherwise} \end{cases}$	
$\begin{cases} \frac{1}{2} & \lvert t \rvert < T \\ 0 & \text{otherwise} \end{cases}$	$\dfrac{\sin 2\pi f T}{2\pi f}$	
$\begin{cases} 1 - \dfrac{\lvert t \rvert}{T} & \lvert t \rvert < T \\ 0 & \text{otherwise} \end{cases}$	$\dfrac{\sin^2 \pi f T}{T\pi^2 f^2}$	
$e^{-a\lvert t \rvert}$	$\dfrac{2a}{a^2 + 4\pi^2 f^2}$	
$sgn(t)$	$\dfrac{1}{j\pi f}$	
$\delta(t)$	1	
1	$\delta(f)$	
$e^{j2\pi f_0 t}$	$\delta(f - f_0)$	
$\cos 2\pi f_0 t$	$\dfrac{1}{2}[\delta(f - f_0) + \delta(f + f_0)]$	
$\sin 2\pi f_0 t$	$\dfrac{j}{2}[\delta(f + f_0) - \delta(f - f_0)]$	
$U(t)$	$\dfrac{1}{2}\delta(f) + \dfrac{1}{j2\pi}$	

Appendix IV

The Error Function

$$\text{erf}(x) = \frac{2}{\sqrt{\pi}} \int_0^x e^{-u^2}\, du$$

x	Erf (x)	Erfc (x)	x	Erf (x)	Erfc (x)
0	0	1	0.05	0.056	0.944
0.10	0.112	0.888	0.15	0.168	0.832
0.20	0.223	0.777	0.25	0.276	0.724
0.30	0.329	0.671	0.35	0.379	0.621
0.40	0.428	0.572	0.45	0.475	0.525
0.50	0.521	0.479	0.55	0.563	0.437
0.60	0.604	0.396	0.65	0.642	0.358
0.70	0.678	0.322	0.75	0.711	0.289
0.80	0.742	0.258	0.85	0.771	0.229
0.90	0.797	0.203	0.95	0.821	0.179
1.00	0.843	0.157	1.05	0.862	0.138
1.10	0.880	0.120	1.15	0.896	0.104
1.20	0.910	0.0901	1.25	0.923	0.0768
1.30	0.934	0.0659	1.35	0.944	0.0564
1.40	0.952	0.0481	1.45	0.960	0.0400
1.50	0.966	0.0338	1.55	0.972	0.0284
1.60	0.976	0.0238	1.65	0.980	0.0199
1.70	0.984	0.0156	1.75	0.987	0.0128
1.80	0.989	0.0105	1.85	0.991	8.53×10^{-3}
1.90	0.993	6.91×10^{-3}	1.95	0.994	5.57×10^{-3}
2.00	0.995	4.59×10^{-3}	2.05	0.996	3.68×10^{-3}
2.10	0.997	2.93×10^{-3}	2.15	0.998	2.33×10^{-3}
2.20	0.998	1.84×10^{-3}	2.25	0.999	1.44×10^{-3}
2.30	0.999	1.13×10^{-3}	2.35	0.999	8.80×10^{-4}
2.40	0.999	6.82×10^{-4}	2.45	0.999	5.26×10^{-4}
2.50	1.000	4.03×10^{-4}	2.55	1.000	3.08×10^{-4}
2.60	1.000	2.34×10^{-4}	2.65	1.000	1.77×10^{-4}
2.70	1.000	1.33×10^{-4}	2.80	1.000	7.46×10^{-5}
2.90	1.000	4.09×10^{-5}	3.00	1.000	2.20×10^{-5}
3.10	1.000	1.16×10^{-5}	3.20	1.000	6.00×10^{-6}
3.30	1.000	3.06×10^{-6}	3.40	1.000	1.52×10^{-6}
3.50	1.000	7.43×10^{-7}	3.60	1.000	3.56×10^{-7}
3.70	1.000	1.67×10^{-7}	3.80	1.000	7.70×10^{-8}
3.90	1.000	3.48×10^{-8}	4.00	1.000	1.54×10^{-8}
4.10	1.000	6.70×10^{-9}	4.20	1.000	2.86×10^{-9}
4.30	1.000	1.19×10^{-9}	4.40	1.000	4.89×10^{-10}
4.50	1.000	1.97×10^{-10}	4.60	1.000	7.75×10^{-11}
4.70	1.000	3.00×10^{-11}	4.80	1.000	1.14×10^{-11}
4.90	1.000	4.22×10^{-12}	5.00	1.000	1.54×10^{-12}

Appendix V
Answers to Selected Problems

CHAPTER 1

1.2. Squared error $= 16$

1.4. (a) $a_n = 0$ for n even; $a_n = 2/\pi^2 n^2$ for n odd.

1.10. (d)

1.11. $mse = 0.0947$

1.15. $\frac{1}{2a} e^{-a|t|}$

1.38. $S(0) = \frac{3}{4}; S(\Omega) = -j/4; S(2\Omega) = \frac{1}{4}; S(3\Omega) = j/4$

CHAPTER 2

2.5. $H(f) = j2\pi f/(1 + j2\pi f); h(t) = -e^{-t} U(t) + \delta(t)$

2.10. (b) $mse = 0.0947$

2.25. (a) $1 \; -1 \; 1 \; -1 \; 1 \; -1 \; 1 \; -1 \ldots$;
(b) $0\ 0\ 0\ 0\ 0\ 0\ 0 \ldots$

2.28. (a) $K/2$; (c) $\frac{1}{4}$

2.31. (a) 1; (b) $1/(4\pi^2 f^2 + 4)$; (c) $\frac{1}{4}$

2.32. 0

CHAPTER 3

3.2. $\frac{5}{6}$

3.4. $\frac{2}{9}$

3.5. $\frac{93}{256}$

3.6. $\frac{5}{32}$

3.8.   0.223

3.10. $x_0 = 6.56$

3.12. $E(y) = \frac{5}{6}$; $\sigma_y^2 = \frac{11}{54}$

3.20.   (b) $E(x^2) = \frac{1}{3}$; $E(y^2) = \frac{25}{3}$; $E(z^2) = \frac{26}{3}$

$\sigma_x^2 = \frac{1}{3}$; $\sigma_y^2 = \frac{25}{12}$; $\sigma_z^2 = \frac{29}{12}$

3.22.   (a) $A = 1$ (b) $1 - e^6$; (c) $\frac{1}{2}$

3.30.   (b), (c)

3.31. $P_{out} = 2f_m$

3.40.   (a) 300; (b) 0.011 dB; (c) 15.18 dB

CHAPTER 4

4.5. $g(t) = (\sin 0.5t)/\pi t + (\sin 0.5t \cos 0.75t)/\pi t$

4.9.   (a) $s(t)$; (b) $P_r/2$; (c) P_r

4.14. $\sigma = 26°$

4.27.   SNR $= 98.17 = 19.9$ dB

4.28.   (b) SNR $= 31.25 = 14.9$ dB

(c) SNR $= 1.67 = 2.22$ dB

4.29.   DSBSC $-$ SNR $= 6.28 = 7.98$ dB;

SSBSC $-$ SNR $= 3.14 = 4.98$ dB;

DSBTC $-$ SNR $= 6.28 = 7.98$ dB assuming a large carrier to noise ratio entering the envelope detector.

4.30.   SNR $= 1/4N_0$

4.31.   SNR $= 10 = 10$ dB

CHAPTER 5

5.9.   (a) $BW = 20,048$ Hz/ (b) $BW = 40,096$ Hz

5.10. $BW = 60,048$ Hz

5.12. $BW = 20$ kHz

5.14. $BW = 3,500$ Hz

5.20.   (a) 22.8 dB

5.24.   0.197 clicks/second

5.26.   (a) $A = 16.89$; (b) SNR $= 553 = 27.3$ dB

CHAPTER 7

7.2.   010; 101; 110; 111

7.7.   36 kbits/sec; minimum bandwidth = 18 kHz

7.8.   Output sequence: 1 1 1 1 0 1 1 1 0 1 1 1 0 1 1 1 ...

7.11. 9.77×10^{-4}; 9.3×10^{-10}

7.12. $A = \dfrac{\sin f_m T}{f_m T}$

$$mse = \frac{N_0}{2\pi} \left(f_m - \frac{\sin^2 f_m T}{f_m T^2} \right)$$

7.14. 0.0013 V

7.15. $mse = 0.012\sigma^2$

7.21. $mse = 0.016$; mse (uniform) = 0.021

7.22. (a) 0.0013; (b) 0.0026

CHAPTER 8

8.4. $BW = 400$ Hz

8.6. $BW = 3$ kHz

8.8. $BW = 5142$ Hz

8.15. $P_e = 7.5 \times 10^{-11}$

8.17. (b) $P_e = \dfrac{1}{2} \text{erfc} \dfrac{1.35}{\sqrt{N_0}}$; (c) $P = \dfrac{1}{2} \text{erfc} \dfrac{1.58}{\sqrt{N_0}}$

8.19. (b) $P_e = \dfrac{1}{2} \text{erfc} \dfrac{4 \times 10^{-4}}{N_0}$

(c) $P_{ND} = \dfrac{1}{2} \text{erfc} \left(\dfrac{1.6 \times 10^{-3} - K}{\sqrt{3.2 \times 10^{-3} N_0}} \right) - \dfrac{1}{2} \text{erfc} \left(\dfrac{1.6 \times 10^{-3} + K}{\sqrt{3.2 \times 10^{-3} N_0}} \right)$

$P_e = \dfrac{1}{2} \text{erfc} \left(\dfrac{1.6 \times 10^{-3} + K}{\sqrt{3.2 \times 10^{-3} N_0}} \right)$

8.20. $P_e = 7 \times 10^{-7}$

Index